TRIBOLOGY AND BIOPHYSICS OF ARTIFICIAL JOINTS

Vol. 27 Dissipative Processes in Tribology (Dowson et al., Editors)
Vol. 28 Coatings Tribology – Properties, Techniques and Applications in Surface Engineering (Holmberg and Matthews)
Vol. 29 Friction Surface Phenomena (Shpenkov)
Vol. 30 Lubricants and Lubrication (Dowson et al., Editors)
Vol. 31 The Third Body Concept: Interpretation of Tribological Phenomena (Dowson et al., Editors)
Vol. 32 Elastohydrodynamics – '96: Fundamentals and Applications in Lubrication and Traction (Dowson et al., Editors)
Vol. 33 Hydrodynamic Lubrication – Bearings and Thrust Bearings (Frêne et al.)
Vol. 34 Tribology for Energy Conservation (Dowson et al., Editors)
Vol. 35 Molybdenum Disulphide Lubrication (Lansdown)
Vol. 36 Lubrication at the Frontier – The Role of the Interface and Surface Layers in the Thin Film and Boundary Regime (Dowson et al., Editors)
Vol. 37 Multilevel Methods in Lubrication (Venner and Lubrecht)
Vol. 38 Thinning Films and Tribological Interfaces (Dowson et al., Editors)
Vol. 39 Tribological Research: From Model Experiment to Industrial Problem (Dalmaz et al., Editors)
Vol. 40 Boundary and Mixed Lubrication: Science and Applications (Dowson et al., Editors)
Vol. 41 Tribological Research and Design for Engineering Systems (Dowson et al., Editors)
Vol. 42 Lubricated Wear – Science and Technology (Sethuramiah)
Vol. 43 Transient Processes in Tribology (Lubrecht, Editor)
Vol. 44 Experimental Methods in Tribology (Stachowiak and Batchelor)
Vol. 45 Tribochemistry of Lubricating Oils (Pawlak)
Vol. 46 An Intelligent System For Tribological Design In Engines (Zhang and Gui)
Vol. 47 Tribology of Elastomers (Si-Wei Zhang)
Vol. 48 Life Cycle Tribology (Dowson et al., Editors)
Vol. 49 Tribology in Electrical Environments (Prashad)

Aims & Scope
The Tribology Book Series is well established as a major and seminal archival source for definitive books on the subject of classical tribology. The scope of the Series has been widened to include other facets of the now-recognised and expanding topic of Interface Engineering.

The expanded content will now include:
• colloid and multiphase systems; • rheology; • colloids; • tribology and erosion; • processing systems; • machining; • interfaces and adhesion; as well as the classical tribology content which will continue to include • friction; contact damage; • lubrication; and • wear at all length scales.

TRIBOLOGY AND INTERFACE ENGINEERING SERIES, 50
EDITOR: B.J. BRISCOE

TRIBOLOGY AND BIOPHYSICS OF ARTIFICIAL JOINTS

L.S. PINCHUK
V.A. Belyi Metal-Polymer Research Institute of the National Academy of Sciences of Belarus, Gomel, Belarus

V.I. NIKOLAEV
Gomel State Medical University Gomel, Belarus

E.A. TSVETKOVA
V.A. Belyi Metal-Polymer Research Institute of the National Academy of Sciences of Belarus, Gomel, Belarus

V.A. GOLDADE
V.A. Belyi Metal-Polymer Research Institute of the National Academy of Sciences of Belarus, Gomel, Belarus

ELSEVIER

Amsterdam – Boston – Heidelberg – London – New York – Oxford
Paris – San Diego – San Francisco – Singapore – Sydney – Tokyo

ELSEVIER B.V.
Radarweg 29
P.O. Box 211, 1000 AE Amsterdam
The Netherlands

ELSEVIER Inc.
525 B Street, Suite 1900
San Diego, CA 92101-4495
USA

ELSEVIER Ltd.
The Boulevard, Langford Lane
Kidlington, Oxford OX5 1GB
UK

ELSEVIER Ltd.
84 Theobalds Road
London WC1X 8RR
UK

First edition 2006

ISBN-13: 978-0-444-52162-0
ISBN-10: 0-444-52162-3
ISSN: 1572-3364 (series)

06 07 08 09 10   10 9 8 7 6 5 4 3 2 1

Printed in Great Britain.

PREFACE

The present book is devoted to endoprostheses of joints, i.e. artificial joints implanted into the human body. Available in literature information on structural endoprosthetic materials complying with the requirements of biocompatibility and wear resistance is reviewed in this book and a retrospective analysis of modern joint endoprosthetic designs is presented. Data on clinical aspects of endoprosthetics are cited. Along with biological methods the approaches of genetic engineering are paid attention to as promising techniques of designing bone and cartilage transplants. Tribological mechanisms of operation *in vivo* of the endoprosthesis are examined as opposed to the natural joint functioning. The analysis is presented of endoprostheses removed at revision operations and tribological test procedures are characterized.

The traditional designs of artificial joints are known to embody the fundamental ideas advanced in the 1960-ies by an English orthopedist J. Charnley and resemble much machine members that are insufficiently adapted for operation in human organisms. The authors put forward a concept on simulation of biological functions of the bone and cartilage tissue, and bioelectret potentials of natural joints in joint endoprostheses. Information is given on developed by the present authors artificial cartilage based on high-molecular weight polyethylene; its structural, physico-mechanical, tribological and medico-biological characteristics are expounded. Experimental evidences obtained in the course of investigations visualize that blood and synovia exhibit spectra of the thermally stimulated current without any electrical treatment. A principal way of improving lubrication of joint endoprostheses via a constant electric (electret) or magnetic field is justified. Novel joint endoprosthesis designs realizing the established tribological and biophysical regularities are described. A forecast of contemporary trends in joint endoprosthetics is set forth.

The book is addressed to specialists in orthopedy, biophysics, immunology and engineers engaged in developing artificial joints.

CONTENTS

LIST OF ABBREVIATIONS

AAOS – American Association of Orthopaedic Surgeons
BMI – body mass index
CAD – computer-aided design
CAM – computer-aided manufacturing
CART – Clinical and Radiographic Terminology
CITO – Central Institute of Traumatology and Orthopedy named after N.N. Priorov (Russia)
CMC – carboxymethyl cellulose
COG – Children's Oncology Group (USA)
COSSG – Cooperative Osteosarcoma Study Group
CT – computer tomography
DEL – double electrical layer
DGOT – Deutshe Gesellschaft fur Orthopedie und Traumatologie
DNA – deoxyribonucleic acid
DTA – differential thermal analysis
EOI – European Osteosarcoma Intergroup
HDPE – high density polyethylene
HUA – hyaluronic acid
ICNIRP – International Commission on Non-ionizing Radiation Protection
IDES – International Documentation and Evaluation System
MCI – morphological cortical index
MDA - Medical Device Agency (UK)
MF – medicinal form
MS – medicinal substance
MTS – macromolecular therapeutic system
MVO – Medical Vaseline oil
NAPM – non-steroid antiphlogistic medications
PE – polyethylene
PEEK – polyether-etheroketone
PETF – polyethyleneterephthalate
PMMA – polymethyl methacrylate
POM – polyoxymethylene
PTFE – polytetrafluoroethylene
PVA – polyvinyl alcohol
PTFCE – polytrifluorochlorethylene
RR – reinforcement ring
RSA – radiographic spectral analysis
SEDICO – Secure Data Integration Concept
SEM – scanning electron microscopy
SICOT – Scientific International Council of Orthopaedics and Traumatology
SSG – Scandinavian Sarcoma Group

TGF – transforming growth factor
THA – total hip arthroplasty
TSC – thermally stimulated current
UHMWPE – ultrahigh-molecular weight polyethylene
WHO – World Health Organization
ZCP – zero charge point

INTRODUCTION

Endoprostheses are implanted into the human organism mechanical appliances that replace lacking organs or parts of the body. They have come into our life as a magnificent achievement of the mankind comparable to the development of the ocean or space exploration. Not many novelties of modern medicine can stand on a par with endoprosthetics in raising quality of human life. Today endoprostheses of almost all organs have found application in clinical medicine (Fig. 1).

Endoprosthetics of joints is considered as a most efficient method of recovering mobility of joints by their partial or total replacement by artificial components. More than 800,000 endoprosthetic operations on different joints are made in the world yearly [1]. This figure may be much higher since according to the World Association of Health Protection there is an objective necessity in endoprosthetics of joints per each thousand of the population [2]. This fact underlines global character of endoprosthetics of the present day.

The production of joint endoprostheses is a specialized science intensive commercial sphere. The materials used for them should strictly meet a series of requirements, namely chemical inertness, biological compatibility, admissible amount of impurities and so on [3, 4]. Endoprostheses of joints are related as a rule to high-precision products whose friction surfaces are machined very thoroughly [5]. Tolerances are also paid much attention to along with a given accuracy of the conjunction [6]. Members of endoprostheses manufactured at different enterprises should conform to international standards and be interchangeable [5, 6]. The production process of endoprostheses employs high-accuracy machine tools, automatic machines, robots and repeated control of process regimes. The rooms where joint endoprostheses are manufactured, packed and sterilized should meet strict sanitary and hygienic norms [7]. The degree of biological and physical contamination of air in such rooms can be characterized by the term "*clean room*".

Endoprosthetics of joints is made in the course of surgical operations by highly qualified orthopaedists, traumatologists in specifically equipped operating rooms with available antibiotics of a wide spectrum, preparations for prophylactics of thromboembolic complications and so on. As a result, endoprosthetics of joints has isolated recently into an independent trend in operative orthopaedics. A tendency has been also traced at the end of the 20th century of conducting endoprosthetic operations in specialized regional centres.

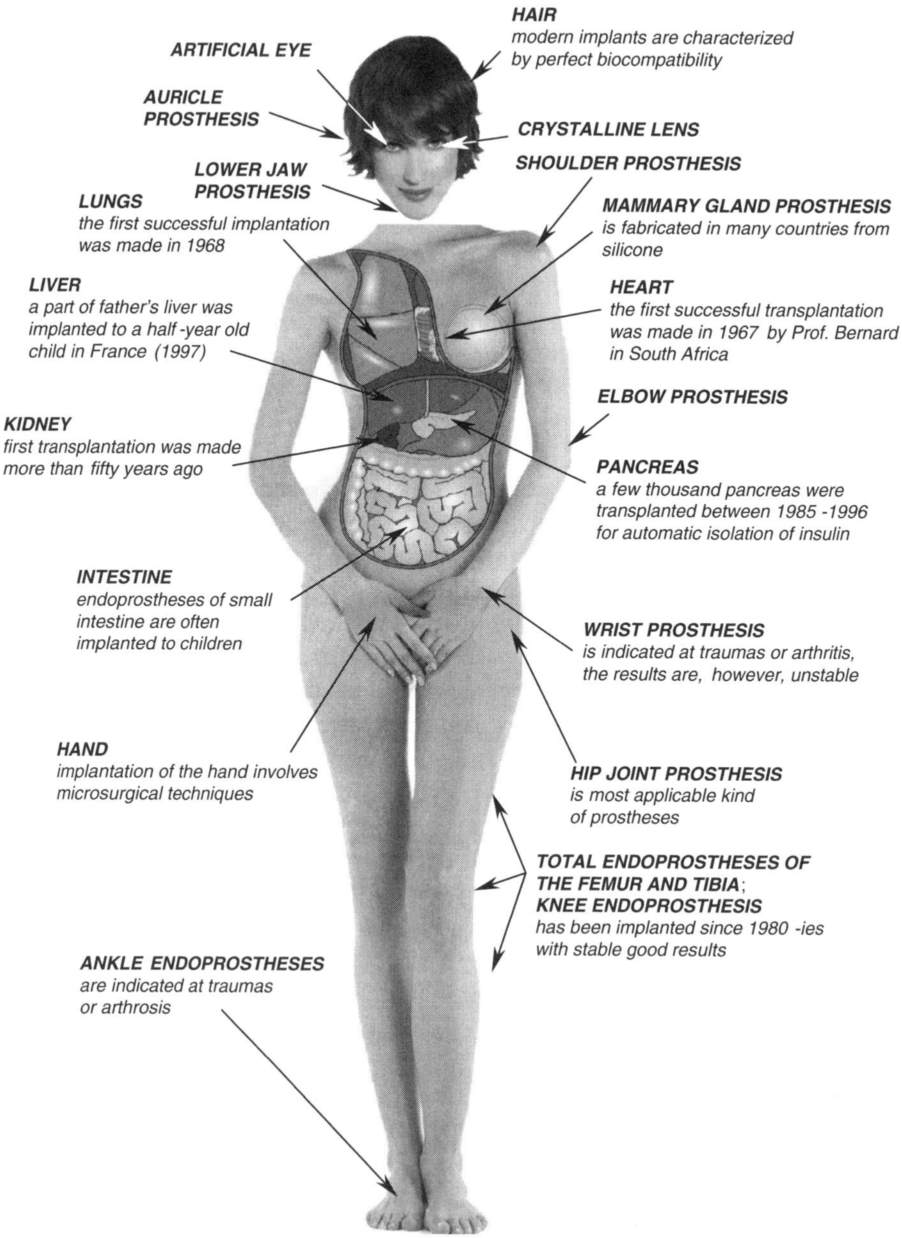

Fig. 1. Potentialities of reconstruction surgery

Within the first years after the endoprosthetic operation the results are commonly good and excellent. Medical observations show that with time there arises a necessity in substitution of the endoprosthesis or its components. The data on the terms of revision operations on e.g. the hip are contradictory [8-10], and the ratio of the initial to revision operations has a dangerous tendency to grow 4:1 and even 3:1. This situation is a subject of anxiety for the developers of endoprostheses as well as orthopaedists and traumatologists.

Human joints consist of biological tissues that are less strong than the modern structural materials of endoprostheses but surpass them much in wear resistance. Most apparent difference of the endoprosthesis from a natural joint is in fundamentally different lubrication mechanisms [11]. The key role in joint lubrication is played by the cartilage, which along with the antifrictional material fulfils the function of a porous reservoir for the synovia. There is no such an element in traditional endoprostheses designs. The comparison of the designs, lubrication mechanisms and functioning of natural joints to endoprostheses suggests that the latter resemble machine joint insufficiently adjusted for operation in human organism.

The analysis of removed during revision operations endoprostheses has proved that the chief problem limiting their stability and durability is inadequate wear resistance of the friction joints. Since this problem is found at the junction of such sciences as medicine, biophysics, triboengineering, materials science, and etc., its solution requires close collaboration of orthopaedists, traumatologists, technicians, biomechanics, immunologists and other professionals.

The present authors have pursued the aim to challenge specialists of different spheres interested in endoprosthetics and its rapid development. Just in cooperative work a new generation of endoprostheses can be created to perform not only mechanical but some biological functions as well, bring them close to natural tissues by their physico-chemical structure and so on. This noble aim is undoubtedly attainable from the standpoint of modern achievements in science and engineering. It requires attraction of the knowledge accumulated in biology, medicine, along with biophysical regularities of functioning of the locomotor apparatus, and high techniques of material preparation and processing.

The book contains both experimental and clinical results obtained at V.A. Belyi Metal-Polymer Research Institute of National AS of Belarus (MPRI), Belarussian Research Institute of Traumatology and Orthopaedics (BelRITO) and Gomel State Medical University.

The authors are grateful to Yu.M. Pleskachevsky and N.K. Myshkin, the former and the present directors of MPRI – for their meticulous attention to this work, E.D. Beloenko, Director of BelRITO, S.I. Boltrukevich, Head of Grodno Regional Traumatology-Orthopedic Center and Clinics, Zh.V. Kadolich, researcher of MPRI – for creative participation in elaboration of the ideas, V.A. Struk, Head of Chair of Grodno State University – for the ideological support and realizing of new types of endoprostheses in practice, L.S. Pushkina, I.I. Kekukh, E.A. Sementovskaya and S.V. Zotov, collaborators of MPRI – for invaluable contribution in preparation of this work. We are sincerely grateful to Professor of the University of Leeds, Duncan Dowson whose fruitful ideas assisted in the development of the artificial cartilage.

References:

1. Neverov V.A., Zakari S.M. Revision endoprosthetics of the hip. St. Petersburg, Education Publ., 1997, 112 p.
2. Grady-Benson J. Revision hip replacement surgery. *Current Orthopaedics*, 1995, No. 9, p. 9-20.
3. ISO 5832–12, 1996. Implant materials for surgery. Metallic materials. Part 12. Deformed alloys based on cobalt, chromium and molybdenum.
4. ISO 5834–1: 1985. Implantation materials for surgery. High-molecular mass polyethylene. Part 1. Powdery form.
5. ISO 7206–2: 1996. Implantation materials for surgery. Partial and total prosthesis of joints. Part 2. Joint surfaces of components of metal, ceramic and polymer materials.
6. ISO 5834–2: 1985. Implantation materials for surgery. High-molecular mass polyethylene. Part 2. Molded parts.
7. ISO 6018: 1987. Implantation orthopaedic materials. General requirements to marking and labelling.
8. Charnley J. The long-term results of low-friction arthroplasty of the hip performed as a primary intervention. *J. Bone Joint Sur.*, 1972, V. 54B, p. 61–76.
9. Morscher E.W. Endoprosthetic. Berlin Springer-Verlag, 1995, 431 pp.
10. Petty W. Total joint replacement. Philadelphia, W.B. Saunders Co, 1991, 811 pp.
11. Kupchinov B.I., Ermakov S.F., and Beloenko E.D. Biotribology of synovial joints. Minsk, Vedy, 1997, 272 p.

Chapter 1. ARTHROLOGY AND JOINT ENDOPROSTHETICS

Arthrology (*arthrologia*) is a section of medicine studying joints and their diseases. Modern arthrology is closely interrelated with endoprosthetics as one of most efficient means of treating joint pathologies that result from traumas or degenerative-dystrophic, inflammatory, oncological or other injuries. This intimate interdependence between arthrology and endoprosthetics is disclosed in the present chapter. A brief account of human joints and their pathology precedes the discussion on the main clinical indications for joint replacement. It is followed by the analysis of symptoms for surgical treatment, preoperative planning and delayed results of joint endoprosthetics. Named problems are described in evolution and estimated from the viewpoint of professionals in this field in the prehistory essay.

1.1 HUMAN JOINTS AND THEIR PATHOLOGY

A simplified anatomic scheme of a human joint illustrated in Fig. 1.1 shows the principal structural elements with articulating surfaces covered by a hyaline cartilage and enclosed in a joint capsule. The capsule is formed of a fibrous shell covered by a synovial backing on the inside and filled with a synovial fluid, which serves as a lubricating medium. The hyaline cartilage thickness as about 0.2–6.0 mm depending on the distribution of loads over the contact surface [1].

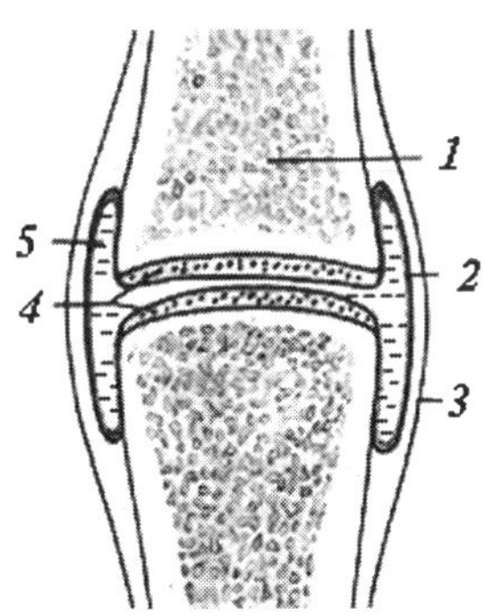

Fig. 1.1. Anatomic structure of joint: 1 – bone, 2 – synovial backing, 3 – fibrous capsule, 4 – joint cartilage, 5 – synovial fluid

The joint cartilage consists of chondrocytes and a collagen backing. The chondrocytes are oval or spherical cartilage cells having small processes and enclosed in the collagen cavities (lacunas).

The physico-chemical and contact interactions of the structural elements create an optimum biophysical basis for the exchange processes between the joint cavity and blood vessel. These interactions are intrinsic for the normal long-term functioning of joints.

The outline of articulating surfaces complies with geometrical bodies like the cylinder, ellipsoid, sphere, and other (Fig.1.2), which defines the number of axes for articulation and sliding of the joints. The cylindrical configuration makes possible rotation only about a single axis, the ellipsoidal – about two axes, and spherical – about three mutually perpendicular axes. This is why the biomechanical classification of joints envisages their subdivision into a uni-, bi- and triaxial types. The cylindrical and block-shaped joints are related to the uniaxial types (Fig. 1.2, *a*). Ellipsoidal (*b*), saddle-like (*c*), and condylar (a transient from the block to the ellipsoidal form, e.g. the knee) belong to biaxial types of joints. Triaxial joints represent a ball conjugated with a socket whose depth predetermines either a spheroid (shoulder, 3) or a cup-shaped (hip) joint [2].

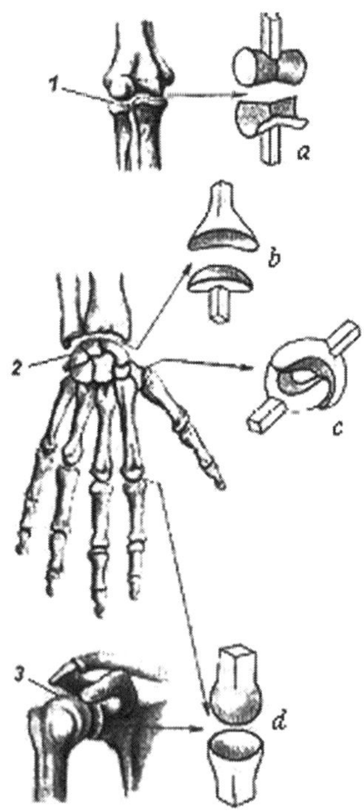

Fig. 1.2. Joints (1–3) and movable contact schemes of articular surfaces (*a*–*d*): 1 – elbow joint, 2 – wrist joints, 3 – shoulder; *a* – block-type, *b* – ellipsoidal, *c* – saddle-shaped, *d* – spheroid

The joints of the upper and lower extremities presenting most interest for endoprosthetics are considered below in this chapter.

The upper extremity includes the shoulder, elbow, wrist and hand joints.

The shoulder (*articulatio humeri*) is formed by a humeral head and a glenoid, which is rather shallow and small. The humeral head looks like a ball, while the glenoid presents a flattened spherical socket. The head is covered by a cartilage whose area exceeds roughly three times that of the glenoid. The head is retroverted approximately 30° relative to the longitudinal axis of the humerus and the glenoid is retroverted 7–10° relative to the scapular body. The described anatomic peculiarities ensure larger range of motion in the shoulder.

The pathological states indicated for endoprosthetics of the shoulder are the next. Total replacement (substitution of all components forming a friction pair) is indicated in case of degenerative changes in the cartilage surface of the head and socket resulted from rheumatoid arthritis or other kinds of arthritis, aseptic necrosis of the head, suppurative infection, acute fractures, dislocations and other. Unipolar endoprosthetics (replacement of one component of the friction pair) is indicated for multicomminuted fractures, fracture and dislocation, dissected head, compression fracture of the head when 50 and more p.c. of the joint surface is lost, fracture of the humeral neck when the broken fragment is disconnected with the soft tissues and blood vessels. The humeral head is replaced only if the glenoid cartilage remains undamaged.

The elbow (*articulatio cubiti*) presents a conjunction of three bones: humeral, ulnar and radial. Three joints are found between them, namely the humeroulnar, humeroradial and proximal radioulnar (Fig. 1.2).

The humeroulnar joint (*articulatio humeroulnaris*) is formed by a conjunction of the humeral block and a block-shaped part of the ulna. The hyaline cartilage covers the humeral block not fully but over the arc of ~300° in the sagittal plane (passing vertically from the front backwards along the body).

The humeroradial joint (*articulatio humeroradialis*, spheroidal) presents a junction of the humeral capitate and the articular cavity of the radial head.

The proximal radioulnar joint (*articulatio radioulnaris,* cylindrical) is formed by the articular circumference (cylindrical generatrix) of the radial head and a corresponding part of the radius of the ulna.

To restore the range of motion after endoprosthetics in the elbow an orthopedist should estimate the mutual orientation of articulating bones. This problem is complicated by the presence of several planes of revolution in the joint and a positional specificity of the axes of revolution in each junction.

Total replacement of the elbow is indicated for anchylosis (immobility of a joint because of symphysis), deformation arthrosis, false joints of the supracondylar zone of the shoulder, tumour processes in the bones.

Partial endoprosthetics of the elbow (the term “unipolar endoprosthetics” is not used here because the elbow is formed of three bones) presents interest only in historical attitude as the early stage practice. Modern designs of endoprostheses are intended for the total replacement of the elbow.

The radiocarpal joint (*articulatio radiocarpalis*) is formed by the articular surface of the radius, articulation disc and the proximal (found closer to the shoulder) surfaces of the first row of carpal bones (Fig. 1.2). The carpal bones are found between the forearm and metacarpal bones and provide for the variety of motions of the wrist. These bones form the following joints: radiocarpal, medcarpal, intercarpal and carpal-metacarpal. By the outline of its articular surfaces the radiocarpal joint is ellipsoidal, biaxial, i.e. having the frontal (perpendicular to the sagittal plane) and sagittal axes of motion.

Although the carpal joints exercise rather complex motions, their axis of rotation passes during flexion-extension and abduction through a fixed point on the capitate bone. The position of this point is independent of the range of motion of the radiocarpal joint. It is spaced ¼ length of the capitate bone from its proximal part.

Articular surfaces of the carpal bones look like a portion of a torus. They have different curvature radii at flexion-extension (R_1) and abduction-adduction of the radioulnar joint (R_2), $R_1 < R_2$. Thanks to this configuration, the head of the radiocarpal endoprosthesis design is in the form of ellipsoid with a corresponding cavity in the counterbody (Fig. 1.2, *b*).

Total endoprosthetics of the radiocarpal joint is recommended for the expressed painful syndrome caused by extensive degenerative dystrophic changes in the wrist as a result of rheumatoid arthritis or severe post-traumatic arthrosis of the joint. The orthopedists have to address to endoprosthetics also in the event of unsatisfactory traditional orthoplasty or removal of affected bones that may cause a shift of the remaining bones of the hand, violate the congruence of articulating surfaces and induce further degenerative changes [3].

Partial endoprosthetics (replacement of separate bones of the wrist) is exercised at localized pathological processes and non-union.

Joints of the fingers connect the metacarpal bones with the phalanges and finger phalanges with each other (metacarpus is a part of the upper extremity between the carpus and the main phalanges of the fingers). These joints are subdivided into two groups.

Metacarpophalangeal joints (*articulations metacarpophalangeales*) are formed by the heads of metacarpal bones and the bases of proximal phalanges. The articulating surfaces of the heads are rounded and the articular cavities are ellipsoidal. The metacarpophalangeal joints allow for the motion about two axes. Flexion-extension is exercised about the frontal axis within 90° and reminds that of the umbrella. Finger abduction and adduction are exercised in the sagittal plane (45-50°). The metacarpophalangeal joints can also perform circular motions. Modern designs of these endoprostheses ensure the rehabilitation of flexion-extension, but can not add stability for more accurate functions.

The interphalangeal joints of the wrist are formed by the heads and bases of neighbouring phalanges. All articulating surfaces of the joints are typically block-shaped. They are able to move only about the frontal axis by performing the flexion-extension within 90°.

Endoprosthetics of the metacarpophalangeal and interphalangeal joints is indicated for severe destructive changes in the wrist commonly due to the rheumatoid arthritis. The difficulties in attaining stable results of the replacement are attributed mainly to irreversible degenerative changes in the ligaments, muscles and the wrist. Note that in case of degenerative changes of the traumatic character in one or several joints the endoprosthetics gives stable positive results [4].

The joints of the lower extremity include the hip, knee, ankle and foot. Their structure is to provide for running, walking, standing, sitting and balancing actions.

The hip (*articulatio coxae*) is formed by the pelvic acetabulum and femoral head. By its articular surfaces this joint belongs to the cup-shaped ones. When in motion the whole articular surface of the femoral head and a part of the acetabular cartilage surface (semilunar surface) participate in friction. One edge of the acetabulum forms a symphysis with a fibrous-cartilage formation called the acetabular lip, which enlarges the acetabulum volume. This anatomic structure envisages rotation in three axes. Flexion and extension of the hip are possible about the frontal axis. Flexion reaches 118-121° during flexion of the knee, and 84–87° at its extension. Extension of the hip does not exceed 13°. Abduction and adduction of the lower extremity to the body proceeds about the sagittal plane (80–90°). The femoral head rotates about the vertical axis (40–50°). The range of motion in the hip is restricted by strong neighbouring ligaments, muscles, and the femoral head ligament. This cylindrical ligament adheres by one end to the femoral head and to the pelvic bone near the acetabulum by the other.

Replacement of the hip restores close to a normal functioning of the articulation. The artificial limb ensures the patient's everyday activities when walking, going upstairs, sitting, etc. except for athletic activities or other excess loads.

The total hip replacement as a major surgical method, which is indicated for the degenerative-dystrophic changes in the joint accompanied by an expressed painful syndrome limiting physical activities and requiring regular medicines. These changes may occur at primary and secondary arthrosis, aseptic necrosis of the femoral head, rheumatoid or gouty arthritis, posttraumatic state and so on. Sometimes the total hip replacement is a primary method for curing elderly people with medial fractures of the hip neck, Paget's disease, primary or secondary tumours.

The orthopaedists should be extremely careful during total endoprosthetics on young patients. Only a few types of hip endoprostheses are known to offer more than 20 years of useful operating life [5, 6]. The reconstructive surgical interventions on the proximal femoral and pelvic bones are preferable for the patients under 40 [7, 8]. An original revascularization (reconstruction of vessels) procedure is proposed for the femoral head at aseptic necrosis that uses bone autotransplants on a muscle vascular crus [9].

Unipolar endoprosthetics of the hip presupposes the replacement of the femoral head and neck. Among its drawbacks is wear of the acetabular cartilage aggravating with time and leading to inevitable revision operation and implantation of the total endoprosthesis. An alternative solution is the bipolar endoprosthesis. It alleviates frictional load on the acetabular cartilage thanks to rotation of the femoral head in a spherical component movably attached to the acetabulum.

The knee (*articulatio genus*) is the largest human joint consisting of three bones, namely femur, tibia and patella. They form a block-shaped rotary joint with two types of motion, that is, flexion-extension and rotation to a lesser extent.

The articulation surfaces of the tibia and femur are covered with a hyaline cartilage as well as intra-articular cartilages named menisci, the medial (internal) and lateral (external) ones. The menisci improve the congruence between the articular surfaces to ensure a uniform pressure distribution during the femur and ankle articulation. They influence the knee wear rate since contact pressures in the joint during walking, going upstairs and downstairs correspond on the average to as much as 4.5 times that of the body mass. Each meniscus presents a fibrous-cartilage plate of semilunar form in horizontal plane and triangular in cross-section. The thicker edge of the meniscus facing outwards adheres to the joint capsule, while the thinner one is directed inwards. The menisci are linked from the front via a transverse ligament.

The neighbouring muscles and ligaments form a stiffening system for the knee joint. The muscles stabilize the joint within the range of active motion; outside it mainly the ligaments are active. By functioning as an integral synergetic complex, the ligaments strain and inhibit passive motions in the joint exceeding the rate of active motions. This strain of the extra- and intra-articular ligaments governs limiting positions at flexion, extension and rotation of the knee.

Above-mentioned transverse ligament as well as the anterior and posterior cruciate ligaments belongs to the intra-articular ones. The anterior ligament hampers shifting of the femur backwards at extension of the knee, while the posterior one bounds the femur shift forward.

The fibrous (external) membrane of the knee capsule functions as a mechanical stabilizer of the limb. It sometimes spreads out and thickens in some places so that it is possible to single out seven extra-articular ligaments.

Hence, the range of motion in the knee can be characterized in general by the following parameters. Flexion and extension of the knee about the transverse axis proceeds at a 140–150° amplitude. Revolving about the vertical axis is roughly 15° for the active motions and 30–35° for the passive ones.

Prior to planning endoprosthetic operation of the knee the specialists estimate acuteness of the pain and the extent of functional violations resulted from joint diseases. Even though the painful syndrome alone had been soothed and the knee function improved, the surgical operation would be considered justified. Endoprosthetics of the knee occupies today a notable place in the arsenal of reconstructive orthopaedics [10]. In spite of a high level of operations on the knee, the problems connected with the prosthesis design, its ligaments, implantation technique, treating of condyle fractures and posttraumatic arthrosis are still disputable and awaiting their decision [11, 12].

The total replacement of the knee is commonly indicated for the late stages of deformation arthrosis, degenerative changes as a result of chondromatosis or villous synovitis, gouty or rheumatoid injury of the bones, aseptic necrosis and posttraumatic arthrosis. This operation can be made after the corrective osteotomy of the femur and ankle. The total replacement of the knee has strictly limited indications for the young patients. As an alternative usually the corrective osteotomy of the femur, ankle or arthrodesis (creation of immovable joint) can be fulfilled. Arthroscopy (low-invasive penetration of optical or other tools into the joint cavity for imaging) is often used as a method of diagnostics that helps to choose an optimum reconstructive operation [13]. This highly efficient method has for long been used in Russia to cure damaged menisci and intra-articular ligaments of the knee [14].

Such diseases as suppurative arthritis, neuropathic arthropathy and primary arthrodesis have been formerly among strict contra-indications for the total replacement of the knee. Today these pathologies do not hinder total replacement using special endoprostheses designs and modern bactericidal preparations, but the surgeon and patient should estimate the risk in each concrete case.

Partial endoprosthetics of the knee is a disputable problem. Pathological injures of one osteochondrous element of articulations is met rather seldom. The attempts in its local replacement have lead to further total endoprosthetics of the joint shortly after the primary operation. Nevertheless, the endeavours are still underway with the aid of computer navigation systems [15], which stirs up the interest in advancing partial endoprosthetics of the knee.

The ankle (*articulatio talocruralis*) is a typical block-shaped joint formed by articular surfaces of the ankle and talus. The ankle bones envelope the talus block like a fork. The capsule of the ankle joint resembles a cuff attached to the tibia from the front so that forms together with the articular cartilage a 5–8 mm gap. From the rear and sides it adheres the articular cartilage line. The joint capsule is stiffened by auxiliary ligaments on the lateral surfaces of the joint, namely the medial (deltoid) and three lateral ones.

The axis of rotation of the ankle passes over the tops of the lateral and medial malleolus. The amplitude of flexion and extension of the joint about this axis is 60–70°. The lesser amplitude transforms loads on the neighbouring subtalar and medial metatarsus joints. A normal ankle joint sustains at walking a load corresponding to 4–5 masses of a man and generates a shear force equal to 80 % of the mass. Contact stresses in the ankle joint are lower than in the hip or knee.

Total endoprosthetics of the ankle is an optimum method for the sick with systemic arthritis whose low activity is connected with troubles in walking. This group of diseases includes rheumatoid, psoriatic and haemophilic arthropathy, as well as systemic lupus erythematosus. This kind of endoprosthetics is indicated for the patients with posttraumatic arthrosis and avascular necrosis of the talus. An alternative of endoprosthetics for the young, obese and physically active patients is arthrodesis of the ankle. After the operation the pain disappears and the patient acquires the ability to fulfil standing a long-term physical job. Immobility of the ankle is compensated during walking by increased range of motion in small joints of the foot [16].

Endoprosthetics is inexpedient in case of purulent infections or neuromuscular diseases with a spastic component. Partial endoprosthetics of the ankle is impracticable.

Joints of the foot include junctions between bones of the tarsus, tarsus and metatarsus, and toe phalanges. Twelve bones of the tarsus and metatarsus are linked into tight joints and serve as a solid base for the foot. These joints are intricate in structure and functional qualities. Some of them acquire individual capsules and separate articular cavities. Their complex multiple-plane movements are optimized by the muscles and tightly strained ligaments.

Certain experience has already been gained in the world in experimental endoprosthetics of the foot bones and joints. This experience has not been yet adequately adopted in clinical practice [16].

Malignant tumours in bones and joints are a specific pathology of human limbs. Recent successes in chemotherapy of oncological patients have lead to a radical reassessment of acknowledged approaches to surgical treatment of these patients. The survival and lifetime of oncological diseased has raised, so the significance and upgrading of the operations on bones and joints has augmented too. Only recently preference in bone oncology has been given to amputations and alloplastic replacement of bone defects or articular ends. The reasons were i) difficulties in the diagnostics of spreading oncogenic processes before the operation; ii) greater safeness of amputation in contrast to adjuvant (auxiliary) therapy; iii) the need in complex reconstruction of joints and manufacture of individual endoprostheses [17]. Perceptible progress has been attained in the differential diagnostics of bone tumours during the last decades. Named operations are being actively substituted at present by endoprosthetics of joints and bone defects in the event of tumours of the limbs, and have established as a new direction in bone oncology. Recent advances in the development of specific tumour endoprostheses of bones and joints underpined this process. Their designs are considered in chapter 3.

This brief review of joint structures and pathologies is one more evidence that their mechanical function can be reproduced by endoprostheses. Any endoprosthesis design presupposes a requisite presence of a friction pair that simulates a movable junction of bone ends. The operation of any friction joint, the implanted endoprostheses included, subjects the frictional surface to inevitable damage and generation of wear debris. In this connection, the problems of designing endoprostheses with improved wear resistance of the articulating joints and disposal of their wear products are related to most urgent ones. The intricacy of mentioned problems puts joint endoprostheses within a rank of most science consuming, highly technological and costly products.

1.2 PREHISTORY ESSAY

The reconstruction of lost mobility of human joints has been and still remains an actual task of orthopaedics. Numerous attempts in this sphere since the end of 19th century can be subdivided in their historical evolution into five phases:

- arthroplasty with osteotomy (reconstruction of joint functioning with bone resection);
- interpositional arthroplasty, i.e. interpositioning of natural tissues between articulating bone ends;
- reconstructive (restoration) arthroplasty;
- intra-articular (partial) endoprosthetics;
- total endoprosthetics.

The history of endoprosthetics has started when Th. Gluck replaced in 1890 a part of the knee and hip joints by ivory prostheses [18]. Four years later J.E. Pean substituted the proximal shoulder with a platinum endoprosthesis. He used platinum after unsatisfactory previous tests with steel and iron implants subject to severe corrosion.

Hip endoprostheses are at present most refined designs of artificial limbs. In 1919 a French surgeon Deblet used a reinforced rubber endoprosthesis of the femoral head for the surgical treatment of a fractured head neck. In 1927, R. Robineau proposed to use bakelite and ebonite as the materials for the unipolar hip endoprostheses. Isolated instances of their implantation ended in a suppuration.

Late 1930ies were marked by intensive studies of the organism reactions on metals. In 1938, P.W. Wiles tried stainless steel to replace both the acetabular and femoral elements of the hip with fixation accomplished by screws and a buttress plate [19]. At the same time, C.S. Venable had experimented on animals with a new chromium-cobalt alloy and used it for joint endoprostheses. The alloy was called Vitalium and consisted of 30% chromium,

5% molybdenum and 65% cobalt. The alloy proved to be reliable and is still employed in traumatology and orthopaedics. In 1940, A.T. Moore implanted a femoral head endoprosthesis made of Vitalium [20]. The Judet brothers proposed in 1950 a metal-polymer endoprosthesis for the proximal hip. It included an acrylic head and a stainless steel rod [21]. However, their endoprostheses, like Moore's one, lead to intensive wearing of the cartilage (chondrolysis) and in many cases the prosthesis protruded into the acetabulum.

A slightly different prosthesis of the femoral head and neck was designed by F.R. Thompson in 1951. The elements were formed on a curved stem to fit the proximal femur [22]. Some surgeons attempted to place the Thompson or Moore femoral head inside a metal cup but found that the constant friction of metal against metal resulted in accumulation of wear debris in the joint capsule leading finally to a chemical abscess. At the beginning of the 1950ies, E.J. Haboush was first to use an acrylic monomer-polymer blend to fix a unipolar endoprosthesis made of Vitalium [23].

All these implants had the same structural imperfection, namely, their short stem could not provide a reliable fixation in the marrowy canal of the femur and showed unsatisfactory follow-up results. Attempting to resist loosening, J. Jepson in 1948 and L.T. Peterson in 1951, proposed steel endoprostheses designs in which the short stem was screwed via a splint to the femoral diaphysis. L.T. Peterson filled the deficiency of the bone between the artificial head and remainders of the femoral neck by metal washers of different thicknesses.

Another trend in the development of endoprostheses of the femoral head represents the structures with a long stem. Their creators attempted to place the weightbearing function along the femoral axis. These were the stems of two types, i.e. straight and curved. To prevent subsiding of the stem in the medullary canal the prostheses were furnished with a plateau (collar). A series of endoprostheses with a long bent stem were designed to preserve the intertrochanteric section of the femur (a part of the femoral neck). The prostheses created by Moore, Thompson, Judet and associates constituted an important stage in the evolution of reconstruction surgery of the hip. The round stem was substituted in 1951 for a trihedral one to prevent its rotation in the medullary canal, and the plateau was made broadened to hamper subsidence of the endoprosthesis in the femur.

By the mid of the 1950ies the variety of femoral heads ran into 50. Successes in endoprosthetics of joints stimulated growth of indications for this kind of surgical treatment that included also the deforming arthrosis of the hip. It became, however, evident with time that the artificial femoral head undergoes acute changes in the articular cartilage of the acetabulum. As a countermeasure it was proposed to make a freely rotating head of the endoprostheses, but the results were discouraging.

Failures of the femoral head replacement were mainly explained by two reasons: i) insufficiently strong fixation of the stem, and ii) affect of the artificial head on the acetabulum.

The works on the development of the acetabular endoprosthesis have begun in the 1950ies. The fame of being a founder of this direction is ascribed to an American surgeon M.R. Urist. First designs of the artificial acetabulum suffered from premature loosening and degenerative changes in the articular cartilage of the femoral head. This was the reason of simultaneous attempts in designing total endoprostheses of human limbs instead of single-component ones.

At the onset of total endoprosthetics the femoral head was mostly replaced by the Judet, Moore or Thompson prostheses. Urist's cups were used for the acetabulum, though their mismatching often turned to be a cause of dislocations of the femoral components and migration. So, there arose a necessity in the total endoprostheses whose components could be interchangeable. Their development has taken two main paths, i.e. with mating components and split prostheses.

The endoprostheses with split components were essentially perfected by G.K. McKee, UK. The hip prostheses implanted by him in 1951 consisted of a spherical cup screwed to the pelvic and a conjugated spherical head on a short stem.

K.M. Sivash 1924 – 1989

J. Charnley 1911 – 1982

M. Muller 1917

R. Mathys 1921–2000

Fig. 1.3. The developers of classical orthopaedic methods and endoprosthesis designs

K.M. Sivash (USSR, Fig. 1.3) is known to propose an all-metal nonseparable total hip endoprosthesis in 1956 [24], which have found wide application in orthopaedics. In the course of his work on this endoprosthesis, K. Sivash has solved three fundamental problems:

1) he has developed an original endoprosthesis design fit for a prolonged operation *in vivo*;
2) selected metals for the friction pair (CoCrMo alloy) and immovable load-bearing parts of the prosthesis (titanium) proceeding from the criteria of biocompatibility, electrochemical homogeneity, wear resistance and fatigue strength;
3) elaborated and substantiated the operation schedule with a mechanical type of fixation based on experimental studies of reparative processes during regeneration of the osseous tissue.

In 1966, P.A. Ring created a variant of the total metallic endoprostheses of the hip with split components. Its femoral element represented a modified Moore's prosthesis, while the cup was tapered and had a high-molecular weight polyethylene liner.

In spite of a pleiad of bright names in the history of orthopaedics, J. Charnley, an English orthopaedist (Fig. 1.3), is credited with a key contribution in the development of modern endoprosthetics. Two crucial Charnley's ideas have made a revolution in endoprosthetics: i) implantation of the artificial hip joint using bone cement, and ii) provision of a minimum low friction in the prostheses.

Bone cement used to fix the femoral component has made it possible to spread body weight more uniformly on the cancellous bone of the proximal femur. Charnley considered that bone cement could function not only as an adhesive but also as a means of ensuring high primary stability of endoprostheses. By filling the free volume between the implant and cortical bony layer it transfers the mechanical load over a larger area of the osseous bed. By reducing the femoral head diameter to 22 mm, he diminished the friction moment of the endoprosthesis.

In 1958, Charnley was first to implant a total hip endoprosthesis in which the pelvic component was made of Teflon. The nearest results of the operation were impressive, however fast wearing of the cup and negative reaction of surrounding tissues on Teflon debris made him abandon this material. In 1962 Charnley changed Teflon to ultra-high molecular weight polyethylene that still remains a most reliable polymer frictional material for joint endoprostheses.

Proceeding from the results obtained, Charnley has formulated a principle of a low-friction arthroplasty, to which the surgeons adhere up till now in endoprosthetic structures. Its essence is expressed in two postulates: 1) the endoprosthesis friction pair should have a *min* friction coefficient; 2) the friction torque exerted on the pelvic element and loosening it should be lessened by reducing the head diameter and increasing thickness of the artificial acetabulum [25].

The Charnley total hip endoprosthesis was termed a "Gold Standard" and has not been essentially modified since then. Among Charnley's achievements is the creation of extra-sterile operation units, which lowered the frequency of post-operative suppurative infections from 10 down to 1%. He has also formulated the bases for evaluation of the remote results of endoprosthetics by analyzing post-operative complications and defining new means of their prophylactics.

A Swiss orthopaedist, M.E. Müller (Fig. 1.3), has thereafter modified Charnley's endoprosthesis by using a mantle ring for fixing the pelvic component reattached to the acetabulum by screws. Müller has optimized the shape of the stem fixed intramedullary in the femur and increased the prosthesis head till 32 mm [26].

The endoprostheses of the hip with both cemented and cementless fixation are continuously perfected. In the 1990ies, R. Mathys (Fig. 1.3) elaborated an endoprosthesis having metal stems lined with a polymer and operating in resonance with the bone, called Isoelastic and Isotitan. The manufacture of a broad range of unique in their clinical efficiency modern endoprostheses is connected with the name of R. Mathys. These endoprosthetic designs have become a standard in orthopaedic industry.

The latest structures of endoprostheses make use of up-to-date rustproof steels and alloys, various coatings to provide reinforcement or growing of the bony tissue into the implant, wear-resistant grades of UHMWPE, and novel types of impact-proof ceramics. The contemporary endoprosthetic procedures employ the precision electro polishing, laser treatment, numerical control machines, computer design and processing, and robotics.

It turned out in the history of orthopaedics so that greater attention was drawn to the hip endoprostheses. This is attributed, first of all, to higher traumatizing degree and morbidity of the hip, which has put the problems of its endoprosthetics to a forefront in the second half of the 20th century. Nevertheless, endoprostheses of in fact all human joints have been designed and tested in the course of 1940-1970ies.

Endoprosthetics of the knee has come into practice in the 1950ies. An English surgeon, D.L. Mackintosh started to install acrylic endoprostheses on the tibial plateau in 1958. Two years later, D.C. McKeever suggested to make this prosthesis from a CrCo alloy and attach it to the bone by two blades placed at a right angle to each other.

The bearing distal part of the femur was first replaced in 1964 [27]. The prosthesis was a duocondylar cup attached to a long stem fixed intramedullary in the femur. G. Platt designed in 1967 a cup, which he simply placed on the end of the femur without a fixation. A similar cup was described by G. Gariepy, but secured with screws.

B. Bechtol was the first to construct patella prosthesis [28]. In 1947, the Judet brothers made a total knee endoprosthesis but never applied it in clinical practice. In 1949, an Italian surgeon G.M. Majnoni d'Intignano implanted the first total knee endoprosthesis with articulating joint in the form of a door hinge that allowed for extension-flexion in the knee. Beginning with 1951, B.

Walldius started using analogous prostheses with acrylic elements but changed to metal joints before 1958. The hinge endoprosthesis did not require a strong ligamentary system in the knee but there arose a problem of fixing the prostheses joints in the bones with multicentric axes of motion. As a result, they frequently loosened and broke. Eventually, H.H. Young stated that the prosthetic knee joint should no longer be used [29].

The first nonhinged total knee endoprosthesis was designed by F.H. Gunston in 1968 during his work in England with J. Charnley. Gunston tried to retain the collateral and cruciate ligaments with a minimal resection of the bone. H.W. Buccholz and E. Engelbrecht working in Hamburg developed the "St. Georg" prosthesis in 1969. It consisted of four pieces that contoured the femoral condyles and acted as a sledge for tibial components.

Later on the buttress plate became trough-shaped to raise stability of the sledge prosthesis since a requisite condition for its implantation was the presence of strong ligaments and muscles. In 1971, a group of American orthopaedists (M.G. Coventry and co-workers) designed a total knee endoprosthesis called geometrical.

A new prostheses GUEPAR was designed in the 1970ies in Paris that contained polymethylmethacrylate elements. H. Kaufer with associates attempted to stabilize the endoprosthesis and eliminate the drawbacks of the hinge one by using a ball-socket joint. M. Freeman proposed a structure in which stability was supported by an orginal roller mechanism.

So-called rotary total endoprosthesis of the knee was developed in Hamburg in 1979 and is still successfully used. It consists of two metal parts fixed via stems in the medullary canals of the femur and ankle. The rotary prosthesis is said to join the merits of the hinge and sledge ones. Flexion of the knee and rotation of the foot is provided by a special cross coupling placed in the intercondylar unit found in the femoral part of the endoprosthesis. The rotary prosthesis is most convenient in the case of deficient ligamentary system but strong enough muscles of the femur and ankle [30].

Endoprosthetics of the ankle was first undertaken by Glück in 1890. The operation involved resection of a part of the talus and introduction of tapered prongs into the canal to fix the endoprosthesis. Unfortunately, his prosthesis rapidly loosened. It was not until the 1970ies when Engelbrecht, Freeman, Smith and associates developed more perfect designs of ankle endoprostheses, but none has proved very satisfactory. This is why arthrodesis still remains the most widely employed procedure in destructive arthritic conditions of the ankle [16].

The shoulder joint was first replaced by Pean in 1894. Numerous designs of upgraded humeral endoprostheses appeared by the 1950ies. C.S. Neer, a New York surgeon, designed an original humeral prosthesis, for which G. Stillbrink used a high-density polyethylene cup installed into the glenoid. Further modifications of the humeral prostheses attempted by various groups of orthopaedists and being in-process up till now did not, however, bring any principal changes in this structure.

The elbow endoprosthesis was employed for the first time in 1925 by a French surgeon R. Robineau. It was fabricated of metal and covered by dental vulcanized rubber. The first prosthesis to replace both sides of the elbow joint and fixed in the forearm and humerus was utilized by I. Boerma and D.J. de Waard. Acrylic endoprostheses of the elbow were installed in 1947 and nylon ones in 1954. Custom prostheses appeared in the 1950ies for specific deficits, and in the 1970ies hinged endoprostheses of the elbow were developed [28].

In the early 1970ies R.W. Coonrad started working on a constrained total elbow prosthesis that was eventually modified by B.F. Morrey with associates. A constrained endoprosthesis "Stanmore" became widely adopted soon. Semiconstrained prostheses of R.W. Pritchard and A.P. Schlein make rotation in the elbow possible about two and three axes. Several designs of non-constrained endoprostheses of the elbow were reported in the 1980ies. In 1951, C.R. Carr and J.W. Howard used a metallic cap, while A.B. Swanson introduced a Silastic one instead of a fractured radial head.

Endoprosthetics of the radiocarpal joint was preceded by a total arthroplasty of the wrist made by N. Gschwend and P. Lalive in 1969. In 1973, H.C. Meuli implanted his articulated nonhinged prosthesis of the ball-and-socket trunion design. R.G. Volz developed in 1976 an articulated nonhinged prosthesis. A. Swanson utilized a Silastic prosthesis hinge core similar to his metacarpophalangeal design to serve as a spacer for a resection arthroplasty of the radiocarpal joint.

Endoprostheses of the hand appeared first in late 1940ies. Total replacement of the scaphoid by a metallic prosthesis was reported in 1948 by T.R.Waugh. In 1959 E.W. Brannon and G. Klein developed a metal hinged endoprosthesis for the metacarpophalangeal joint. The following year, A.E. Flatt introduced his hinged prosthesis. It soon became evident that endoprostheses of hard metals damage the articular cartilage of contacting joints leading to fast degenerative changes in the hand. This made Swanson to introduce elastic endoprostheses of the metacarpophalangeal and interphalangeal joints fabricated of Silastic. Analogous prostheses were developed by I.A. Movshovich in 1976 [4]. In 1971, J.J. Niebauer implanted similar endoprostheses of Silastic with a Dacron core. A year later the prostheses based on polypropylene appeared. Successes in total arthroplasty held with polymethyl methacrylate have inspired A.D. Steffe to fix endoprostheses of the hand using bone cement. By the mid-1970ies cemented hand endoprostheses fall into disfavour because of loosening and other complications. Complete enough information on the history of joint endoprosthetics the readers may find in monographs [19, 28, and 31].

A number of endoprostheses that were developed early in the history of arthroplasty are still being used, but most have been supplanted by the prostheses of new generations. In spite of the diversity of endoprosthetic devices, which embody the unique properties of novel materials and new ideas, we would like to emphasize once again that today's successes of endoprosthetics proceed from the pioneering efforts of the innovative surgeons of the last century who have contributed to the early development of endoprosthetics as an actual trend of orthopaedics.

1.3 SURGICAL OPERATIONS OF JOINT ENDOPROSTHETICS

Endoprosthetics of joints can be justifiably called today a mass operation. This method of reconstructive surgery has not, however, become a routine procedure but still requires a creative approach to all its stages. As clinical experience has shown, successes of endoprosthetic operations depend on a number of interrelated factors. M. Muller has noted that good endoprosthetics differs from a plain endoprosthetics by a thousand of minor aspects [26]. In this connection, endoprosthetic methods of joints are being developed presently using robotics [32]. The factors accompanying endoprosthetic operations can be subdivided into three groups [33].

1. Engineering factors dealing with endoprosthesis designs, quality of chosen materials and biomechanical parameters of their installation. The endoprosthetic elements forming a friction joint should be secured immovably in the mating bony ends in order to restore the centre of rotation of a natural joint. Such characteristics like technical perfection of endoprostheses as an articulating joint and fixation reliability of its elements are strongly important.
2. The operative intervention at joint endoprosthetics is a concluding procedure on the replacement of pathologic elements by endoprostheses. Its main stages are the choice of the endoprosthesis design, its dimension type and fixative means proceeding from the radiographs of the joint to be operated. Surgical access to the joint is defined in terms of atraumatism and convenience for the operation. The questions of fabricating non-standard tools are analyzed.

   During the operation a surgeon should estimate the quality of primary fixation of the endoprosthesis and strictly keep to the implantation procedure. The success of the operation depends largely on the experience of the leading surgeon, coordination of the operation team including assistants, anaesthesiologists, operation nurses and some other factors.
3. The biological aspects of endoprosthetics include biocompatibility of the implants, biophysical mechanisms of a stiff primary fixation and secondary osteointegration of the prosthesis parts, the kinetics of debris generation and their toxicity. Service life of the implanted endoprosthesis depends on the distribution of mechanical loads in places of fixation, friction joints and the

materials of endoprosthesis elements. To this also belong the responses of the bony body and surrounding soft tissues on the loading.

Most of these factors are set forth in detail in the chapters to follow. In continuation of this chapter the terminology and most important medical aspects of endoprosthetics concerning the preoperative planning, postoperative and follow-up results will be touched upon.

Preoperative planning envisages some variants of the behaviour of a surgeon and his team prior and after the operation as well as certain systems of services and rehabilitation procedures of the patient intraoperatively and after the operation. Besides, it is necessary to schedule preparation of the patient for the operation, and the operation itself. This includes the choice of dimension-type of the prosthesis, the technique of performing the operation in stages, anaesthesia and prescribed medicines. Along with this, a basic variant of the postoperative cure and medical rehabilitation are selected. A system of a long-term guide over the life-supporting systems of the operated patient is also planned.

Preparation for the operation begins with the estimation of the total condition of the sick and the extent of injure of the joint to be operated. The condition of a patient before the operation is estimated on the base of analyzing the results of clinical laboratory investigations of biological fluids, functional state of critical organs and systems, the presence and severity of coexisting diseases.

The radiographic examination of a damaged joint is a major and requisite stage of the preoperative planning for endoprosthetics. Such modern methods of diagnostics as computer tomography, magnetic resonance and ultrasonic examinations, and some other are auxiliary procedures to be employed if required. Radiograms of joints are made in the front and side views, or in other planes, if necessary. The results of radiography visualize the extent of osteoporosis of articulating bone ends or spreading of the tumour in case of malignant processes. Proceeding from the results, the dimension-type and fixation means of the endoprosthesis are chosen.

The complete clinical, laboratory and radiographic examination of the patient also serves to evaluate the operation risk and to plan countermeasures for avoiding possible complications. The preoperative stage includes sanitation of the oral cavity and nasopharynx, removal of carious teeth. To prevent festering, antibiotics of a broad spectrum are indicated, as well as heart and sedative medicines.

Endoprosthetic operations are performed, as a rule, by a regular team of surgeons, which is headed by a most experienced practitioner in reconstructive operations. The assistants and nurses should be aware of all stages of the operation to be of aid for the leading surgeon. Operations on replacement of joints are allocated a special operating room meeting strict aseptic rules.

Anaesthesia of operations on joint replacement utilizes the whole stock of means of modern anaesthesiology and intensive therapeutics. The leading centres of endoprosthetics commonly utilize the following types of anaesthesia [34, 35]:

- general anaesthesia (narcosis) using inhalation anaesthetics and pulmonary ventilation;
- general anaesthesia with guided hypotonia to reduce blood losses during operation;
- less often regional anaesthesia, this means blocking of painful pulses from the operative zone to the central nervous system.

The choice of anaesthesia depends on the anatomic zone to be replaced and anticipated duration of endoprosthetic operation. Regional anaesthesia is attained by rendering anaesthetic to nerve-endings and trunk in the operative site (local anaesthesia) or to the spinal cord (spinal anaesthesia).

Medicamental supply of endoprosthetic operations has its peculiarities since these operations are most traumatic in orthopaedics and are accompanied by a considerable blood loss. The surgeons encounter the problem of how to stop bleeding intraoperatively and replace the blood loss by the autologous blood. The general state of the patients for orthopaedic diseases usually admits preoperative blood donation for further autohemotransfusion, i.e. transfusion of autologous blood. Autohemotransfusion is administered in two variants: in the form of blood conservation a few days prior to the operation or as reinfusion after a special treatment of the blood collected from the wound [36, 37]. Regional anaesthesia is characterized by reduced perioperative blood losses attained by a constant refinement of the fast and low-traumatic operation technique, careful treatment of adjacent tissues. The patient is additionally injected with plasma and other indicated fluids.

The infection of wounds mainly occurs intraoperatively in case the endoprosthesis or the open wound sterility is violated. Yet, perioperative bacteria ingress does not lead obligatory to the wound infection. For the prophylactics of suppurative complications in the operative site a special antimicrobial therapy is exercised. Most often, various schemes of injecting a wide-range of antibiotics are utilized. Today they are the antibiotics of a cephalosporin series of the 3rd and 4th generations.

The prophylactic of thromboembolic complications (connected with artery occlusion by clots) is administered using the heparin-based preparations. They are usually injected 12 h before the operation and are kept on for 5–7 days. The risk of thromboembolic complications rises in the event of abrupt refusal from anticoagulant therapy [38]. An elastic bandage is applied on both the operated and healthy feet as a supplementary measure in endoprosthetics of lower extremities.

The operation technique includes the choice of an optimum access to the operated organ, intraoperative estimate of the pathologic process, final decision-making on the endoprosthesis design, fixation method and implantation methodology. For prophylactics of hematomas, the operative site is actively drained using vacuum systems (24 till 72 h after the operation). A surgeon checks the accuracy of primary fixation of each endoprosthesis element and its stability as a whole intraoperatively.

Postoperative therapy consists of the medicamental curing to avoid suppurative or thromboembolic complications, medical supervision of all crucial organs and systems. The therapeutic physical training with/without special-purpose trainers is an integral part of the rehabilitation period starting from the first days after the operation. The patient is moved into a specialized rehabilitation ward after healing of the operative wound and removal of stitches (9-10 days after the operation) for final adaptation.

The postoperative management is administered by specialists of clinic where the endoprosthetics was performed. The first stage lasts 6 to 12 months at a set periodicity and continues in the course of patient's life with specified intervals. During this period the clinicians traditionally evaluate the functions of replaced joints to reveal the signs of instability by clinical or radiological means, computer or magnetoresonance tomography. Modern approaches presume in addition the estimation of tribological aspects of the endoprosthesis functioning (see Ch. 5). Further in this chapter an information is given on the methods of estimating the results of endoprosthetics of various joints.

1.4 RESULTS OF JOINT ENDOPROSTHETICS

The hip undergoes endoprosthetics using various types of prostheses depending on the pathologic state of the joint. Various rating methods of estimating short-term and long-term endoprosthetic results enable comparisons of the efficacy of any operation.

The rating scale for hip functions was developed first by M.D. Aubigne and M. Postel in 1954 and modified by Charnley in 1972 (Table 1.1). According to Charnley, most patients for whom hip endoprosthetics is indicated have Grade 3 or 4 pains [39]. Grade 1 and Grade 5 are too mild to consider for operative treatment. Postoperatively, Grade 6 and 5 are considered good and excellent results. These rating scales are used now throughout the world.

Analogous 10-point systems were developed in the USA [40, 41]. In 1969, W.H. Harris reported in his work an original rating system [42] in which the critical criteria were pain and functional capacity of the hip (Table 1.2).

Additionally, functional deformities and range of motion in the joint are estimated. In the years to follow, Harris' system underwent various modifications and was much complicated [43].

Table 1.1
Numeric classification of the clinical state of the hip [39]

Grade	Pain	Motion range, degrees	Walking
1	Severe and spontaneous	0–30	Few steps or bedridden Two sticks or crutches
2	Severe on attempting to walk	60	Time and distance very limited without sticks
3	Tolerable, permitting limited activity	100	Limited with one stick Difficult without a stick Able to stand long periods
4	Only after some activity	160	Long distances with one stick Limited without a stick
5	Slight or intermittent. Pain on starting to walk but less with normal activity	210	No stick but a limp
6	No pain	260	Normal

Radiologic evaluation of loosening of the total hip prosthesis was proposed by T.A. Gruen for the femoral stem [44] and by Charnley for the acetabular component [45]. Gruen singled out 7 zones in the proximal femur round the endoprosthesis stem (Fig. 1.4). Charnley has divided the acetabulum round the pelvic component into three zones (Fig. 1.5).

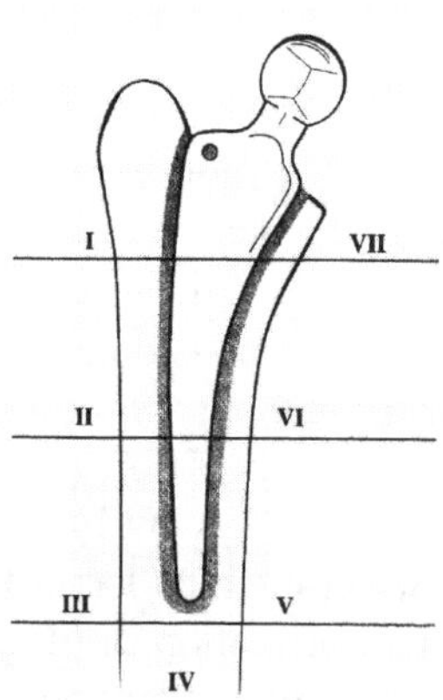

Fig. 1.4. Gruen zones of the femoral stem

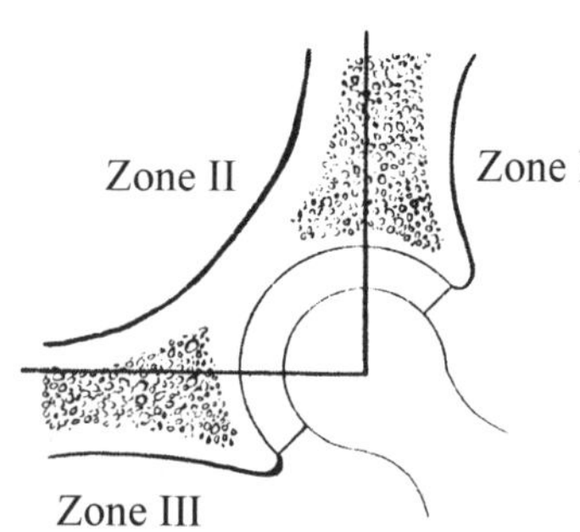

Fig. 1.5. Division of circumference of acetabulum into three zones

Table 1.2
Harris' hip rating system [42]

I. PAIN (*max* 44 points)		
Category	Symptoms	Points
A	None or ignores it	44
B	Slight, occasional	40
C	Mild pain, some limitation of ordinary activity	30
D	Tolerable, but makes concessions to pain; limitation of ordinary activities	20
E	Marked pain, serious limitation of activities	10
F	Totally disabled, crippled pain in bed, bedridden	0
II. FUNCTION (*max* 47 points)		
A. Gait (*max* 33 points)		
1. Limp		
a	None	11
b	Slight	8
c	Moderate	5
d	Severe	3
2. Support		
a	None	11
b	Cane for long walks	7
3. *Max* distance at walking		
a	Unlimited	11
b	Six city blocks	8
c	Two or three building blocks	5
d	Able to reach the door	2
e	Unable to walk	0
B. Activity (*max* 14 points)		
1. Stairs		
a	Normally without a rail	4
b	Normally using a rail	2
c	In any manner	1
d	Unable to do stairs	0
2. Shoes and socks		
a	With ease	4
b	With difficulty	2
c	Unable	0
3. Sitting		
a	Comfortably in ordinary chair, 1 hour	5
b	On a high chair for ½ hour	3

c	Unable to sit	0
4. Enter public transportation		1
III. DEFORMITY		
Absence of deformity 4 points are given if the patient demonstrates flexion contracture <30°, adduction <10°, internal rotation <10°, and limb length discrepancy <3.2 cm		
IV. RANGE OF MOTION (*max* ~ 5)		
Index values are determined by multiplying the degrees of motion possible in each arch by the appropriate index and summation.		
A. Flexion:		
a	0–45° × 1.0	
b	45–90° × 0.6	
c	90–110° × 0.3	
B. Abduction:		
a	0–15° × 0.8	
b	15–20° × 0.3	
c	>20° × 0	
C. External rotation in extension:		
a	0–15° × 0.4	
b	>15° × 0	
D.Internal rotation in extension:		
a	any × 0	
E. Adduction:		
a	0–15° × 0.2	
Overall rating for range of motion equals to a sum of points (A + B + C + D + E) × 0.05		

Loosening of the acetabulum is characterized by a shift and fracture of the cup or its fixtures, or shifting of the prosthesis head centre relative to the cup due to wear. Evidence of probable loosening for cemented prostheses is a continuous radiolucent line (up to 2 mm wide) at the bone-cement mantle interface. Lines at the bone-cement interface that are progressive in either width or extent are considered to represent possible decreasing of stability of cup's fixation. This method is recommended as a universal estimate of instability in hip endoprostheses of all kinds [46].

As an example of total hip endoprosthetics using Charnley's "Gold Standard" we may cite the results 15 to 21 years after surgery [47]. Postoperatively the functions of the limb improved essentially in all observed patients. In a follow-up 85% of them had no pain at walking, another 11% had only occasional discomfort. Deterioration of the joint function was associated with advancing age of the patients whose mean age constituted 53 years at the time of operation. In spite of this fact, 68% of the patients had nearly full range of motion and another 10% had normal motion.

The knee joint is the second in frequency of endoprosthetic operations after the hip one.

Rating systems for total knee arthroplasty are not as sophisticated as for the total hip replacement. The most popular is the first rating scale HSS (Hospital for Special Surgery, New York) introduced in 1976 [48] (Table 1.3). In 1989, the American Knee Society introduced a new total knee rating system (Table 1.4). Its important features include separation of the artificial knee rating by pain sensation from functional assessment. Besides, it envisages deduction of points in both parts of the system for overall sums. The Knee Society has developed also a scoring system for radiographic evaluation of knee endoprosthetics [55]. Wide adoption of these two systems is able to facilitate the comparison of reports on total knee arthroplasty from different clinics.

The results reported by a leading surgeon of ENDO clinics in Hamburg, G.W. Baars, were summarized for 1837 patients within the period between 1981 till 1989. The system used was a rotary knee endoprosthesis "Endo" implanted with preserved patella [30]. The follow-up observation time made up 2 till 12 years with patients 22–99 years (mean age 66 years). The postoperative functional results were satisfactory: 98% of the operated joints showed full extension, the mean index of flexion was 110°, being an 18° improvement. During 7 years of follow-up observation aseptic loosening of the endoprostheses scored 5.8 %, 3 % of which were replaced and 2.8 % were limited to surgical corrections without the replacement. Various infections (1.9 %) were eliminated by a suction-lavage drainage or replacement of the endoprosthesis with fixation by a cement containing antibiotics.

The implanted endoprostheses remained serviceable (survival rate) in 97% for 7-year follow-up results, including the patients above 66 years it constituted 98% and below 66 – 93%.

Table 1.3
Knee rating scale - HSS [48]

PAIN (30 points)	
None	30
No pain at walking	15
Mild pain at walking	10
Moderate pain at walking	5
Severe pain at walking	0
No pain at rest	15
Mild pain at rest	10
Moderate pain at rest	5

Severe pain at rest	0
FUNCTION (22 points)	
Walking and standing unlimited	22
Walking distance of 5–10 blocks and standing ability intermittent (< 30min)	10
Walking 1–5 blocks and standing ability up to 30 min	8
Walking less than 1 block	4
Cannot walk	0
Climbing stairs	5
Climbing stairs with support	2
Transfer activity	5
Transfer activity with support	2
RANGE OF MOTION (18 points)	
1 point for each 8° of arch of motion to a maximum of 18 points	
MUSCLE STRENGTH (10 points)	
Excellent: cannot break the quadriceps power	10
Good: can break the quadriceps power	8
Fair: moves through the arch of motion	4
Poor: cannot move through the arch of motion	0
FLEXION DEFORMITY (10 points)	
No deformity	10
Less than 5°	8
5–10°	5
More than 10°	0
INSTABILITY (10 points)	
None	10
Mild: 0–5°	8
Moderate: 5–10°	5
Severe: more than 15°	0
SUBTRACTION	
One cane	1
One crutch	2
Two crutches	3
Extension lag in 5°	2
Extension lag in 10°	3
Extension lag in 15°	5
Each 5°of varus	1
Each 5° of valgus	1

Table 1.4

Knee score – the Knee Society [49]

Patient category

A. Unilateral or bilateral (opposite knee successfully replaced)
B. Unilateral, other knee symptomatic
C. Multiple arthritis or medical infirmity

PAIN	Points
None	50
Mild or occasional	45
• stairs only	40
• walking and stairs	30
Moderate	
• occasional	20
• continual	10
Severe	0
RANGE OF MOTION	
5° = 1 point	25
STABILITY (*max* displacement in any position)	
Anteroposterior	
• < 5 mm	10
• 5–10 mm	5
• 10 mm	0
Mediolateral	
• < 5°	
• 6–9°	15
• 10–14°	10
• 15°	5
	0
Subtotal:..	
DEDUCTION (minus)	
Flexion contracture	
• 5–10°	2
• 10–15°	5
• 16–20°	10
• < 20°	15
Extension lag	
• < 10°	5
• 10–20°	10
• >20°	15

Alignment	0
• 0–4°	3 points each degree
• 5–10°	3 points each degree
• 11–15°	0
Other	20
Total deduction:..	
Total score:.. *(If total is a negative number, score is 0)*	
FUNCTION	
Walking	
• unlimited	50
• 10 blocks	40
• 5–10 blocks	30
• < 5 blocks	20
• housebound	10
• unable	0
Stairs	
• normal up and down	50
• normal up and down with rail	40
• up and down with rail	30
• up with rail; unable down	15
• unable	0
Subtotal:..	
Deduction (minus)	
Cane	5
Two canes	10
Crutches	20
Total deduction:..	
Function score:..	

The ankle joint is most successfully replaced by endoprostheses in elderly patients who are sedentary or suffer from systemic arthritis with advanced destruction of the tibiotalar joint.

The quality of gait after the total ankle arthroplasty depends on many factors including the condition of the patient, strength of the muscles and the implant type and alignment, its fixation method, and surgical approach [51]. The ideal arthroplasty provides: (1) a stable pain-free ankle offering 10° of dorsiflexion and 20° of plantar flexion; (2) satisfactory durability of the prosthesis for normal activities in daily living; (3) *min* bony resection; (4) stability of the joint under normal mechanical loading conditions [16]. Total ankle endoprosthetics commonly gives most satisfactory early follow-up results.

The experience of ankle endoprosthetics can be summarized in the following recommendations. In contrast to endoprosthetics of other joints of lower extremities, total ankle arthroplasty requires more careful choice of the prosthesis design, its dimension type, thorough preoperative planning, and a patient who will comply with the postoperative regime. The endoprosthesis structure (preferably custom-made) is selected so as to minimize bone stresses round the fixation. Cemented fixation is undesirable in this case. The biological type of fixation is preferable to provide regulation of the prosthesis centre of rotation intraoperatively.

Worsening of the state of artificial limbs in 2–6 years the follow-up period (radiographic translucency at the bone-cement interface and endoprosthesis loosening in the talus) is a reason of pessimistic approaches to this operation [52].

The shoulder joint subjected to endoprosthetics discharges its functions the better the closer the normal biomechanics of the humerus is simulated by the prosthesis.

The state of the shoulder before and after endoprosthetics can be estimated using a system (Table 1.5) devised by the American surgeons [53]. It gives an estimate of the pain, range of both active and passive motion, strength of the muscles brining the humerus in motion and its functions.

As observations have shown, loosening of the glenoid component presents a major problem of shoulder endoprostheses. This can be traced on radiograms of the implanted endoprostheses. The appearance of translucent lines at the bone-cement interface has been recorded in 33–80% patients according to the reports of American surgeons [53]. As for the clinical examinations, they support loosening of the glenoid component only in a limited number of cases. Therefore, the cause of the early appearance of translucent lines is often considered as imperfection of surgical techniques, in particular, of the cementing method.

Long-term investigations have defined the next mechanism of failure of the implanted shoulder prostheses during functioning [54]. The prosthesis component with a spherical head fixed in the humerus migrates proximally while wearing the glenoid cavity. This changes essentially the kinematics of the shoulder. Under the upward force developed by the deltoid and meeting no resistance, the eccentrically directed compressive strains affect the glenoid. They evoke loosening of the cup fixed in the glenoid cavity and augment the sign-variable stresses at the bone-cement interface.

Postoperative rehabilitation is just as important as the surgery itself. For a successful outcome the patient should start active motions with the artificial limb two weeks after the operation and keep on under a supervision of a physiotherapist for not less than two months [53]. The best results in reconstruction of the functions are evident 1–2 years of exercises after the shoulder replacement.

The elbow joint as an object of total endoprosthetics is characterized by less evident successes as compared to the hip and knee mainly because of a limited number of such operations.

The efficacy of elbow arthroplasty is related to the optimal choice of the endoprosthesis, on the first place.

Table 1.5
Shoulder examination data form [53]

I	*Pain* (5 – none, 4 – slight, 3 – after unusual activity, 2 – moderate, 1 – marked, 0 – complete disability)	
II	*Shoulder motion*	
	A.	Patient sitting 1. Active total elevation of arm, degrees 2. Active external rotation with arm at side, degrees 3. Active rotation – 90 degrees abduction, degrees 4. Passive internal rotation (segment reached is recorded)
	B.	Patient supine 1. Passive total elevation of arm, degrees 2. Passive external rotation with arm at side, degrees 3. Passive external rotation – 90 degrees abduction, degrees 4. Passive internal rotation – 90 degrees abduction, degrees
	C.	Impingement 1. Painful arc of motion 2. Relieved by subacromial xylocaine injection 3. Crepitus
III	*Strength* (5 – normal, 4 – good, 3 – fair, 2 – poor, 1 – trace, 0 – paralysis)	
	A.	Anterior deltoid
	B.	Middle deltoid
	C.	External rotation
	D.	Internal rotation
	E.	Trapezius
	F.	Triceps
	G	Biceps
IV	*Stability* (5 – normal, 4 – apprehension, 3 – rare subluxation, 2 – recurrent subluxation, 1 – recurrent dislocation, 0 – fixed dislocation)	
	A.	Anterior
	B.	Posterior
	C.	Inferior
	D.	Superior

V	*Function* (4 – normal, 3 – mild compromise, 2 – difficulty, 1 – with aid, 0 – unable)
	1. Use back pocket 2. Rectal hygiene 3. Wash opposite underarm 4. Eat with utensil 5. Comb hair 6. Use hand, arm, shoulder 7. Carry 10-15 lb/arm at side 8. Dress 9. Sleep on shoulder 10. Pulling 11. Use hand overhead 12. Throwing 13. Lifting 14. Do usual work 15. Do usual sport
VI	*Patient response* (3 – much better, 2 – better, 1 – same, 0 – worse)

For the rheumatoid arthritis, two types of endoprostheses are used commonly to ensure rotation of the forearm in three mutually perpendicular axes: (1) unconstrained, and (2) semiconstrained. The former types are implanted with rather strong medial and lateral ligamentary system preserved in the elbow joint. Whatever strict requirements to the surgical technique may be, they can not avert the probability of postoperative complications, subluxation and dislocation of the implants. There can be, besides, a haphazard of injuring the ulnar nerve during lateral surgery. As an example, we will cite the results of a survey of 202 patients 2 to 15 years after the total endoprosthetic operation (prostheses with a capitellocondylar component) [55]. The mean range of motion at flexion was 30° till 138° with supination (external rotation of the forearm with palm up) 64° and pronation (reverse rotation) 72°. Radiographic translucent lines were recorded in 8 humeral and 23 elbow components of the prostheses. The causes of the revision operations were 4 because of aseptic loosening, 2 due to instability, and 3 of infections. 24 cases failed because of temporary and 10 persistent paralysis. Instability of the implant was observed in 17 patients, 7 of which had evident dislocation and 10 subluxation. In spite of a high rate of early complications, follow-up results of this series of operations are considered good.

The second type of endoprostheses (partially constraining rotation) has proved to show good early postoperative results. These implants allow soft tissues evenly distribute the load in the artificial joint. The reduced stresses at the bone-cement interface diminish the probability of loosening and wear rate of polyethylene components. 137 patients examined within the period of 2–13 years after the endoprosthetics have shown the following results. The mean range of motion at flexion in elbow was 103°. The results of primary operations were satisfactory in 125 patients. Three of them underwent revision surgery. 7 deep suppuration cases were recorded in this series. No revision operations were recorded because of loosening, whereas 8 were due to dislocations necessitating the replacement of polyethylene liners or axial components of the endoprostheses. Among other complications 3 cases of persistent neuropathy of the ulna nerve were reported and 7 supracondylar fractures of the humerus. The translucent line was recorded in 10 patients [56].

The operations for the post-traumatic arthrosis of the elbow present a serious problem for the cases of failed surgery because of bony and soft tissues deficiency, strangulation or injury of the ulnar nerve. Semiconstrained endoprostheses or the custom-made ones are recommended for these operations. The operation should be thoroughly planned. It is reported in [57] that out of 54 total elbow replacement operations for the post-traumatic arthrosis 47 were a success. 13 revision operations were made within 2 to 14 years of observation (11 were successful), 13 implants showed loosening and one was infected. The latter case and two revision replacements necessitated a resection arthroplasty. The analysis has proved that total endoprosthetics at post-traumatic arthrosis may be indicated for the patients above 60 as a salvage procedure.

The radiocarpal joint is replaced rather seldom. When the endoprosthetics is recommended the surgeon should consider the patient's age, occupation, leisure activities, area of bone and joint involvement, the cause and severity of destructive changes in the wrist, and associated injury of muscles, tendons and joints. When the primary destructive changes are confined to the radiocarpal and midcarpal joints, and resection or fusion of the wrist bones are no longer useful, there remain three alternatives: resection arthroplasty, endoprosthetics, and total wrist arthrodesis.

Endoprosthetics of the radiocarpal joint can be performed by the implantation of elastic interpositional spacers between the bones or substitution of the joint by a total endoprosthesis.

Implantation of elastic spacers is a variant of the resection arthroplasty. It reconstructs the configuration of the radiocarpal joint, assists in aligning and formation of a joint capsule round the implant. As radiographic investigations have demonstrated, not only elastic spacers alone contribute to the motion of the artificial joint but sliding of its unattached to bone ends in and out of the intramedullary canal as well. Early results of such operations display a significant pain relief, increased range of motion and correction of the deformed joints. However, radiographic examination of 71 patient 5 years after the operation showed injury of the joint in 70%, although in practice only 20% of the implants turned to be broken. 60% of the patients had good clinical follow-up results [58]. This proves that like in the situation with other joints, radiological evidences of the impaired state of the radiocarpal joint with implanted flexible spacers do not correlate directly with the data of clinical examinations.

One of the problems of the total wrist arthroplasty is how to preserve the centre of rotation in the artificial limb like that in the natural one. Their misalignment, especially at the outset of the total wrist endoprosthetics, has lead to shifting and deformations at flexion of the ulnar joint. Early results of the total wrist endoprosthetics are analogous to those reached at implanting flexible spacers. With time almost half of the patients start to experience various problems associated with loosening of the endoprosthesis stems in their fixation sites and overloading on the distal radius. Subluxation of the distal endoprosthesis component results in strangulation of the medial nerve and breakage of the flexor tendons. The two major signs of good total wrist endoprosthetic results are the normal functioning of the extensor, as well as correction of the palmar and elbow contracture (stable bounding of motion). Good results are reported in roughly 60% of this type of operations.

The metacarpophalangeal joints are indicated for endoprosthetics at persistent pains and unresponsive to medical management destruction of the metacarpophalangeal and interphalangeal articulations. The alternative operations are the ones on soft tissues of fingers and arthrodesis. Endoprosthetics of the finger joints employs either flexible implants or hinged total metacarpopghalangeal and interphalangeal endoprostheses.

The flexible implants present an auxiliary means for the resection arthroplastics serving to enlarge the range of motion in the phalanges and improve their aesthetics. The flexible core is not fixed in the medullary canal, therefore flexion-extension in the joint is realized through both flexion of the interphalangeal part of the core and its sliding intramedullary. The available clinical data reports 70–80% overall subjective satisfaction of such operations. 2.5–5 years after the endoprosthetics the range of active motion in the patients made up 40%. Increased power grip has been recorded in 70% of patients. The probability of fatigue-induced failures of the rods does not surpass 5%. The follow-up radiographic studies have identified cortical erosion in 40–100%, and bone production (ossification) on articulating bone ends in 35–50 % of patients [59].

Total endoprosthetics of the metacorpophalangeal joints is often associated with the problems of fracture and shifting of the loaded components of prostheses, osseous erosion and loosening of fixed in the bones parts. A report of up to 20 years of follow-up data on Flatt's metallic hinged prosthesis revealed stem perforation of the metacarpal bone (44%) and perforation of the proximal phalanx (59%) [60]. Later results of total endoprosthetics using the metal-UHMWPE as a friction pair did not show improvements but lead to unpredictable results. There exists an opinion that until total metacarpophalangeal endoprosthetics provides stable results comparable to those of flexible implant arthroplasty, it should be better entrusted to specialized medical centres with staffs having broad experience in joint arthroplasty of the hand.

Implantation of hinged interphalangeal endoprostheses suffers from the same problems as the total arthroplasty of the metacarpophalangeal articulations. They are high failure probability, loosening and migration of the prostheses. To avoid these, it is necessary to strictly observe indications for the operation and we believe that the interphalangeal hinged endoprostheses may become similarly successful like the operation with implantation of flexile spacers.

It should be noted in conclusion that biomechanical observation of the operated joints is highly important for the objective estimation of endoprosthetic results. In view of today's tendency of elevated responsibility for the patient it will be incompetent to examine solely the operated joint in the post-operative period. A long-term supervision of the patients with implanted joints should include a qualified control over their rehabilitation, which involves orthopaedists along with other medical professionals.

◆◆◆

Endoprosthetics of joints is a science-intensive interdisciplinary sphere of arthrology. A success of each operation is interrelated with numerous factors including engineering, medical, biological and other. Modern advances in endoprosthetics proceed from the ideas and experience of the pioneering orthopaedists of the last century. The basic designs of endoprostheses and their implantation evaluation procedures undergo a continuous refinement. Just these approaches have made possible the estimation of operation efficacy by comparing the endoprosthetic operations performed with different kinds of endoprostheses, medical techniques and condition of the joints.

References:

1. Human anatomy update. Ed. by E.N. Marieb, and J. Mallat, 3rd ed., New York, Benjamin Cummings, 2002, 844 p.
2. Biomechanics. New York, SAE Int., 2003, 170 p.
3. Fundamentals of hand and wrist Imaging. Ed. by C.Van Kuijk, G. Guglielmi, and N.K. Genant. Heidelberg, Springer-Verlag, 2001, 500 p.
4. Movshovich I.A., and Vilensky V.Ya. Polymers in traumatology and orthopaedics. Moscow, Medicine, 1978, 320 p.
5. Neverov V.A. and Zakari S.M. Revision endoprosthetics of the hip. St. Petersburg, 1997, 112 p.
6. Wroblewski B.M. 15–21 year results of the Charnley low-friction arthroplasty. *Clin. Orthop.*, 1986, V. 211, p. 30–35.

7. Sokolovsky A.M. and Kryuk A.S. Surgical treatment of hip diseases. Minsk, Navuka i Tekhnika, 1993, 248 p.
8. Bombelli R. Osteoarthritis of the hip. Berlin, Springer-Verlag, 1976, 450 p.
9. Divakov M.G. Aseptic necroses of bones and substantiation of their curing methods. Dr. Med. Sci. Thesis, Moscow, 1991.
10. Scuderi G.S. and Tria A.J. Surgical techniques in total knee arthroplasty. Berlin, Springer-Verlag, 2000, 785 p.
11. Voronovich I.R. Intraarticular injuries of the knee (condylar fractures, breakage of tendons, menisci). Dr. Med. Sci. Thesis, Minsk, 1968.
12. Knee surgery. Complications, pitfalls, and salvage. Ed. by M.M. Malek. Heidelberg, Springer-Verlag, 2001, 507 p.
13. Strobel M.J. Manual of arthroscopic surgery. Berlin, Springer-Verlag, 2001, 1074 p.
14. Mironova Z.S., and Falex F.Yu. Arthroscopy and arthrography of the knee. Moscow, Medicine, 1982, 108 p.
15. Perlick L., Bathis M., Tingart M., et al. Minimally invasive unicompartmental knee replacement with a nonimage-based navigation system. *Int. Orthop.*, 2004, V. 28, p. 193-197.
16. Bone and joint disorders of the foot and ankle. Ed. by M. Bouysset. Heidelberg, Springer-Verlag, 1998, 357 p.
17. Szendroi M. Advances in orthopaedic oncology. *Int. Orthop.*, 2002, V. 26, p. 195-196.
18. Gluck Th. Die Invaginationsmetode der Osteo- und Arthroplastic. *Berlin Klin. Wochenschr.*, 1980, No. 33, S. 732 – 757.
19. Shans A.R. Historical milestones in the development of modern surgery of the hip joint. In: The Hip. St. Louis, C.V. Mosby Co., 1976, p. 12–21.
20. Moore A.T. Hip joint surgery: an outline of progress made in the last forty years. Columbia, S.C., 1963, 402 p.
21. Judet J. and Judet R. The use of an artificial femoral head for arthroplasty of the hip joint. *J. Bone Joint Surg.*, 1950, V. 33 B, p. 166–173.
22. Thompson F.R. Two and half years experience with a Vitallium intramedullary hip prosthesis. *J. Bone Joint Surg.*, 1954, V. 36 A, p. 489–502.
23. Haboush E.J. A new operation for arthroplasty of the hip based on biomechanics, photo elasticity, fast-setting dental acrylic, and other considerations. *Bull. Hosp. Joint Dis.*, 1953, No. 14, p. 242–251.
24. Sivash K.M. Alloplastics of the hip. Moscow, Medicine, 1967, 150 p.
25. Charnley J. Low friction arthroplasty of the hip. Berlin, Springer-Verlag, 1979, 355 p.
26. Muller M.E. Total hip prostheses. *Clin. Orthop.*, 1970, V. 72, p. 46-68.
27. Jones W.N., Aufrance O.E., and Kermond W.L. Mould arthroplasty of the knee. *J. Bone Joint Surg.*, 1967, V. 49A, p. 1022–1030.

28. Mc Elfresh E. History of arthroplasty. In: Total joint replacement. Ed. by W. Petty. Philadelphia, W.B. Saunders Co., 1991, p. 3–18.
29. Young H.H. Use of a hinged vitallium prosthesis (young type) for arthroplasty of the knee. *J. Bone Joint Surg.*, 1971, V. 53A, p. 1658–1659.
30. Engelbrecht E. The tibial rotating knee prosthesis "Endo" model: surgical technique. *J. Orth. Surg. Techn.*, 1987, V. 3, No. 2, p. 83–98.
31. Dowson D. History of tribology. Leeds, Leeds Univ., 1997, 720 p.
32. Jerosch J., Peuker E., von Hasselbach C., et al. Computer-assisted implantation of the femoral stem in THA – an experimental study. *Int. Orthop.*, 1999, V. 23, p. 224-226.
33. Morscher E.W. Endoprosthetic. Berlin, Springer-Verlag, 1995, 432 p.
34. Surgery. Ed. by J.A. Norton, R.R. Bollinger, A.E. Chang, et al. Heidelberg, Springer-Verlag, 2000, 1281 p.
35. Kustov V.M. Tactics of anaesthetic provision of endoprosthetic operations on the hip. *Traumatology and orthopaedics*, 1994, No. 5, p. 17–26.
36. Earnshaw P. Blood conservation in orthopaedic surgery: the role of epoetinalfa. *Int. Orthop.*, 2001, V. 25, p. 273–278.
37. Sinha A., Sinha M., and Burgert S. Reinfusion of drained blood as an alternative to homologous blood transfusion after total knee replacement. *Int. Orthop.*, 2001, V. 25, p. 257–259.
38. Borghi B. and Casati A. Thromboembolic complications after total hip replacement. *Int. Orthop.*, 2002, V. 26, p. 44–47.
39. Charnley J. The long-term results of low-friction arthroplasty of the hip performed as a primary intervention. *J. Bone Joint Surg.*, 1972, V. 54B, p. 61–76.
40. Wilson P.D., Amstutz H.C., Czerniecki A., et al. Total hip replacement with fixation by acrylic cement. *J. Bone Joint Surg.*, 1972, V. 54A, p. 207–236.
41. Amstutz H.C., Thomas B.J., Jinnah R., et al. Treatment of primary osteoarthritis of the hip. A comparison of total joint and surface replacement arthroplasty. *J. Bone Joint Surg.*, 1984, V. 66, No. 2, p. 228–241.
42. Harris W.H. Traumatic arthritis of the hip after dislocation and acetabular fractures: treatment by mold arthroplasty. *J. Bone Joint Surg.*, 1969, V. 51A, p. 739–755.
43. Petty W. Results of primary total hip arthroplasty. In: Total Joint Replacement. Ed. by W. Petty. Philadelphia, W.B. Saunders Co., 1991, p. 315–348.
44. Gruen T.A., Mc Neice G.M., and Amstutz H.C. "Model of failures" of cemented stem-type femoral components. *Clin. Orthop.*, 1979, V. 141, p. 17–27.
45. De Lee J. and Charnley J. Radiological demarcation of cement sockets in total hip replacement. *Clin. Orthop.*, 1976, V. 121, p. 20–32.

46. Jonston R.C., Fitzgerald R.H., Harris W. H., et al. Clinical and radiographic evaluation of total hip replacement. *J. Bone Joint Surg.*, 1990, V. 72A, p. 161–168.
47. Wroblewski B.M. 15–21 year results of the Charnley low-friction arthroplasty. *Clin. Orthop.*, 1986, V. 211, p, 30–35.
48. Insall J.N., Ranavat C.S., Aglietti P., et al. A comparison of four models of total knee-replacement prostheses. *J. Bone Joint Surg.*, 1976, V. 58A, p. 754–765.
49. Insall J.N., Dorr L.D., Scott R.D., et al. Rationale of the Knee Society clinical rating system. *Clin. Orthop.*, 1989, V. 248, p. 13–14.
50. Ewald F.C. The Knee Society total knee arthroplasty roentgenographic evaluation and scoring system. *Clin. Orthop.*, 1989, V. 248, p. 9–12.
51. Unger A.S., Inglis A.E., Mow C., and Figgie H.E. Total ankle arthroplasty in rheumatoid arthritis: A long-term follow-up study. *Foot Ankle*, 1988, No. 8, p. 173–179.
52. Newton S.E. Total ankle arthroplasty: Clinical study of fifty cases. *J. Bone Joint Surg.*, 1982, V. 64A, p. 104–111.
53. Turner A.J. Shoulder replacement arthroplasty. In: Total Joint Replacement. Ed. by W. Petty. Philadelphia, W.B. Saunders Co., 1991, p. 601 – 658.
54. Walch G. and Boileau P. Shoulder arthroplasty. Heidelberg, Springer-Verlag, 1999, 444 p.
55. Simmons E.D., Sullivan J.A., and Ewald F.C. Long-term review of the capitellocondylar total elbow replacement presentation. *6-th Open Meeting of the American Shoulder and Elbow Surgeons*, New Orlean, 1990, p. 21–28
56. Figgie M.P., Inglis A.E., Figgie H.E., and Mow C.S. Semiconstrained total elbow replacement in rheumatoid arthritis. *57-th Ann. Meeting American Ac. of Orthopaedic Surgeons*, New Orleans, 1990, p. 311–314.
57. Morrey B.F. and Bryan R.S. Total elbow arthroplasty for post-traumatic arthritis. *Orthop. Trans.*, 1988, No. 12, p. 675–685.
58. Brase D.W. and Millender L.H. Failure of silicone rubber wrist arthroplasty in rheumatoid arthritis. *J. Hand Surg.*, 1986, V. 11A, p. 175–183.
59. Blair W.F., Shurr D.G., and Buckwalter J.A. Metacarpophalangeal joint implant arthroplasty with a silastic spacer. *J. Bone Joint Surg.*, 1984, V. 66A, p. 365–370.
60. Chidgey L.K. and Dell P.G. Arthroplasty in the hand. In: Total Joint Replacement. Ed. by W. Petty. Phyladelphia, W.S. Saunders Co., 1991, p. 725–745.

Chapter 2. MATERIALS FOR JOINT ENDOPROSTHESES

Numerous factors have contributed into convincing successes of modern endoprosthetics, among which the polished mastery of surgeons and better understanding of the relationships between the organism and implants are most decisive. Joint endoprostheses of today are manufactured from such materials that neither they themselves nor their corrosive or wear debris may cause rejection in the living tissue. Biological influence of the organism on implants resulting in aging and corresponding impairment of strength and triboengineering characteristics of the joints, do not however, inevitably lead to an unscheduled refuse of the endoprosthesis. In the present chapter we are going to discuss the requirements imposed on endoprosthetic materials and analyze a variety of metallic, ceramic, polymer and composite materials employed in joint endoprosthetics. In conclusion, the authors endeavoured to describe the future trends in the materials science of endoprostheses.

2.1 REQUIREMENTS TO MATERIALS

The living tissues contacting some foreign matter form an interface that may result in a hostile reaction towards the implant. Various fluids in the organism contain biologically active substances, including enzymes, which are the biological catalysts based on proteins present in all living cells. They can generate the radicals able to destroy polymer implants. Complex electrolytes incorporated in biological fluids may initiate electrochemical corrosion of metals and dissolve ceramics [1]. This is why there are only a few materials that remain intact after implantation. In severe conditions the organism may initiate certain complex reactions that attack the implant and may cause inflammation.

Recent representations on the mechanism of the living tissue response to the implant prove that they bear a dynamic and intricate interface (Fig. 2.1). The implant's surface layer acquires a new structure as a result of machining, exposure to air, sterilizing and biological environment. Its composition is quite different from that of the implant bulk. The surface of metallic implants is gradually covered by an oxide layer and the first to interact with it is the serum whose proteins are adsorbed on the oxides. The composition of adsorbed protein layers may vary in a wide range with time [2].

The adsorbed protein molecules enter into a donator-acceptor interaction with the implant oxides generating complex chemical compounds, called ligands. The receptors of the cell membranes interact selectively with these ligands. A cascade of intercellular chemical reactions "notifies" the cell nucleus that a sufficient amount of the receptor-ligand complexes has been formed. These reactions regulate the cellular functions in contact with the implant, including adhesion, changes in organoids (special intracellular structures), matrix deposition (a substance filling the organoids) and so on.

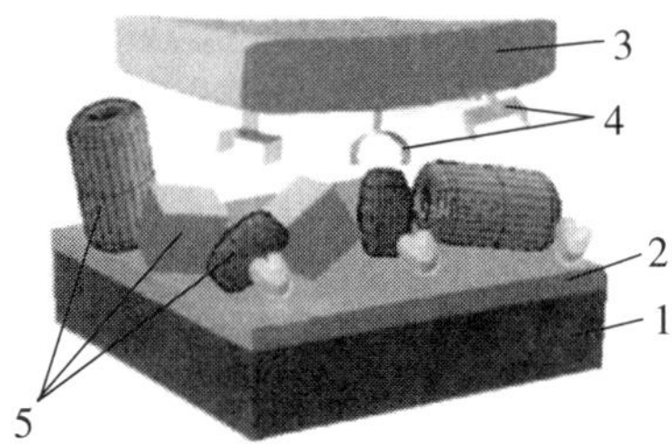

Fig. 2.1. Schematic presentation of the cell-implant interface: *1* – metallic implant, *2* –oxide surface layer, *3* – cell, *4* – receptors, *5* – adsorbed protein molecules

In response to foreign inclusions the macrophages adhere (connectivum cells) to the implant and merge in gigantic cells. This is dependent on the orientation of peptides (synthesized by the amine acid cells) [3] or the length of hydrocarbon chains immobilized on the implant surface [4]. Regulating of the tissue response may avert chronic and unfavourable phenomena in the healing wound and promote safe growing of the implant into the organism.

The response can be guided by the choice of implantation materials. Four main types of reactions on the implant are given in Table 2.1 [5].

Table 2.1.

Types of local reactions occurring at tissue-implant interface [5]

Type	Implant-tissue reaction	After-effects
1	Toxic	Toxic damage of the tissue Necrosis, destructive inflammation, dystrophy and atrophy, degeneration
2	Bioinert	The tissue forms a fine in adherent fibrous capsule round the implant
3	Bioactive	The tissue is biologically linked with the implant at the interface
4	Implant dissolution	The tissue substitutes the implant

Type 1. The implant should be non-toxic for the tissue cells and restrain from liberating chemical compounds in blood, lymph and tissue fluids that may inflict damage to the organism.

Type 2. The formation of a fibrous capsule round the implant is a most common reaction of the living tissues. The capsule tends to isolate and force the foreign body out of the organism. This protective reaction occurs as a result of inflammatory processes initiated by the chemical peculiarities of the implanted material [6]. The implants shown in Figs. 2.2 *a* and *b* are made of polycarbonates that differ only in the structure of alkyl ether groups grafted to the backbone chain. The mute reaction of the osseous tissue to the implant (*a* – with grafted ethyl groups) changes for the unfavourable leading to formation of a thick fibrous capsule between the bone and implant (*b* – with grafted acrylic groups). The capsule forms in the following manner.

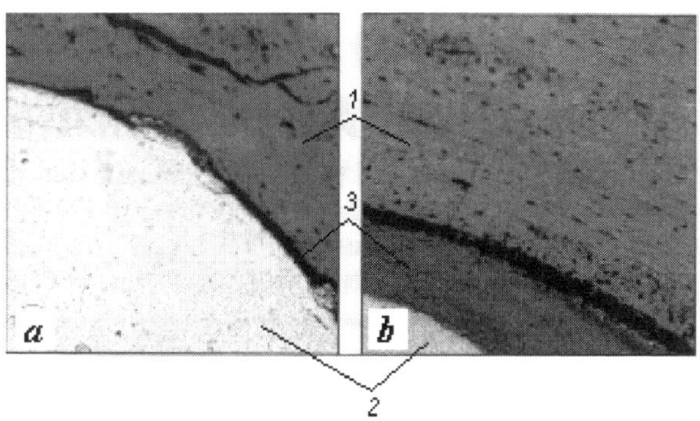

Fig. 2.2. Interfaces between polymer implants and rabbit bone [5]: *1* – bone, *2* – implant, *3* – fibrous capsule (see explanation in the text)

Any inflammation may lead to accumulation of macrophages (representing chiefly white blood cells and leukocytes) at the tissue-implant interface. In case the implantation was not complicated by infections, this inflammation would end roughly in three days. 4–5 days after the operation the macrophages start to dominate round the implant, though lymphocytes (types of leukocytes), plasmatic cells and fibroplasts (the main cellular form of connective tissue of animals and man) are also observed here. The population of macrophages decreases abruptly in case of the friendly reaction of the tissues in 1–2 weeks and the fibroplasts form a collagen-like capsule round the implant. Its thickness changes with time (Fig. 2.3) [7].

Type 3. A bond called a bioactive interface is formed at the interface between the implant and the tissue, which prevents mass-transfer between them. As a result, the initial tissue structure is restored at the interface. Since the implant-tissue system is found in the dynamic equilibrium state, this reaction depends on the rate of electrochemical and biological processes occurring in it.

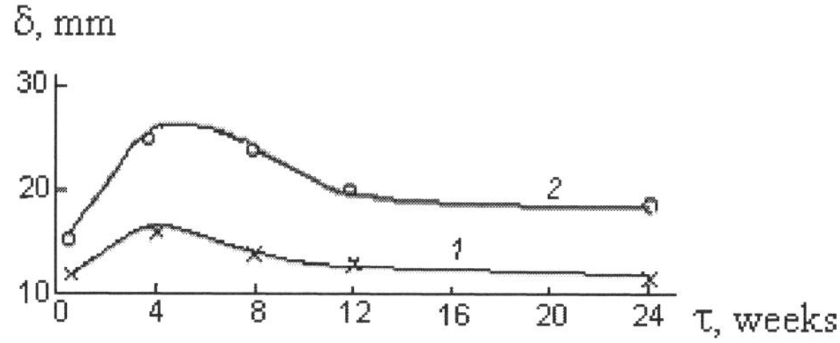

Fig. 2.3. Thickness of the capsule versus implant life in rat's organism. Implant materials: *1* – polydimethyl siloxane, *2* – polystyrene with polymethyl siloxane coating

Type 4. Biologically active interfaces vary so quickly that the implant material resorbs and are substituted by the tissue. In other words, the resorbed material degrades under the action of blood, lymph, tissue fluids and is readily digested by the microphages. Their degradation products are non-toxic and assimilate in the cells.

Here and further we will adhere to the terms and definitions used in medical materials science and recommended by the 5th and 6th International Congresses on Biomaterials (1998 and 2000), as well as ISO/TR standard No. 9966 and Russia State Standard No. 511480-98.

Biocompatibility is the ability of an implant to stimulate a proper response of the organism (host) under specific conditions of interactions [8]. This is not a property of the material the implant is made of but a characteristic of the implant-organism system corresponding to certain conditions of their contact. Biocompatibility does not mean that the implant is absolutely non-toxic and is devoid of other negative features. This is a property of the implanted material to stimulate a response of the organism, which will solve the stated problem. A biomaterial is a nonviable material intended to bear medical functions in contact with the living tissues. A biomaterial should be biologically compatible and may be biodegradable. Biocompatible materials interact smartly and consistently with the organism so as to avoid any disease.

As we have already noted above, the implant surface layer undergoes fast changes due to settling of protein molecules. Their molecular mass grows as the settled layer thickness increments (Vroman's effect) [1]. Biocompatibility of the implant is a measure of the extent the organism accepts the settled protein layer. Biocompatibility tests serve the base for estimating potentialities of using materials in endoprostheses. In Nottingham's University (UK) the first methods of quantitative estimates of the reaction of cells on implants, as well as biochemical and microscopic means of determining their compatibility have been developed using numerous cell cultures and types of materials. These methods have become a standard procedure today [1].

The lack of a generalized theory on the mechanism of organism responses to the implant agitates animated discussions and anxiety of specialists studying biocompatibility. Unfortunately, this problem is still far from being solved. As an example [9], three specimens of one and the same non-toxic polymer with different surface layer porosities have been implanted into the soft tissues. It appeared that the tissue together with blood vessels grew into the sample with spherical pores 10–20 μm in diameter. In contrast, the nonporous specimen and the one with elongated pores (10×200 μm) induced a classical reaction of the tissues to a foreign body with encapsulation. After healing all three specimens can be called biocompatible, although their biocompatibility degrees are different. In spite of the fact that the mechanism of response of the tissues to surface roughness of the implants has not received any adequate explanation in literature, the technique of microtexuring is used to stimulate mobility of the cells and regulate their functions. The cells of osseous tissues are migrating over the surface grooves of a corresponding size (about μm). This proves that the bone tissue formation is dependent on the micro topography of the implant surface [10].

In reality, only relatively biocompatible materials may exist and cause no negative reactions in the organism during some limited time. The standard ISO/TR 9966, 1989 (E) recommends to estimate biocompatibility of materials on a case by case study and treat the results obtained very thoroughly. It is in fact impossible to predict the reaction of the organism on some implant; this is why the criterion of biocompatibity still remains the experiment *in vivo*.

Biodegradable materials may dissolve fully or partially when contacting the living organism (are adsorbed by microphages and enter in metabolic and biochemical reactions) and the living tissue substitutes them. *Bio stability* of materials is a property opposite to biodegradability, meaning the ability of materials to resist the affects of the biological medium and preserve their functional properties.

To characterize biocompatible materials implanted into the bone tissue we use such terms as bio tolerant, bioinert and bioactive.

Bio tolerant materials interact with the bone via a follow-up osteogenesis, i.e. they are separated from the bone tissue by an ingrowing fibrous layer.

Bioinert materials do not actually interact with the tissues in the contact and do not result in any explicit fibrous layer (the capsule thickness is below 50 μm) or osteogenesis. So, the bone may be formed near the bioinert implant. It has, as a rule, a protective layer that permits no free ions out and biological fluids in. The bioinert materials get integrated with the bone by penetration of the connectivum into the implant pores. There are not evidently absolutely bioinert materials since the biological response is always present at the tissue-implant interface and its intensity depends on a number of factors (Table 2.2).

Table 2.2.
Factors effecting biological response of bone tissues in contact with orthopaedic implants [5]

Factors affecting tissue response	Factors affecting implant response
Tissue type	Implant composition
Presence of damages	Amount of phases
Tissue age	Presence of interfaces
Circulation of blood and lymph in tissues	Surface structure
Tissue shifting at interfaces with implant	Electrochemical reactions
Compliance of implantation place to anatomic and physiological criteria	Accuracy of implant assembly
Mechanical load on tissue	Loading intensity
General reactions of organism (immunological, allergic, neuroendocrinous)	Organism functioning

Bioactive materials stimulate biological reactions in the organism leading to integration of this material with the organism. They are intended to form bonds with the biological systems and thus improve the treatment, form or substitute a tissue or an organ, and to perform some function in the organism. All bioactive implants used in orthopedy and traumatology are united by a common feature, namely, a carbonate layer whose composition and structure are analogous to the bone mineral phase. This layer is composed of polycrystalline agglomerates incorporating collagen fibrils in contact sites with the bone.

Reliability of the implants is an actual problem of materials science in medical engineering. It arose in the 1950s when endoprosthetics of the hip has acquired a mass character and become an every-day practice in many clinics. In contrast to the usage of medicines, the mechanical endoprosthesis can be adjusted only by a surgical intervention. Therefore, joint endoprostheses system must operate without failure and maintenance for many years. This means that endoprosthetic materials, their strength and wear resistance should not change perceptibly in the hostile biological environment with time.

Modern endoprostheses employ practically all types of structural materials: metals, polymers, ceramics and composites on their base. Composites have been used as bio imitative materials simulating the structure and properties of the natural tissues since early 1990s, which was initiated by the far-famed centres engaged in the development of biomaterials. They are the Toronto University, Japanese Institute of Medical and Dental Engineering and London

IRC. They represent a new generation of materials developed purposefully for medical applications as implants. We may now ascertain with certainty that the period when the implants were made of the best materials borrowed from electronics, precise instrument-making and military spheres has passed into history.

The life span of joint endoprostheses recorded by the observations *in vivo* is, as a rule, much less than that determined by the mechanical testing machines. None the less different testing systems for simulating work of implanted endoprostheses are used in the countries producing this kind of products. Along with other eminent institutions, the databases on tribological characteristics of hip endoprostheses have been created in the Central Institute of Traumatology and Orthopedy named after Priorov (CITO, Russia). In Great Britain the life of all implanted in this country endoprostheses is controlled by the Department of Health Medical Devices Agency. This Agency performs simulation tests of all joint endoprostheses designs used in orthopedy. Beginning with the 1990s the Department of Trade and Industry in UK has been backing the programs dealing with mechanical tests and prediction of the useful operating life of joint endoprostheses. To this end, the European Community has adopted three basic documents called "European Medical Devices Directives" acting as the laws in the participating countries. They contain the requirements to endoprosthetic materials aimed at minimizing risk for the health and safety of patients. The materials and endoprostheses meeting these demands have a "CE" mark meaning the best quality and approved status. The top-level requirements to endoprosthetic materials seem unattainable for the traditional engineering. Therefore, the biomaterials are more often custom-made today, including individually designed composites, functional polymers, bio imitative and bioactive coats on metallic implants and so on [11]. Just these materials predetermine today is level of endoprosthetic implants.

2.2 METALS AND ALLOYS

Metals and their alloys are employed in endoprosthetics in the form of cast or wrought products. Their biomechanical characteristics and chemical properties define the designs of joint endoprostheses.

Native gold was a starting point in the long-standing experience of metallic implants. Among its numerous advantages are purity, absence of oxides, and simplicity of processing into rods, wire, sheets or plates thanks to a perfect ductility. The absence of any oxides has made it possible to apply cold welding for the endoprosthetic elements. Owing to its high bioinertness golden implants are readily biocompatible. Other noble metals like platinum, iridium and palladium show similar merits but did not find the market because of, firstly, their low strength, and secondly, high cost. The transfer to noble alloys Au+Ag+Cu has somewhat improved the situation. Nevertheless in the follow-up experiments, copper alloys showed unfavourable biological reactions in animals due to the elevated corrosion.

There were numerous attempts to employ alloys based on copper, nickel, iron and cobalt that appeared durable for the marine equipment. In contrast to noble metals, their bioinertness depends on the properties of passivating films forming in saline environments. The experiments on laboratory animals have visualized that commercial brass, bronze, chrome-nickel and carbon steels are inappropriate for implantation due to poor biocompatibility, and low strength impaired by corrosion [12].

Cast products of the CoCrMo alloy, termed Stellite, and the wrought stainless steel alloy FeCrNi were used in implants in the 1940s. Their bioinertness was conditioned by the formation of the passivating surface film. In the long run, the implants based on titanium, zirconia and their alloys appeared in early 1950s [13].

The chemical and electrochemical criteria of biocompatibility of metallic materials involve next parameters.

So-called vital chemical elements that are valuable in biological respect, or in other words, intrinsic for viability are 11 chief (C, H, O, N, S, Ca, P, K, Na, Cl, Mg) and 15 trace ones (Fe, Zn, Cu, Mn, Ni, Co, Mo, Se, Cr, I, F, Sn, Si, V, As). They participate in metabolism or may be present in enzymatic systems, and are found in optimal concentrations in tissues and organs. When the concentration of some vital element is either surplus or scanty, various biochemical defects or physiological dysfunctions may occur and even become fatal [14].

Owing to a permanent relationship with the environment, the human organism commonly incorporates 20 to 30 inessential trace elements (Al, Sb, Cd, Hd, Ge, Rb, Ag, Pb, Au, Bi, Be, Ti, Zr, Nb, Ta, etc.). Some of them are toxic even in low quantities (Cd, Hg, Pb, Be), others are indifferent physiologically (Al, Ti, Zr, Nb, Ta) [15].

The chemical elements introduced from endoprostheses omitting natural barriers, may excel some critical concentration and disturb the natural biochemical processes in the cells. From this standpoint, the titanium implants turn to be preferable to aluminium ones. The experiments in animals have shown that aluminium specimens are more toxic for the osseous tissue than titanium and accumulated round the implant titanium does not impose any functional changes on the cells [16].

Biocompatibility of the metallic materials depends on their position within the voltage series as well. The products of electrochemical reactions of the implants in electrolytes effect negatively the surrounding tissues. Therefore, the implants from Ti, Zr, Nb or their alloys are considered biocompatible as they resist corrosion in chlorine-containing fluids of the organism (blood, lymph, secretion, etc.). Only 13 metals suit optimally this criterion, among which are 8 noble metals (Au, Ag, Pt, Pd, Ir, Ph, Ru, Os) and 5 passive ones (Ti, Ta, Nb, Zr, Cr) forming a protective oxide film on the surface [17].

The authors of [5] keep to the opinion that a trend of substituting steel implants by the titanium, niobium and zirconia ones appeared at the beginning of the 21st century. Note that the titanium alloys heat treated in vacuum can be further subjected to the magnetic-resonance tomography after an orthopaedic operation with minimal artefacts.

Modern metallic materials for joint endoprosthetics are subdivided into two main groups: 1) cast cobalt-based alloys, and 2) wrought titanium, cobalt or stainless steel-based alloys. Their properties are described in Table 2.3 [13].

Cast cobalt-based alloys originate from the Stellite group of materials. Most promising type of their processing into components is the lost-wax casting in air. Being subjected to the initial industrial cleaning, only 1% of nickel is left in cobalt. Cobalt is the main component of the alloys forming a matrix that contains chromium and molybdenum-based phases. Chromium adds strength to the alloys, and chemical inertness as it forms a passivating oxide film on the implant surface. Molybdenum ensures resistance to corrosion (crevice, pitting or local), durability, and reliability of the implants.

Table 2.3.
Properties of metals and alloys used for joint endoprostheses [13]

Composition (wt. %)	Elasticity modulus, GN/m^2	Breaking strength at tension, MN/m^2	Relative elongation at rupture, %	Surface layer composition
Ti (99)	97	240–550	> 15	TiO_2
Ti (90) + Al(6)+V(4) wrought	117	860–896	> 12	TiO_2
Ti (90) + Al(6)+ V(4) cast	117	860	> 8	TiO_2
Stainless steel Fe (70)+Cr(18)+Ni(12)	193	480–1000	> 30	Cr_2O_3
Co(66)+Cr(27)+Mo(7) cast	235	655	> 8	Cr_2O_3
Co(55)+Cr(20)+W(15)+Ni(10)	235	860	> 30	Cr_2O_3
Co(45)+Ni(35)+Cr(20)+Mo(10)	235	793–1793	50–8	Cr_2O_3
Co(52)+Ni(20)+Cr(20)+Mo(4)+W(4)	235	600–1310	50–12	Cr_2O_3
Zr(99)	97	552	20	ZrO_2
Au(99)	97	207–310	> 30	Au

Iron and other admixtures enter in interactions with the main components of cobalt alloys that form carbides and other secondary phases adding stability to the matrix in abrasive wearing. Carbon content in the alloy should be so low as to avoid excess growth of the carbide phase, since it may worsen strength and ductility of the alloy. Cast components undergo annealing to impart spherical structure to the carbide phase and improve elasticity of the alloys.

Cobalt alloys are noted for high stiffness and wear resistance, and furnish perfect surface polishing. Therefore the majority of the heads, as well as spherical and conical cups of hip endoprostheses are cast from Co (66%) – Cr (27%) – Mo (7%) alloys following ISO 5832-4. In the 1950-1960s this alloy has become popular as stems for the Moore, Thompson and Muller endoprostheses. They, however, frequently broke because of insufficient fatigue strength and were later substituted by the wrought stems.

Wrought alloys are processed into ready products by rolling, forging, stamping and drawing.

Wrought components of cobalt alloys surpass cast ones in strength and ductility. The analogous regularity is characteristic for the iron-based alloys. Stainless steel is subjected to corrosion (intercrystalline or pitting) when the passivating film on the implant surface dissolves or spreads under cyclic loads. To hamper stress concentrations in the passivating film, the steel implants with rough-turned or porous surfaces should be rejected.

Titanium alloys combine high strength and ductility with perfect corrosion resistance. When exposed to air the titanium implants form an oxide film in contact with the living tissues. In contrast to cobalt and stainless steel-based alloys, they are susceptible to fretting-corrosion (in response to oscillating motion of conjugated elements in corrosive medium). This is why numerous scores may appear on the implant surface, while the tissues contacting titanium start to decolorize [18]. For a long time wrought alloy heads of Ti (90%)–Al (6%)–V (4%) have been used in the USA but were abandoned due to severe wearing of the heads in pair with the polyethylene cups. Both titanium alloys (ISO 5832-3, -10, -11) and pure titanium (ISO 5832-2) have been used in wrought endoprosthesis stems and cups since 1970s. They offer high strength, biocompatibility and osseointegration of the implants.

Since then, vivid comparative evidences have been accumulated on serviceability of the alloys *in vivo*. The problems of the aseptic loosening rate of femoral stems (design of M.E. Muller) manufactured from three alloy grades have been studied in [19]. Namely, CoCrNi (SS77), Ti-6Al-7Nb, and SLS Ti alloys. Greater elasticity of Ti alloy was established to be a factor responsible for loosening. This was supported by the observations [20] of Mc. Kee-Arden's femoral stems made from either Co-Cr alloy or Ti alloy. Time to failure was significantly shorter in the titanium group. Loosening and peri-prosthetic osteolysis occurred with significant frequency in the titanium group as compared to the CoCr group.

Wrought steel of the composition Fe(65%)–Cr(18%)–Ni(14%)–Mo(3%) (ISO 5832-1) has been successfully employed for the spherical heads of joint endoprostheses. In the 1980s J. Charnley used a stronger and more rustproof wrought steel Fe(61%)–Cr(20%)–Ni(10%)–Mn(4%)–Mo(3%)–Nb(1%)–N(1%), ISO 5832-9 for 22 mm in diameter heads.

Properties of metal implants are conditioned in many respects by their technology and design.

The polished cobalt, titanium or stainless steel-based implants are similar in colour. The steel and cobalt specimens have close densities (8–10 g/cm^3) but are heavier than the titanium ones (3–4 g/cm^3). The cobalt-based components of endoprostheses are harder and more wear resistant, while titanium implants are highly responsive to stress concentrations. Of course, sharp angles, notches and surface features such as scratches and cracks must be avoided for all kinds of metallic prostheses.

The potentials of the porous and non-porous titanium implants differ. Therefore, their corrosion rate *in vivo* depends upon the smooth to porous surface area ratios (see 5.4). This regularity is typical for the cobalt implants as well [13].

The tissues adjacent to the titanium implants, and their wear debris react biologically more peacefully as compared to cobalt and stainless steel. Since the corrosion products of both groups of prostheses contain nickel, they are not implanted to hypersensitive to these metal patients.

The size and configuration of endoprostheses is believed to define macro distribution of the loads, whereas roughness and properties of the material effect most strongly stress distribution at the bone-implant interface. The interfacial stress concentration arises from a great difference between the mechanical properties of the bone and implant. The elasticity modulus of titanium alloys is 5.7 times as high as the compact osseous tissue. As for the stainless steel and cobalt alloys, this value is even higher 9.3 and 11 [13]. In certain cases the soft titanium dioxide layer on the implant surface may integrate with the bone, while chromium oxide on the cobalt implant is incapable of the integration. Ingrowth of the tissues into the implant effects significantly stress distribution within the endoprosthesis and adjacent bones.

To initiate the ingrowth, the contacting metal implant is covered with a porous coating by sintering of spherical or irregular in shape particles, wire chops, by plasma or by the fire treatment. Although the distant follow-up clinical data on such coats are still disputable, the nearest experimental results in animals have shown their doubtless efficiency. A fibrous tissue and bony mixture are seen to grow into the porous coating. The reduction of stresses in the implants depends mostly on the coating processing type and pore size [13, 16].

It is evident that perceptible advances in the development of metallic implants require the long-term, many-sided, and laborious investigations. Available for the time being alloys have been optimized by the criteria of their biocompatibility, strength, corrosion and wear resistance. These facts as well as their inherent process ability, and vast experience in the adoption of metal implants underlie their long-term domination over other materials in this field.

2.3. POLYMERS

Abundant experience has been gained in the application of practically all structural polymer materials offered by the modern technology [11, 12, 16, and 21]. Some of them (polyamide, polystyrene, polyvinylchloride) meet but insufficiently the criteria of biocompatibility because of the migration of the process additives (plasticizers, low-molecular compounds, stabilizers and other) into the adjacent tissues. Others, though admitted as biocompatible (the majority of polyolefin's, fluoroplasts, silicone polymers) are unfit for application in the endoprosthetic friction joints due to their low wear resistance and unfavourable effect of the wear debris (observed in different periods). At the threshold of the 21st century three main types of polymers have found wide application in endoprostheses, namely ultrahigh-molecular weight polyethylene (UHMWPE), polymethyl methacrylate (PMMA), and silicone rubber. UHMWPE offered for the use in endoprostheses by J. Charnley has remained unsurpassed up till now as a frictional material for articulation with metals, and ceramics. This is because of its high wear resistance, inertness of the wear products, low friction coefficient and a self-lubrication property *in vivo*. Being a base of the bone cement, PMMA was also first used in endoprosthetics by J. Charnley. The implants in the form of flexible rods functioning as minor joints of the hand and foot are made at present of silicone elastomers.

Ultrahigh-molecular weight polyethylene is obtained via the polymerization of ethylene monomers by a specific technology [22]. The molecular chain lengthening is accompanied at polymerization by branching and formation of side groups. Low-density polyethylene (of 30,000-40,000 molecular mass) acquire numerous side units and a corresponding large free volume. High-density polyethylene (50,000-60,000) and UHMWPE (3,000,000 and higher) get a few side units. UHMWPE fabricated for endoprostheses in the USA has, according to the standards [23], density 0.930-0.944 g/cm^3, yield strength 19 MPa, breaking strength at tension 27 MPa, and relative elongation at rupture 200%.

Some process and physico-mechanical characteristics of UHMWPE fabricated by Hoechst AG, GmbH (Germany) are presented in Table 2.4. These materials used for joint endoprostheses are stronger and more resistant to impact loading than the previous ones [24]. Named properties are preserved within a wide temperature range since permolecular formations of UHMWPE get linked at melt crystallization by the transit macromolecules and physical nodes (macromolecular engagement).

Table 2.4.
Physico-mechanical characteristics of Hoechst UHMWPE used for endoprostheses

Characteristics	Unit	Standard	*Chirulen* grade		*Hostalen GUR* grade	
			1020	11120	2122	4120
Flow index under stress (150/10)	H/mm^2	DIN 53493	0.23	0.2	0,22	0.22
Density	g/cm^3	DIN 53479 ISO/R 1183	0.935	0.93	0,93	0.936
Yield point at tension (23 °C)	N/mm^2	DIN 53455 ISO 527	22.8	23	21	22
Breaking stress at tension (23 °C)	N/mm^2	DIN 53455 ISO 527	39.6	49,5	43	44
Relative elongation at rupture (23 °C)	%	DIN 53455 ISO 527	335	512	450	427
Impact toughness	mJ/mm^2	DIN 53453 ISO 179	195	197	140	145

Polymer components of artificial limbs should be adjusted for operation under loads for a prolonged time. Table 2.5 presents the creep factor of UHMWPE at extended compressive tests [22]. The creep factor is seen to depend insignificantly on the load and temperature but shows a tendency to reduce with temperature rise (within the physiological temperatures).

By its triboengineering characteristics UHMWPE is close to the antifrictional polymers like fluoroplastics and polyamides. Its wear resistance is twice as high as that of all other polyethylene brands. As it is been mentioned previously, UHMWPE wear debris does not exert any expressly negative reaction on the soft tissues as, e.g. fluoroplast debris do. Nevertheless, they may lead to the appearance of a granuloma round the foreign body. Different aspects of the wear resistance of implanted in pair with metals or ceramics UHMWPE will be considered in chapters 5 and 6.

Table 2.5.
Creep factor of Hostalen GUR-grade UHMWPE versus compressive strength and temperature [22]

Load, MPa	Temperature, °C	
	23	40
3	1.6	1.5
4	1.6	1.6
5	1.6	1.6
6	1.8	1.7
8	1.8	1.7
10	2.0	1.9

In 1997 Internet sites placed an information that a Swiss company Sulzer Orthopaedics, Ltd. - the world largest manufacturer of biomaterials and medical implants, had developed and commercialized jointly with the Massachusetts General Hospital and the Massachusetts Institute of Technology (USA) a new type of UHMWPE for the endoprosthesis friction joints. This material is produced as follows. The rods, discs or plates for endoprostheses elements from, e.g. Hostalen GUR grade, are subjected to a high-energy treatment in electron beams (10 MeV), and annealing under 125 °C temperature (melting point of UHMWPE is 137 °C). As a result, the macromolecules impart the material a cross-linking structure. This structure is i) more resistant to aging *in vivo*, ii) has better deformation and strength parameters, and iii) ameliorated wear resistance [24]. The triboengineering testing of hip endoprostheses on imitators (20 mill. loading cycles equal to 20 years of operation *in vivo*) has shown negligible wear of the cross-linked UHMWPE liner [25]. The affiliated Sulzer Medica, USA advertises the cross-linked UHMWPE as the best of all existing for today polymer friction materials for joint endoprostheses.

The problems arising in UHMWPE components are related mainly to sterilization of the implants. The methods of killing micro organisms may often damage the implanted material [26]. Using IR spectroscopy, the *max* oxidation degree of UHMWPE has been recorded after the autoclave sterilization by vapours. The oxidation of the implant surface layer subjected to γ-sterilization in air (2.5 Mrad) reduced essentially at exposure in argon. Ethylene oxide treatment did not show any UHMWPE oxidation at all. The analysis of SEM images of the implant surfaces has proved that γ-sterilization results in micro cracking (10-30 μm in width) in the surface layer, which limits service life of UHMWPE implants.

Investigations of γ-sterilized in air UHMWPE liners have shown that the long-term durability varied significantly depending on the shelf-life before implantation [27]. The liners with a shelf-life time 3 years and more displayed significantly higher volumetric wear than those with a shelf-life time less than 3 years. IR spectroscopy and scanning calorimetry showed that all explanted implants underwent substantial *in vivo* oxidation and crystallization. The oxidation aging of UHMWPE renders the liner susceptibility to severe wear.

Polymethyl methacrylate forms the base of the orthopaedic bone cement.

It is supplied in clinics in the form of powdery or liquid components (2:1 by mass) further mixed for preparing the cement paste. The powder (30-150 μm particle size) consists of PMMA or its copolymer with styrene and a slight amount of the polymerization initiator (benzoyl peroxide), and, if necessary, rentgenocontrast additives (barium sulphate). The liquid includes a monomer (methyl methacrylate), inhibitor (hydroquinone) and a polymerization activator (dimethyl-*p*-toluene). The inhibitor keeps the monomer from polymerization at storage, while the activator initiates its polymerization after mixing with the powder. The typical compositions of bone cement are cited in Table 2.6 [28].

Table 2.6.
Bone cement components (content in wt %) [26]

Liquid	Powder	
	radio parent	radiopaque
Methyl methacrylate 97.5 Dimethyl-*p*-toluene 2.5 Hydroquinone 0.0075	PMMA or copolymer 88 Barium sulphate 10 Benzoyl peroxide 2 Monomer residues - traces	PMMA or copolymer 98 Benzoyl peroxide 2 Monomer residues - traces

The monomer softens polymer particles during mixing and forms a gel. Dimethyl-*p*-toluene reacts in the liquid phase with the powder of benzoyl peroxide and forms free benzoyl radicals. When interacting with the monomer the radicals break double bonds in methyl methacrylate and initiate polymerization. The molecular mass of PMMA formed during this process surpasses that of the PMMA as powder. Some quantity of the monomer is evaporated at mixing of the cement components, and the *min* of evaporation corresponds to a 2:1 powder-liquid mass ratio. About 2.5% of unreacted monomer may stay in the cement for 200 days upon setting. The monomer exerts a slight toxic effect on blood serum, and impedes its bacteriostatic property (hampering of bacteria multiplication).

The polymerization process is accompanied by the heat generation (130 cal/g of the monomer) with the *max* exothermal reaction at the final stage of the polymer curing. The highest recorded curing temperature of a 10 mm thick cement block is 107 °C. With temperature elevation the polymerization speeds up and the cement molecular mass of the final stage drops leading to impaired mechanical strength of the cement. A thin cement layer not strongly heated up during polymerization is always stronger than the thick one. Bone cement undergoes volume shrinkage to about 8% during the polymerization. As a result, compressive stresses appear in the cement mantle encircling the prosthesis element. So the cement layer should not be too much thick to avoid cracking from shrinkage.

Figure 2.4 illustrates the parameters of the bone cement processing.

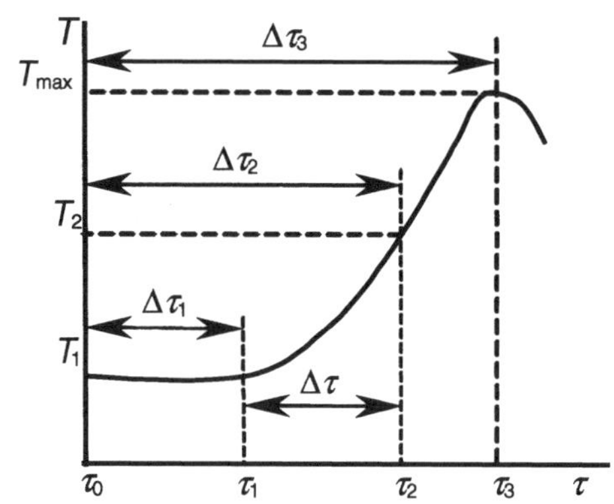

Fig. 2.4. Temperature versus cement setting time (see explanations in the text)

The cement stays in a dough state for time $\Delta\tau_1$ from the moment τ_0 of powder mixing with the liquid, till moment τ_1 when it no longer adheres to a gloved surgeon's hand. The setting period $\Delta\tau_3$ or $\Delta\tau_2$ runs from the beginning of mixing (τ_0) till the temperature peak (T_{max}, τ_3) or the middle of the temperature interval between T_{max} and ambient temperature T_1 (T_2, τ_2). A surgeon may operate with the cement within the time interval $\Delta\tau$, which is a difference between the cement setting period $\Delta\tau_2$ or $\Delta\tau_3$ and its dough time $\Delta\tau_1$, i.e. $\Delta\tau = \Delta\tau_2(\Delta\tau_3) - \Delta\tau_1$ [28].

The cured cement is a composite consisting of solid powder components bonded together by the polymerized monomer with entrapped air bubbles. The composition and porosity depend upon the cement brand and its preparation technology. Centrifuging of the cement dough reduces the porosity and improves considerably its strength.

Viscosity of the dough defines the extent of filling the gaps between the implant and bone. The influence of cement viscosity on the migration of femoral stem (Exeter design) has been studied using RSA [29]. The cements of various grades (Simplex, CMW1, CMW3) were examined in a total of 46 patients over a 12-month period. *In vitro* studies demonstrate that low-viscosity cement forms a more stable bone-cement interface. Several groups have examined the *in vivo* effect of cement viscosity on stem longevity with conflicting results. The cement dough is a non-Newtonian liquid whose viscosity depends on the shear rate, i.e. it is more flow able if forced into the clearance under high rates and pressures. The dough flows quicker when injected by a special tool (cement gun) and fills the pores better. The strength of adhesive joints bone-cement formed after flashing the marrowy canal with a salty solution was compared. In the first variant the cement was finger-pressed in the canal, in the second it was injected by a gun. Almost identical results of the breaking stress at tension were recorded for both variants and much higher breaking stress at shear was recorded for the second variant [30]. In [31] three commercially available cement mixing systems were tested (Howmedica, Summit, and Stryker). After mixing the cement was allowed to cure in metal molds under standard conditions. The fatigue behaviour of the cement blocks were tested until failure. A wide scattering of results was observed in all the three systems. Uneven mixing of the polymer and monomer was observed in the Howmedica system. Significantly stronger cement was produced with the Stryker system.

The cements of different grades differ in their resistance to wear. They can be arranged into a series in the order of their wear resistance reduction: CMW (UK) > Implast (Germany) > Acryloxide (Russia) > Sulfix-6 (Switzerland) > Simplex Surgical Plain (UK) > Palacos (Germany) [32].

Silicone elastomer is an organic polymer with rubber-like properties and a structural formula

$$\begin{array}{c} R \\ | \\ [-Si-O-]_n \\ | \\ R' \end{array}$$

In a low-molecular form it is a silicone liquid. The base of used for endoprosthetics silicone elastomers constitute polydimethyl siloxane macromolecules ($R = R' = -CH_3$, of 750,000 molecular mass) with slight additions of methylvinylsiloxane units ($R' = -CH = CH_2$) able to form cross-links. This binder is filled with a foamed silicone of a large specific area (400 m^2/g) and cured in the presence of catalysts (rare metals and peroxides). The cross-link density between macromolecules after curing is 1 link per 325 silicone atoms. The final product takes the form of a giant 3D molecule, wherefrom the volatile residues are removed [33].

The typical mechanical characteristics of the silicone elastomer of Dow 372 Silastic (USA) are: breaking strength at tension 9.57 MPa, elasticity modulus 350 MPa, relative elongation at rupture, not less than 400%.After a 1.5-year exposition *in vivo* the implant strength impaired by 7%, elongation by 10% and elasticity modulus raised by 8% [21].

Failure of the silicone endoprostheses was first associated with the adsorption of lipids (fats and fatty matter) by the silicones from the ambient tissues accompanied by their decolouration. To clear up the causes, the following experiment was performed [34]. Silicone prosthesis was adequately mounted in the imitator and tested for flexion-extension (7 mill. cycles) in the serum. The experiment showed neither damages on the prosthesis nor any adsorption of the lipids by silicone. After which, the prosthesis was mounted so that its flexion axis did not coincide with that of the imitator (the case of incorrect implantation). After the tests both decolouration of the tissues and lipid adsorption were evident. Certain modifications in the design of the flexible artificial limb have eliminated the problem of adsorption. The main reason of failure of silicone endoprostheses of interphalangeal joints of the hand is considered the rheumatoid arthritis leading to degeneration of the phalange. The resultant sharp edges of the phalange may cut the silicone implant during flexion or extension of the fingers [35].

The sustained challenges to broaden the range of polymer materials employed in modern endoprosthetics and their main directions will be discussed in chapter 6.

2.4 CERAMICS

Although some types of ceramics have been used in medicine for more than 25 years, their wide adoption in endoprosthetics has begun no sooner than in the 1990s [36]. Certain positive experience has been gained in applying some types of ceramics, including alumina oxide Al_2O_3, and zirconia oxide ZrO_2, as well as calcium phosphate $Ca_3(PO_4)_2$ and $Ca_5(PO_4)_3OH$ (hydro apatite). Among their advantages are high chemical stability, stiffness, insignificant deformation under loading, high wear resistance, durability and resistance to aging in biological environment. The phosphate ceramic is related to biologically active and slowly dissolving *in vivo* materials allowing for the bone tissue ingrowth. Its main drawback is brittleness, which makes their behaviour unpredictable under loading [16, 37-39].

Alumina ceramics seems to be most extensively used in medicine. It is characterized by a high compressive strength (breaking point 4,500 MPa, elasticity modulus 380 GPa); whereas it is bending strength (breaking point 550 MPa) and tensile strengths are much lower. Alumina ceramic implants make the least impact on the biological environment *in vivo* [37]. Impairment of mechanic characteristics recorded when modelling the effect of biological environment in vivo on ceramic strength is attributed to the penetration of saline solutions into insufficiently dense implant pores [38].

From the standpoints of the dislocation theory of strength, brittle failure of ceramics is referred to the propagation of cracks whose tips are devoid of the local zones of plastic deformation. In contrast to metals, which display plastic flow at tension and the crack tips get rounded, ceramics is not in fact subject to deformation, so the crack tips remain sharp, and the crack propagates lengthwise under loading [39].

The implants of aluminium ceramics are obtained by the powder metallurgy methods during sintering of compacted Al_2O_3 blanks fewer than 1600 °C. MgO may be used as an additive to improve sintering. Fine powder of Al_2O_3 should contain *min* of contaminants since only the pure ceramic provides the appropriate biocompatibility [40]. Such contaminants as silicates, alkaline oxides and calcareous inclusions form a glassy phase with low chemical stability round the grain boundaries. This phase impairs both breaking and fatigue strengths of the implants. One of the best alumina ceramic brands for endoprostheses is considered Biolox® manufactured by CeramTec (Germany) [36]. The processing technique of ceramic endoprosthetic components requires exceptionally clean rooms.

The typical hip endoprosthesis components made of ceramics are the femoral ball and pelvic liner. Their spherical surfaces undergo finish polishing using diamond tools. Such friction joints are well wetted by the synovia and should show low friction coefficient. Triboengineering characteristics of the ceramic prosthetic components are considered in more detail in chapter 5.

Some properties of the alumina and other structural ceramics of medical purposes are cited in Table 2.7 [36].

Table 2.7

Mechanical properties and formulations of ceramic materials [36]

Characteristics	Ceramics		
	Alumina	Zirconia	
	BIOLOX forte	Y-TZP ZIOLOX forte	Mg – PSZ
Standard	ISO 6474	ISO/DIS 13356	–
Chemical composition	Al_2O_3 + MgO	ZrO_2 + Y_2O_3	ZrO_2 + MgO
Description	Polycrystalline corundum	Polycrystalline tetragonal ZrO_2	Partially stabilized ZrO_2
Elasticity modulus at compression, GPa	380	210	210
Wickers hardness	2000	1250	1250
Breaking strength at bending, MPa	>500	>900	>500
Grain size, μm	<2	<0.5	30

The ceramic components of endoprostheses are commonly sterilized by γ-radiation or vapour treatment in autoclave. There are no hitherto any evidences of aging of alumina ceramics or the reduction of ceramic implants service life. According to standard [40] the implants from alumina ceramics can be subjected to sterilization as many times as necessary. Their white or ivory colour acquires brownish hue upon γ-sterilization. The colour depth depends on the temperature and sterilization time. It is confirmed experimentally [36] that the phenomena effecting colour change are not related to impairment of the mechanical or chemical characteristics of alumina ceramics.

Zirconia ceramics are obtained by sintering the ZrO_2 powder. The grain size (0.5 μm) is less than the alumina ceramic has, which presupposes its better polish and high wear resistance.

The crystalline structure of zirconia ceramics is stabilized by the introduction of 3–9% magnesium MgO or yttrium oxide Y_2O_3. Thanks to its perfect mechanical properties the ZIOLOX® forte trademark ceramics as well as Y-TZP (polycrystalline ceramic with a tetragonal lattice stabilized with Y_2O_3) are applicable at a broad scale in artificial limbs [41]. Its precursor is the ceramic of Mg-PSZ brand (partially stabilized zirconia by magnesium dioxide). Both trademarks are produced by the CeramTec, Germany. The initial ZrO_2 can be present in three crystalline phases, i.e. monocline, tetragonal or cubic. Phase transformations start with cooling of the sintered implant and are accompanied by variations in the volume and nucleation of cracks. The stabilizers are introduced to initiate an optimum phase transformation process. As a result, local compressive stresses arise in the crack tips that hinder their propagation.

According to Table 2.7, the zirconia ceramic concedes the alumina one in compressive strength but is stronger at bending. Its elasticity modulus at compression is higher than Co-Cr alloys have. The hot isostatic pressing elevates density and the mechanical properties of the implants [42]. The biological response to zirconia implants is similarly quite as to alumina ones. *In vivo* tests have visualized that zirconia ceramic prostheses preserve biocompatibility for tens of years [16, 37]. According to the statistical data, adequately manufactured ceramic parts of endoprostheses are reliable enough in spite of the ceramic brittleness [36] and they undergo breaking far rarely because of other reasons (Table 2.8).

Table 2.8

Probability of revision operation of hip endoprostheses with a ceramic head [36]

Probability	Cause of revision operation
~ 10%	Aseptic loosening of endoprosthesis elements
~ 1%	Septic loosening
0.01 – 0.02%	Breakage of ceramic head

In 1996 the UK Medical Device Agency (MDA) informed that the sterilization raises the probability of the ball head fracture made of zirconia ceramics [43]. It was also underlined that worsened quality of finishing of the heads has lead to elevated wear of the conjugated UHMWPE friction parts. In MDA's opinion, vapour sterilizing should be eliminated for preparation of ZrO_2 implants.

Gray colour of zirconia ceramic implants changes for the pinky-gray after γ-radiation. As spectroscopy has shown, this change is due to the radiation-induced electron transitions on the local defects of the ZrO_2 crystalline lattice [44].

The raw stock for zirconia ceramics is chiefly excavated in South Africa and Australia. Some types of this material contain radioactive contaminants (uranium and thorium). As soon as it was established, the manufacturers started to use for endoprostheses only the purified raw materials and the problem was relieved [36]. The raw stock for the alumina ceramic does not contain any radioactive admixtures.

Calcium phosphate-based ceramics (CaP) are distinguished by the highest biocompatibility out of all other engineering materials. Its chemical structure is very similar to the endogenous bone structure (formed inside the organism), so the fibrous capsule is not formed round the implant, there is no inflammation or any reaction to a foreign body, nor any toxic effects, and the implant contacts directly the bone. The CaP-based ceramic interacts with the bone tissue and forms chemical bonds [42].

Two CaP ceramic materials have found extensive application in orthopedy, namely the ceramic based on calcium phosphate $Ca(PO_4)_2$ and hydroxyapatite $Ca_5(PO_4)_3OH$. Both ceramics are biologically active and are perceived calmly by the organism. The chemical and crystalline structure of hydroxyapatite shows strongest affinity to the natural bone tissue. The properties of the apatite raw stuff produced by the chemical industry for medical purposes are presented in Table 2.9.

The dissolution rate of phosphate ceramics in biological liquids *in vivo* is rather high and depends on the implant specific area. $Ca(PO_4)_3$ dissolves in acidic media 12.3 times and in alkaline 22.3 times faster than hydroxyapatite [45]. Since the phosphate ceramic block samples are not very strong, they are employed chiefly for coating metal parts of endoprostheses. In this respect, hydroxyapatite is preferable because it is friendly to the bone, and dissolves slower *in vivo*, which ensures favourable conditions to move the load from the prosthesis onto the musculoskeletal system.

Table 2.9
Properties of hydroxyapatite [37, 39]

Characteristics	Unit	Notes
Density	3.16 g/cm^3	Porous material
Specific area	50 m^2/g	
Breaking strength at compression	100–200 MPa	Depends on porosity
Breaking strength at bending	<100 MPa	Depends on porosity
Elasticity modulus	100 GPa	Depends on porosity
Wickers hardness	500	Similar to a window glass
Thermal expansion coefficient	$11 \cdot 10^{-5}$ K^{-1}	
Melting point	1650 °C	Exceeds decomposition point, Sintering temperature is below 1350 °C.
Colour	White, bluish	Depends on raw stuff and processing regimes

The phosphate ceramic coatings are applied on endoprostheses by the plasma spraying in adjustable gaseous media, e.g. Ar/H_2 or Ar/N_2 [46]. The ceramic powder particles are entrapped by the gas flow, are molten in the plasma moving closer to the substrate, and deposit as a coat. The particles partially decompose under the plasma temperature, so far the microstructure, phase composition, crystallinity, content of hydroxyls, and molar ratio of the Ca/P coating differ essentially from the initial ceramic, and depend on spraying regimes [47]. In the absence of water hydroxyapatite decomposes under 1400 °C:

$$2Ca_5(PO_4)_3OH = 2Ca_3(PO_4)_2 + Ca_4P_2O_9 + H_2O.$$

The coating thickness on metal components should be optimized in terms of strength and time of the phosphate ceramic dissolution *in vivo*. Thin coatings (till 10-15 μm) display perfect adhesive and mechanical strength values, although dissolve shortly after implantation. The thicker coats are capable of a prolonged interaction with the bone tissue, but may fail under shear loads leading to loosening of the endoprosthesis. The optimum coating thickness makes up 50-100 μm [42].

Regretfully, there is little information in literature about physico-chemical characteristics of just as the original phosphate ceramic powders, so the final biocompatible coatings, and their dependence on the process regimes in particular. These data are, as a rule, confidential and contain know-how of the manufacturers. In a number of cases these secrets help to prevent from failure of endoprostheses because of insufficient biocompatibility of their coatings or fixation tightness between the components.

2.5 COMPOSITES

Composite materials or composites present multiphase systems consisting of two and more components that preserve their intrinsic structure and properties within the composite formulation. A composite structure is schematically illustrated in Fig. 2.5. The continuous component across the composite bulk is called a matrix or a binder. The stiffening or reinforcing components are arranged within the matrix in a certain order. The transient surface layers are found at the interfaces between the matrix and other components. The transient layer properties (the third phase) differ from the other phases and designate adhesion of the matrix to other components, service characteristics of the composite, and their stability in time. Under the mechanical loading of the composite, stresses reach their maximum at the interfaces between the components.

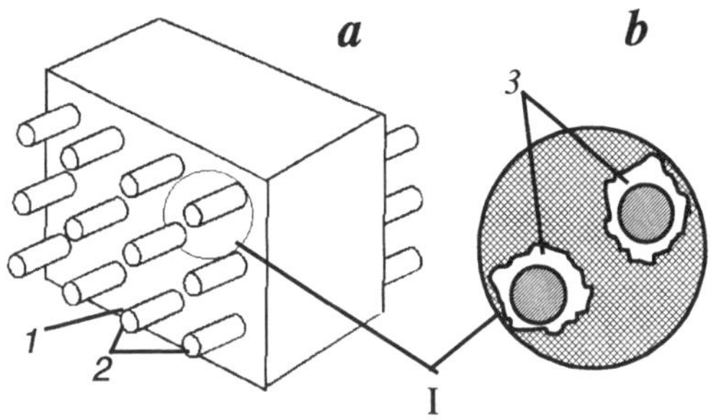

Fig. 2.5. Scheme of a composite material: 1 – matrix, 2 – reinforcing components, 3 – transient layer at the interface between components

Composites are used in endoprostheses of joins to add elasticity to the implants comparable to the osseous tissue. Despite a recognized progress in the development and usage of the metallic implants, all the attempts to impart them adequate elasticity turned to be insolvent. As the clinical experience has proved, all available for today elastic metal parts of endoprostheses fixed in bones (stems in the form of the converging or diverging rods of specific profile, dissected stems, etc.) undergo early failure. Besides, metals are not accepted by the organism as biocompatible materials. Even in the absence of the initial acute reaction to the foreign body, the organism responds by the dystrophy of the osseous and soft tissues contacting the metal implant, negative reactions on the wear debris and corrosion products of the prostheses from the base metals in the follow-up period.

Carbon composites used for endoprostheses contain carbon fibres as the reinforcing elements of carbon matrices.

The endoprosthesis elements are most often formed from the high-modular carbon fibres or threads by weaving, layer-by-layer packing of pierced carbon fabric, strips or bundles of carbon fibres. The matrix composition is introduced into the blank by the impregnation or deposition on fibres as coatings.

As the initial impregnation compositions the carbonified phenolic or furfurol resins and coal-tar pitch can be employed (carbonization means increase of carbon content in the organic substance by heat, light, ionizing radiation or other treatment). Thus impregnated and cured blank is subjected to carbonization by heating in vacuum or inert (nitrogen) gaseous medium up to 650–1100 °C. The output of carbon (coke) from the synthetic resins makes up 70-75%, from coal-tar - 50-65%. After this the coke matrix undergoes graphitizing under 2600–2700 °C. To enhance density of the composite, the technological cycles (impregnation, setting, carbonization, and graphitizing) are repeated. The final thermal treatment is carried to remove the volatile products, complete the structural formation, and reduce residual stresses in the blank.

The matrix material is deposited on the carbon fibres by a gas-phase method. With this aim, methane or natural gas is forced through the heated to 1500 °C carbon skeleton. Up to 1 mm thick carbon coatings are deposited on the skeleton fibres and further subjected to graphitizing. The greatest density and strength of the composite are attained in the case of combined gas-phase precipitation and resin-impregnation. A production layout of the carbon elements of endoprostheses is shown in Fig. 2.6. Depending on the process regimes, initial material of the matrix, packing arrangement of the carbon fibre skeleton, and some other factors, the mechanical properties of endoprostheses can be varied within a wide range.

The carbon matrix of composite endoprostheses possesses high fatigue strength and can achieve a high-grade polish. The wear resistance of the prosthesis components enhances with increasing density of the carbon matrix. However, low resistance to abrasive wearing is an acute problem of carbon endoprostheses, which is in part solved by the application of the gas-phase coatings on the friction surfaces. The biological response to carbon components is independent of the production method. Carbon wear debris accumulated in the glands does not affect the organism. Observations of the carbon implant-bone contact for 8 to 12 weeks have shown that the fibrous capsule is not formed, while the bone tends to ingrow into the implant pores (50-150 μm) [42].

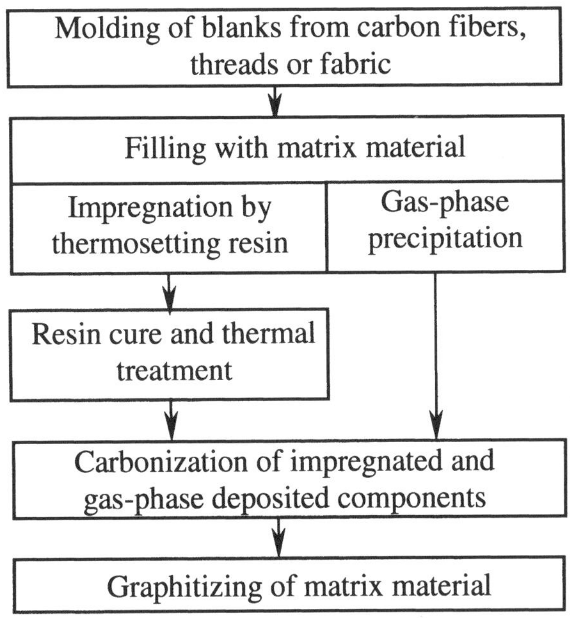

Fig. 2.6. Production layout of endoprostheses parts from carbonic composite materials

The State Research Institute of Graphite-based Structural Materials (NIIGraphite, Moscow) has developed and produces the INTOST-type of carbon materials for endoprostheses. INTOST-3 is obtained by the gas-phase deposition of carbon coatings on the fine-dispersed particles of a non-baked coke kept in the fluidized bed followed by their compaction in moulds and carbonization of the semis.

INTOST-4 has 18-100 μm in diameter through pores. This carbon material is used as a substitute of bone defects and as components of knee prostheses to be fixed in the marrowy canal. After the implantation, the osseous tissue ingrows in the micropores and thus fixes the prosthesis.

Polymer composites consist of the reinforcing carbon long or short fibres encapsulated in a polymer matrix. Short fibres add less strength to the matrix as compared to long ones, but the composites with shorter reinforcements can be processed into products by highly productive methods like extrusion or injection moulding. The composite endoprostheses parts can be hot-pressed into the sheets of polymer-impregnated carbon fibres, called prepregs, or alternating carbon and polymer fibres. The carbon fibres in each layer may be parallel or directed at some angle to the formed part axis (0 till 90°). The blanks are formed as a result of the polymer binder fusing in different layers.

The polymer composites reinforced by carbon fibres produced in the USA are based mainly on polysulfone or polyetheroketone matrices. The latters acquire a polycrystalline structure and are more stable chemically, being on a par with polysulfones in the mechanical properties. A polysulfone prosthesis part structure reinforced by carbon fibres is illustrated in Fig. 2.7. The endoprosthesis stem consists of a composite core made as a bundle of unidirectional carbon fibres within the polysulfone matrix. The core is encased in a braid of carbon fibres, which is enveloped, in turn, in a polysulfone shell. Cross dimensions of the stem are proportional to the marrowy canal to fit well the endoprostheses [48].

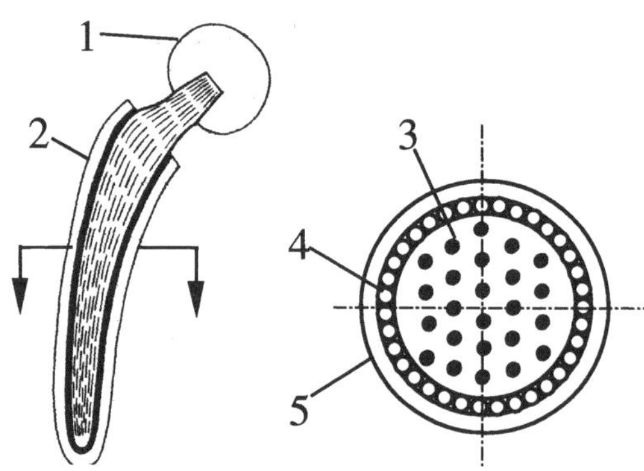

Fig. 2.7. Scheme of a composite hip endoprosthesis: 1 – head, 2 – stem, 3 – core, 4 – braid, 5 - shell

INTOST-1 produced by NIIGraphite (Russia) consists of the short carbon fibres and a polyamide matrix. Its elasticity modulus is twice as low as that of the osseous tissue, but its bending strength is rather high - 300 MPa. Thanks to its low elasticity modulus, the contact stresses at the bone-composite interface distribute evenly and do not lead to resorption of the neighbouring tissue in case of cementless fixation. INTOST-1 is employed for the hip, metacarpophalangeal, interphalangeal stems, and fixatives for treating fractures of the femur neck and long bones.

This review of the materials used for joint endoprostheses is summarized in Table 2.10, which gives a comparison of strength parameters of the artificial materials and biological tissues. Although flow values are cited here only for metals and UHMWPE, it does not mean that other materials undergo exclusively the brittle mode of failure. The biotissues possess sufficiently stable structure kept undamaged almost till the very failure. This is one more evidence to the principal differences between the biotissues and metals.

Most hard of the materials analyzed are the ceramics and carbon. All of them, together with metals are most tough materials, being by far stronger than the biological tissues. This, however, does not belittle the merits of artificial materials since the structural strength is never superfluous. Being inferior to artificial materials in strength, biotissues form such natural limbs whose tribological parameters are incomparable to any man-made friction joints. This problem of endoprosthetics is considered to be most intricate, and the ways of its solution are discussed in chapter 5.

The analysis of Table 2.10 shows imperfection of engineering materials as the analogues of biotissues. The formers are able to bear and transfer the mechanical loads, while the latters represent a controllable biological system whose properties depend upon the distribution of bio potentials and filling with biological fluids. The creation of such systems as components of smart technical materials presents a lofty challenge for the developers.

Table 2.10

Strength of biological tissues and materials for joint endoprostheses [42]

Materials	Yield stress, MPa	Breaking strength, MPa	Relative elongation at rupture, %	Elasticity modulus, GPa
Ceramics				
Alumina		550		380
Zirconia				
– hot-isostatic pressing		1200		200
– sintered		900		200
Metal alloys				
Protasul-10, forged	1000	1200	10	200
Stainless steel				
– cold working	750	1000	9	200
– annealed	170	400	45	200
Ti (6%) – Al (4%) – V alloy	890	1000	12	105

Carbon and composites				
Gas-phase deposited carbon		350–700	2–5	14–21
Carbon fibres				
– low-modular		1720	0.75	380
– high-modular		2760	1.4	240
Carbon-carbon composite				
– parallel fibres		1200		140
– mutually perpendicular		500		60
Polysulfone-carbon composite, parallel fibres		2130	1.4	134
Biological tissues				
Hydroxyapatite		100	0.001	114–130
Bone		80–150	1.5	18–20
Collagen		50		1.2
Polymers				
PMMA bone cement		75	3.5	2.8
UHMWPE	23	40	500	0.5
Silicone rubber		7–10	600	0.0003

◈ ◈ ◈

Biocompatibility and aging in the biological environment of organism still remain the main problems of the endoprosthetic engineering materials. In view of ultimate responsibility assumed by the implantation of artificial organs, there exist and are developing special state control bodies for the biomaterials and implanted endoprostheses. The overwhelming predominance of metals in this field is loosing its absolutism in the 21st century. Composites are believed to simulate the physico-mechanical properties of the osseous tissues and are more affined to the biological environment than other materials. It is a general tendency of the modern medical materials science to cease moving high-grade materials from engineering for their adoption in medicine, but to start a targeted creation of biotissues analogues intended for application in endoprostheses.

References:

1. Nicholson J. Current trends in biomaterials. *Materials Today*, 1998, V. 1, No. 2, p. 6–8.
2. Dee K.C., Puleo D., and Bizios R. Engineering of materials for biomedical applications. *Materials Today*, 2000, V. 3, No. 1, p. 7–10.

3. Kao W.J. Evaluation of protein-modulated macrophage behaviour on biomaterials: designing biomimetic materials for cellular engineering. *Biomaterials*, 1999, V. 20, p. 2213–2221.
4. Jenney C.R., and Anderson J.M. Alkylsilane-modified surfaces: inhibition of human macrophage adhesion and foreign body giant cell formation. *J. Biomed. Mater. Res.*, 1999, V. 46, p. 11–21.
5. Karlov A.V., and Shakhov V.P. Systems of external fixation and regulating mechanisms of optimum biomechanics, Tomsk, STT, 2001, 480 p.
6. James K., Levene H., Parsons J.R., and Kohn J. Small changes in polymer chemistry have a large effect on the bone–implant interface: evaluation of a series of degradable tyrosine–derived polycarbonates in bone defects. *Biomaterials*, 1999, V. 20, p. 2203–2212.
7. Gumargalieva K.Z., Zaikov G.E., and Moiseev Y.V. Macrokinetic aspects of biocompatibility and biodegradability of polymers. *Successes in Chemistry*, 1994, V. 63, No. 10, p. 905–920.
8. Definitions in biomaterials. Ed. by D.F. Williams. Amsterdam, Elsevier, 1987, 280 p.
9. Ratner B.D. Biocompatibility: the convoluted path to a working definition. *Trends in Polymer Science*, 1994, V. 2, No. 12, p. 402–403.
10. Brunettle D.M., and Chehroudi B. The effect of the surface topography of micromachined titanium substrata on cell behavior *in vitro* and *in vivo*. *J. Biomech. Eng.*, 1999, V. 121, p. 49–57.
11. Structural Biological Materials: design and structure – property relationships. Ed. by M. Elices. Amsterdam, Pergamon, 2000, 362 p.
12. Implants in Surgery. Ed. by D.F. Williams, and R. Roaf. Philadelphia, W.B. Saunders Co., 1973, 360 p.
13. Lemons J.E. Metals and alloys. In: Total Joint Replacement. Ed. by W. Petty. Philadelphia, W.B. Saunders Co., 1991, p. 21–27.
14. Underwood E.J. Trace elements in human and animal nutrition. New York, Academic Press, 1977, 255 p.
15. Williams D.F. Corrosion of implant materials. *Annual Review of Material Science*, 1976, V. 6, p. 237–266.
16. Handbook of materials for medical devices. Ed. by J. Davis. Ohio, ASM Int., 2003, 340 p.
17. Pourbaix M. Electrochemical corrosion of metallic biomaterials. *Biomaterials*, 1984, V. 5, p. 122–134.
18. Solar R.J., Pollack S.R., and Korostoff E. *In vitro* corrosion testing of titanium surgical implant alloys: an approach to understanding titanium release from implants. *J. Biomed. Mater. Res.*, 1979, V. 13, p. 217 – 229.
19. Maurer T.V., Ochsner P.E., Schwarzer G., and Schumacher M. Increased loosening of cemented straight stem prostheses made from titanium alloys. An analysis and comparison with prostheses made of cobalt-chromium-nickel alloy. *Int. Orthop.*, 2001, V. 25, p. 77-80.

20. Sott A.H., and Rosson J.W. The influence of biomaterial on patterns of failure after cemented total hip replacement. *Int. Orthop.*, 2002, V. 26, p. 287-290.
21. Movshovich I.A., and Vilensky V.Y. Polymers in traumatology and medicine. Moscow, Medicine, 1978, 320 p.
22. Andreeva I.N., Veselovskaya E.V., Nalivaiko E.I., et al. Ultra-high molecular weight polyethylene of high density. Leningrad, Khimiya, 1982, 80 p.
23. American Society for Testing and Materials (ASTM) F648-84. Standard specification for ultrahigh molecular weight polyethylene powder and fabricated forms for surgical implants. Philadelphia, ASTM Press, 1989.
24. Cross-linked and thermally treated UHMWPE for joint replacement. Ed. By S. Kurtz, R. Gsell, and J. Martell. Philadelphia, ASTM Int., 2004, 270 p.
25. USA patent 6242507, C 08 J 3/28. Process for medical implant cross-linked ultrahigh molecular weight polyethylene having improved balance of wear properties and oxidation resistance. K.A. Saum, W.M. Sanford, W.G. DiMaio, and E.G. Howard, 2001.
26. The effect of sterilization methods on plastics and elastomers. NY, Plastic Design Library/ William Andrew, 1994, 470 p.
27. Puolakka T.J.S., Keranen J.T., Juhola K.A., et al. Increased volumetric wear of polyethylene liners with more than 3 years of shelf-life time. *Int. Orthop.*, 2003, V. 27, p. 153-159.
28. Petty W. Fixation methods. In: Total Joint Replacement, ed. by W. Petty. Philadelphia, W.B. Saunders Co., 1991, p. 61–74.
29. Glyn-Jones S., Hicks J., Alfaro-Adrian J., et al. The influence of cement viscosity on the early migration of a tapered polished femoral stem. *Int. Orthop.*, V. 27, p. 362-365.
30. Surgical adhesives and sealants. Current technologies and applications. Ed. by D.H. Sierra, and R. Saltz. Lancaster, PA, Technomic Publ. Co., 1996, 255 p.
31. Yan W.P., Ng T.P., Chiu K.Y., et al. The performance of three vacuum-mixing cement gun – a comparison of the fatigue properties of Simplex P cement. *Int. Orthop.*, 2001, V. 25, p. 290-293.
32. Gavryushenko N.S. The effect of various physico-mechanical factors on the fate of joint endoprosthesis and its functionality. *Bull. of Traumat. & Orthop. named after N.N. Priorov*, 1994, No. 4, p. 30-34.
33. Frisch E.E. Technology of silicone in biomedical applications. In: Biomaterials in Reconstructive Surgery, ed. by L.R. Rubin. St. Louis, C.V. Mosby, 1983, p. 73–90.
34. Rose R.M., Paul I.L., Weightman B., et al. The role of stress enhanced reactivity in failure of orthopedic implants. *J. Biomed. Mater. Res. Symp.*, 1973, V. 4. p. 401–418.

35. Swanson J.W., Poitevin L.A., Swanson G.G., et al. Bone remodelling phenomena in flexible implant arthroplasty in the metacarpophalangeal joints. *Clin. Orthop.*, 1986, V. 205, p. 254–267.
36. Willman G. Hip–joint replacement – still a challenge to orthoplaedists, tribologists and designer. In: Industrial and Automotive Lubrication, 11th Int. Coll., Esslingen, V. 1, 1998, p. 7–20.
37. Black J. Biological performance of materials. New York, Marcel Dekker, Inc., 1992, 470 p.
38. Clarke I.C., and Willman G. Structural ceramics in orthopedics. In: Bone Implant Interface, ed. by H.A. Cameron. St. Louis, C.V. Mosby, 1994, p. 203–252.
39. Hench L.L., and Wilson J. An introduction to bioceramics. Singapore, World Scientific, 1993, 350 p.
40. ISO 6474. Implants for surgery – Ceramic materials based on high-purity alumina. 2-nd ed. 1994-02-01.
41. ISO 13356. Implants for surgery – Ceramic materials based on yttria-stabilized tetragonal zirconia (Y-TZP). International Standard, 1997.
42. Brown S.A. Ceramics. In: Total Joint Replacement, ed. by W. Petty. Philadelphia, W.B. Saunders Co., 1991, p. 35–42.
43. Zirconia ceramic heads for modular total hip femoral component: Advice to users on resterilization. Safety Notice MDA SN 97617, 1996.
44. Dietrich A., Heimann R.B., and Willman G. The colour of medical – grade Zirconia. *J. Material Sci.: Materials in Medicine*, 1996, No. 7, p. 559 –565.
45. De Groot C.P., Wolke J.G., and De Blieck-Hogervorsat J.M. Chemistry of calcium phosphates bioceramics. In : CRC Handbook of Bioactive Ceramics, ed. by T. Yakamuro, L.L. Hench, and J. Wilson. Vol. II: Calcium Phosphates and Hydroxyapatite Ceramics. Boston, CRC Press, 1990, p. 3 – 16.
46. Liu D.M., Chou H.M., Wu J.D., and Tung M.S. Hydroxyapatite coating via amorphous calcium phosphate. *Mater. Chemistry and Physics*, 1994, V. 37, No. 1, p. 39–44.
47. Wang B.-Ch., Chang E., and Yang Ch.-Y. Characterization of plasma – sprayed bioactive hydroxyappatite coatings *in vitro* and *in vivo. Mater. Chemistry and Physics*, 1994, V. 37, No. 1, p. 55–63.
48. Koeman J.B., Magee F.P., Longo J.A., et al. Design and testing of a carbon fiber – thermoplastic composite artificial hip prosthesis. *6-th Int. Conf. on Polymers in Med. and Surg.*, 1989, p. 10/1 – 10/4.

Chapter 3. DESIGNS OF JOINT ENDOPROSTHESES

Orthopedy is rapidly moving ahead thanks to upgraded mastery and creative work of surgeons closely coordinated with engineering sciences and industry. Prospect studies of orthopaedists have encouraged recent comprehensive investigations in biomechanics and related technical domains. Advanced software and computer design methods serve the base for developing modern endoprostheses from specific materials using refined procedures. Endoprosthetic operations on joints are inseparably linked not only with the knowledge and experience of surgeons but with technical perfection of endoprostheses as science-intensive commercial products as well.

The present chapter gives an overview of general joint endoprosthesis designs available in orthopedy today. We mean that these structures contain the elements distinguished by the principal features of advanced endoprostheses. We are not going to give a retrospective view of the path the endoprostheses have passed through in spite of the evident appeal and usefulness of such analysis for both orthopaedists and engineers. There are the ones among these designs that have made invaluable contribution to maturing of orthopedy and became a thing of the past.

The chapter begins with a paragraph devoted to the problems and challenges to be solved by endoprosthetics. Meditations over these philosophical questions show how huge is the combination of social, economic, technical, biological and medical aspects that should be embraced by the designers of endoprostheses. Further consideration of endoprosthetic designs gives a survey of artificial joints and the means of their upgrading not yet embodied in engineering solutions. The chapter is concluded by a brief review of the test methods and presents the main trends in perfecting endoprosthetic designs.

3.1 THE PHILOSOPHY OF DESIGNING ENDOPROSTHESES

Several dozens of companies in different countries of the world offer today joint endoprostheses of different designs and technological parameters. The manufacturers of each brand try to advertise their products underlining the advantages and high-minded aims they serve to. To all appearance, it seems possible to create the advanced generation of endoprostheses by a simple summation of the achievements of today's structures. However, this will not bring us nearer to an ideal. When designing joint endoprostheses one should bear in mind so many medical, social, engineering and many other aspects, solve the problems in various fields of knowledge intricately entangled in a cause-and-effect relationship that it is impossible to correlate them in this simplest manner. Let's name most important factors accompanying the development of endoprostheses.

The social and economic aspects influencing designing of endoprostheses are connected, first of all, with a great number of sick people that are in need of the operations on replacement of joints. They amount to 6-7 million per year. These are potentially able-bodied people that can return to a full-fledged work after endoprosthetic operations. The cost of this treatment per patient is related to the mean payment for disablement full or partial as 1:42 [1]. This is why the sick people in want of endoprosthetic operations should be supported just as in moral, so social and economic respects. This support creates certain difficulties, to which we relate the following.

Modification of endoprostheses presupposes an essential limiting of the number of intraoperative, early and remote post-operation complications, and soothing of the pain. From the other hand, improved endoprostheses become more expensive. For example, the hip endoprosthesis may cost from one till five thousand USD depending on the material used and type of machining. So, what are the reasonable criteria of relating the cost to the level of medico-engineering perfection of endoprostheses?

In this connection, we should always remember about the feelings of the operated patients because those expecting a miracle after the operation are unfortunately mistaken. After endoprosthetic operations the patient has to refuse from some habitual feelings and adapt to new ones, should thoroughly and unhurriedly master the prosthesis. One of the most common complications encountered within the first months after the operation on the hip, which is especially peculiar for the elderly people, is dislocation of the endoprosthesis head. The number of additional factors contributing to instability of artificial limbs is growing with age. Therefore, the sick are trained to acquire the skills helpful after the operation. The patients should know the structure of artificial joints, probable complications in the post-operative period and the precaution measures.

It was at the SICOT Symposium (Scientific International Council of Orthopaedics and Traumatology) in 1966 when J. Charnley and K. McKey declared that the total hip replacement had become an accessible and low risk method of the medical aid, which necessitates revision endoprosthetics only in special cases. M. Muller noted that there still remained unsolved problems of instability of endoprosthesis elements and remote infections. He also underlined that there were grounds for developing preconditions of a qualitatively new pathological state, which he called the instability of endoprosthesis. Wherefrom he expressed a conviction that the total joint replacement was an operation of the reserve and the surgeons may resort to it only with patients after 65. These approaches have remained the subject of animated discussions up till now.

Above excerpts give an idea of the social and economic facets of the problems, their global character and urgency. The engineering tasks in modification of endoprostheses are similarly if not more complex than the previous ones.

Wear resistance of joint endoprostheses has been and still remains an actual problem of orthopedy. This is because all kinds of motion in prostheses are exercised via frictional joints fit for operation *in vivo* for many years without failure and renewal. Note that this problem is at least three-sided.

First, the useful operating life (we mean the life till failure due to limiting wear) of some joint endoprostheses can be shorter than the patient's life. Since the implanted joint can be substituted only in the course of the revision operation, wear resistance of the friction joint turns to be critical for both durability of the prosthesis and patient's health.

Second, the debris accumulating in the neighbouring tissues and lymphatic glands bring about complications and sometimes intoxication of the organism.

Third, there is a firmly established opinion on the aseptic origin of endoprostheses instability (looseness) as a result of the tissue clogging by the debris. Wear debris initiate bone lysis and loosening of the prostheses [2, 3]. There arises a question, whether the tissues are capable of removing adequately the particulates from the organism or their amount will grow with wearing. Evidently, the answer to this question is individual for each patient, type of prosthesis and operating conditions.

The choice of materials presents an important stage of designing endoprostheses. In the previous chapter we have pointed out the following chief criteria of meeting the requirements for artificial materials used in implants:

- biocompatibility supported by experimental investigations;
- requisite static and dynamic strength at compression and bending;
- resistance of the cement-implant joint to biological fluids of the organism;
- wear resistance of the materials based on above-named criteria within the endoprosthesis friction *in vivo*.

The endoprosthetic materials are commonly selected based on the experience, which is still rather contradictory. The materials should undergo to laboratory and bench tests adequately simulating service conditions of endoprostheses. The only adequacy criterion of a chosen material is operation results of the endoprosthesis *in vivo*. It is seen from the definitions of the terms of endoprosthetic materials cited in section 2.1 the biocompatible materials represent a combination of bioinert and bioactive materials.

Designing of endoprostheses is a creative process where all aforementioned factors are arranged so as to consider just as force and frictional loading of the artificial joint, so its size, cost and certain special requirements, such as wear debris localizing, specificity of its biological environment and many other.

The main design stages are the following:

- determination of the endoprosthesis kinematic scheme,
- construction of friction joints and fastening elements,
- design for statistical and dynamic strength, and wear resistance.

Modern achievements in biomechanics, the discipline treating regularities of the locomotive system from the standpoint of theoretical and experimental mechanics, form the scientific base for defining the kinematic scheme of endoprostheses [4].

The parameters of endoprosthetic structures are bounded by the corresponding characteristics of the constituent materials, such as breaking stresses, flow, creep, fatigue, brittleness, and other. As for the strength design, it is more reliable in respect to metals and less trusty for ceramics because its unpredictable brittleness may mislead computations. The biological interactions between the structural materials and living tissues complicate designing of endoprostheses still more.

Application of composites makes it possible to predict the required strength of the structures based on the optimum distribution of reinforcing elements in the matrix. The nature itself prompts this solution. A young bone tissue presents a structure naturally self-optimizing its strength criteria. Its biological sensors measure the mechanical stresses over different portions of the bone and suggest the organism where to built-up the bone tissue for levelling its stress field. Analogously, the organism stimulates lysis of the bone tissue on the bone areas perceiving little loads. To make the bone better perceive the loads, fibres in the bone tissue are oriented in direction of stress tensors. The cylindrical bone has its fibres arranged in-plane, whereas on the joint end they are distributed in different directions according to the load vectors.

A strength calculation method of the composite specimens developed in Cambridge (UK) simulates the bone [5] and is based on the Michel theorem, according to which the composite structure with the minimum mass under a given strength can be always presented as a skeleton formed by three orthogonal sets of fibres. The fibrous structure simulated on the base of biomechanical notions is then optimized by the balanced life criterion to reach the minimum mass of the product. This optimization is predetermined by the following reasons. In contrast to the reinforced concrete where the reinforcement is evenly prestressed, the stresses in the composite fibres are distributed randomly. Therefore, the first to break under the loading are the stressed fibre, which augment stresses in less loaded fibres and initiates a cascade of fibre ruptures. If the fibres are distributed proceeding from the balanced life criterion, the load on the composite can be significantly increased, its mass reduced and the structure of endoprosthesis components brought most closely to the bone.

The design approaches for the endoprostheses friction units are based on a vast theoretical and experimental foundation disclosed in chapter 5.

The structure of endoprostheses depends much on the fixation technique of its parts in the bone. Either a cement, press-fitting or hybrid method (as well as tissue grafting to the porous implant) is chosen depending on a number of interrelated factors, such as patient's age, the character of pathological processes in the joint, bone tissue density and other.

Let's examine the effect of the mechanical factors of endoprosthesis fixation on the operating life of artificial joints by way of example of the hip as most elaborated and widely adopted kind of endoprosthesis.

The endoprostheses of cementless fixation are designed largely for the patients under 50 whose bone tissues are still dense and strong. The stem structure presupposes the conical core and the presence of sharp edges. This ensures the immediate mechanical stability of the stem upon locking in the marrowy canal of the femur. The pelvic element is fixed in the acetabulum by screws, threading or dowels. Osseointegration presents a secondary fixation of the components, i.e. growing of the bone tissue in the implant's surface. The rough surface of the stem locked in the cylindrical bone guarantees a 5 times higher strength of the press-fit joint at shear in contrast to a polished one [6]. During walking the parts of endoprostheses undergo microshifting so; the fixing elements should be thoroughly designed to avoid contortion of the bone tissue, wear and perforation of the bone.

The endoprostheses with a cement fixation of components are recommended at osteoporosis with the evident loss of the bone tissue. The cement mount makes it possible to approach the contact area in the bone-cement and implant-cement joints to the geometrical contact area. This is the reason why the cement joints of the stem-bone type show high reproducibility of the breaking strength at shear in a very short period after fixation in contrast to the cementless joints characterized by a large scatter of the results [6]. To optimize the roughness value of the stem surface one should proceed from the full-strength criterion of the stem-cement and cement-bone joints.

Stimulation of the biological fixation of the cementless endoprostheses has taken shape as a direction only recently. This presupposes sintering of the spherical titanium granules to the bone contacting the implant surface to form a porous coating. The pore size is 100-200 μm, which corresponds to the cancellous (piercing the bone tissue) blood vessels. The coating does not form a fibrous capsule in contact with the bone, so the bone grows into the coating pores within 2-3 months [7]. Modern trends in biological fixing of endoprostheses involve the application of hydroxyapatite coats on the surfaces in contact with the bone. Work [8] cites the clinical and radiographic results of two matched series of total hip arthroplasties, one with hydroxyapatite-coated femoral stems, and the other with a similar but porous-coated femoral stem. Circumferential hydroxyapatite coatings reduced the occurrence of osteolysis and eliminated distal osteolysis at 5-10 years of follow-up. In addition, hydroxyapatite coatings of the stems not alter the wear rate.

It is preferable to make joint endoprostheses detachable or interchangeable. A modular design of the femur developed and tested in the

1970s had a spherical head press-fit by its conical hole into the cone of the prosthesis stem [9]. Unification of both cones makes possible the application and combination of the endoprostheses produced by various companies. An example of such structure is BiCONTACT hip endoprosthesis system with the characteristics geared towards providing uniform, optimal treatment under economically favourable conditions for the widest possible stock of patients [10].

For a long time the only method of estimating endoprosthesis designs was an engineering or clinical test. The clinical tests lasted according to WHO for not less than 6 years. This is why the leading manufacturers of endoprostheses started to employ the computer techniques in early 1980s, including computer-aided design (CAD) and computer-aided manufacturing (CAM). To these also belonged various social programs evaluating the stress state of the implanted endoprostheses, the zones of stress concentrations in the bone-implant systems able to promote resorption of the bone tissues and other advanced means (Fig. 3.1).

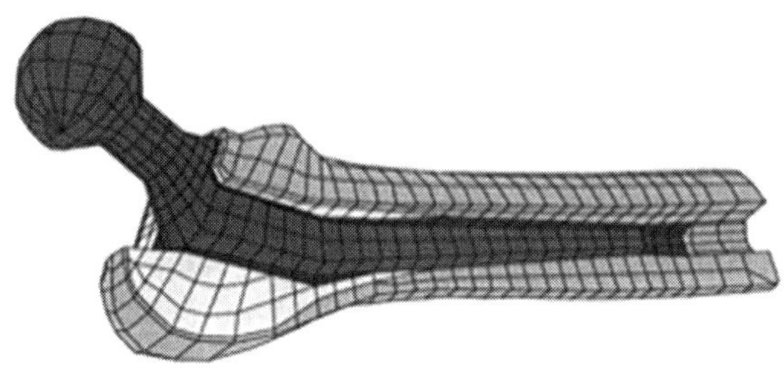

Fig. 3.1. A computer model (the finite element method) of cement fixture of the hip endoprosthesis stem. The lines represent planes of equal stresses

Using the finite element analysis the pelvic bony socket had been studied in [11] and compared with radiological imaging using acetabular cups of three different shapes. This model uses computer simulation to predict bone changes with different designs of implants. The ability to simulate biological conditions is a valuable addition to the testing of mechanical strength.

Focusing our attention on designing joint endoprostheses we would like to underline once again its immense and multifactor character. It is hard to present any unequivocal algorithm for solving this problem, so the best for the existing structures of endoprostheses seems to be a heuristic basis, which involves a creative approach, intuition and talent of orthopaedists and designers.

Rehabilitation of a patient upon the endoprosthetic operation is interconnected in a certain way with the endoprosthesis structure. The rehabilitation includes physiological recombination of the operated bones and surrounding muscles effecting engraftment of the prosthesis. The hydrodynamic effects of the elastic strains occurring at loading of the operated joint regulate the intrabone blood circulation, metabolism in the tissues and stimulate ingrowing of the bone tissue into the endoprosthesis [12]. The rehabilitation of the bone was simulated using a computer model [13]. The mechanical loads on the bone are optimized by the criteria of its strength and speed of response (adaptivity), and makes allowance for the degree of operative damage to correlate it with the endoprosthesis design.

Evaluation of the functional results of joint endoprosthetics is based mainly on the subjective feelings of the patient. The reliable feedback should be strongly adequate to promote upgrading of the endoprostheses. It should include [14]:

- the information accumulation methods on the shifts in joints;
- the evaluation criteria on the functional state of joints (see section 1.4), presupposing their further elaboration and substantiation;
- the database on the remote results of curing the patients with implanted endoprostheses.

Addressing to the natural objects has been for a long time adopted in designing various medical facilities [15, 16]. Either in the virtue of the protracted period allotted to Nature for the work at her objects or thanks to high qualification of the Creator, the biological designs strike our minds by their graceful rationality. The outstanding functional properties of the superior biostructures are related to their capability of adapting to varying external conditions. So, it is not surprising that there is a clear tendency in designing materials for the implants and joint endoprostheses by imparting the properties of cybernetic systems [17, 18]. Just a few examples of designing endoprostheses and their members by repeating natural objects are cited in Ch. 6.

How to know, which endoprosthesis design is better? We may receive an unambiguous answer to this philosophical in its essence question through the cut-and-try approach. With this aim, the results of prolonged treating large groups of people should be analyzed. The world experience in endoprosthetics has proved the necessity of the state control over the fate of each implant and storage of all, clinical, radiological and biomechanical results of observations in the information centre. There are such special independent and partly supported by the government laboratories in a number of countries. They check all available in the country endoprostheses designs in close to real conditions and give the official approval to their use. Analogous tests in Russia are conducted in the Central Institute of Traumatology and Orthopedy named after N.N. Priorov (CITO) and Scientific and Production Amalgamation "Energy" [6, 12]. Recommendations on the choice of the endoprostheses are given to medical institutions proceeding from a coordinated generalization of the ultimate results on the name of the governmental bodies [19].

It is understandable that there are the patients unable to pay a few thousand dollars for a joint endoprosthesis. Therefore, it is convenient to produce the endoprostheses in local conditions. Their designs are, however, far from being perfect and concede in majority of cases to their foreign analogues that have accumulated a longstanding experience and traditions.

We would like to note with regret that the system of numerous aspects participating in designing endoprostheses still suffers from the ambiguity of criteria. The probability theory treats the hypothesis of solving such systems as strongly dubious. Though being based on the trustworthy physical, chemical, biological and medical regularities, the above reasoning still bears the features of a philosophical discourse. Nevertheless, we have got by now a series of durable, high-quality joint endoprostheses and their designs are continuously modified. It would not be out of place to emphasize once again that the new generation of endoprostheses is a result of unconventional solutions, creative thinking, and a deep insight of orthopaedists, engineers, biologists and other involved specialists.

3.2 THE HIP

The biomechanical model of the hip (Fig. 1.2, *d*) presupposes its rotation about three axes. To this end, the functional scheme of endoprostheses is constructed in the form of a stem fixed in the femur and provided with a spherical head (Fig. 3.2). The head is found in a movable contact with a spherical cavity made in the cup and fixed in the acetabulum.

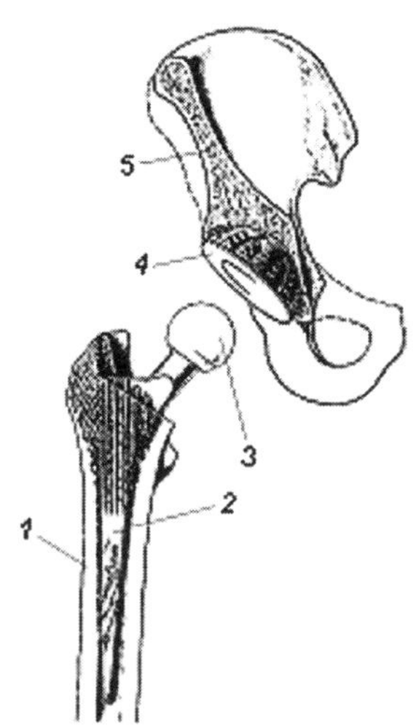

Fig. 3.2. Functional diagram of hip endoprosthesis: 1 – femur, 2 – stem, 3 – ball head, 4 – cup, 5 – pelvic bone

Basic designs of the total hip endoprosthesis are distinguished by a specific manner of fixing and configuration of the stem. A classical curved stem (called "banana") has been first proposed by J. Charnley. Charnley's endoprosthesis with a cement fixation has been adopted in endoprosthetics worldwide as a golden standard. The life span of based on it "Elite^{TM}" and "Elite^{TM} Plus" endoprostheses makes up 30 years in 92.5% of cases. An alternative biomechanical concept of endoprostheses developed by M. Muller has lead to the creation of a cement endoprosthesis "Straight stem". It is considered a standard in respect to the survival and simplicity of implantation. All other used today hip endoprostheses are modifications of above-named two basic structures. Some structural peculiarities of the main members of endoprostheses will be discussed in the sections to follow.

The endoprosthesis stem should be immovably locked in the femoral marrowy canal to ensure a stiff support for the implant.

The cementless fixation is exercised by press-fitting of the stem into the canal. The stems of the early Sivash's endoprostheses were either round or oval in cross-section to allow for the full filling of the marrowy canal. This resulted in tensile stresses in the femur and reduced substantially (from 2.5 till 1.4 t) its bearing capacity. The femoral canal was studied using model systems. With this aim, the cadaver femora were used [20]. Epoxyplastic replicas of cementless stems were used. Then the femora with implanted stems were radiographed.

The cross-sectional area of the stems in the best for today "Spotorno" and "Zweymuller" cementless endoprostheses is less than the full length of the canal of the stem. The "Spotorno" stem fills almost fully the canal in the proximal hip, whereas "Zweymuller's" – in the distal part (distant from the body) (Fig. 3.3). Both stems are straight and are made in the form of a taper with the apex in the distal part. The "Spotorno" stem is provided with the sharp conical ribs parallel to the taper axis to fix the stems intraoperatively in the femur canal. Both stems are popular with the orthopaedists in Europe.

It is widely accepted that the design and geometry of the implant impacts its ability to transfer loads to the femur [21]. Consequently, the geometry of the femoral component plays a pivotal role in explaining periprosthetic bone loss due to stress shielding [22]. Because of the lateral flare geometry, loads in the proximal femur are more evenly distributed. The lateral flare components rest upon an additional lateral column of cortical bone, stabilizing it against subsidence (Fig. 3.4). This additional lateral contact area creates a wider base of support and has been proven to help provide a more physiological distribution in the proximal femur and diminish the stress shielding [23].

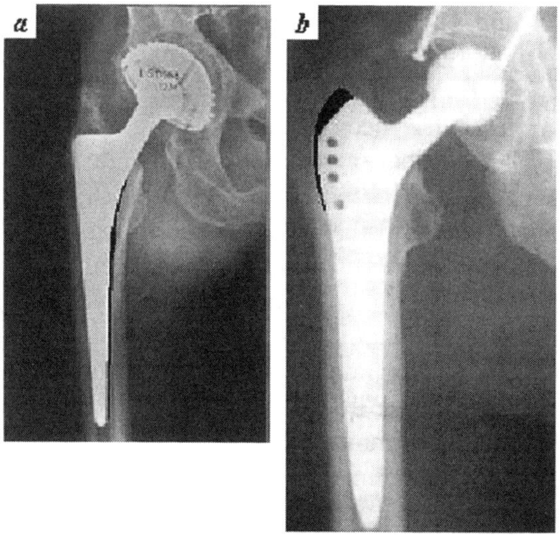

Fig. 3.3. Radiographic photos of implanted femoral endoprosthesis stems: *a* – "Spotorno", *b* – "Zweymuller". Clearances between the implant and the marrowy canal are blackened

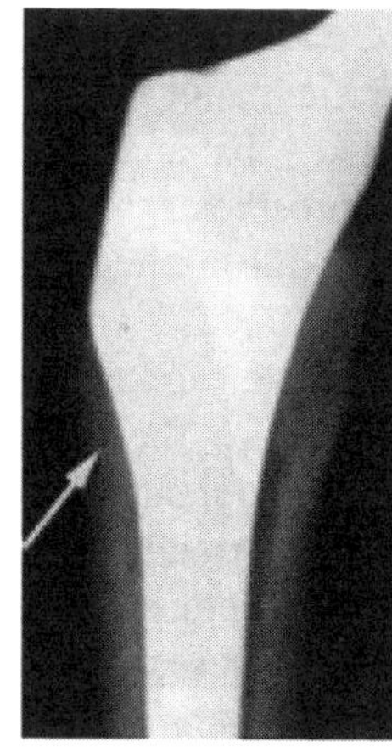

Fig. 3.4. View of the lateral flare feature resting on top of the lateral cortex, and the bone response elicited

The stems with the cement fixation are implanted mainly to elderly patients or to those with explicit osteoporosis of the proximal hip. The distribution schemes of the mechanical stresses round the implanted Charnley curved stem and Muller's straight one are illustrated in Fig. 3.5.

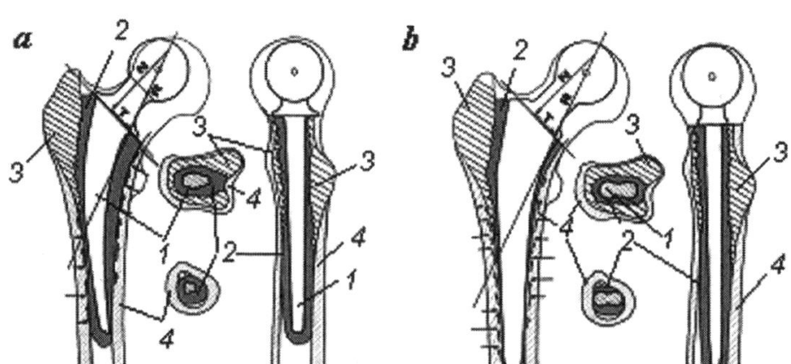

Fig. 3.5. Distribution schemes of compressive and shear stresses in joints with implanted Charnley's (*a*) and Muller's (*b*) stems [26]: 1 – stem, 2 – cement mantle, 3 – cancellous bone, 4 – cortical layer of the femur. Arrows show the direction of stresses in the bone

The stress fields in the contact zone between the stem and cement stem and bone, and bone-cement arise under the action of radial *R* and tangential *T* constituents of load *N* on the endoprosthesis. The straight Muller's stem (*b*) contacts the bone where the cement mantle is broken and perceives less loads. The role of cement mantle thickness in femoral stem survivorship is poor understood [24]. To form a continuout cement mantle and regulate its thickness the modern cementing technology uses medullary plugs. They are made of a bone tissue, PE, bio absorbable gelatine and other materials. The material of the plug does not in fact influence the quality of the cement mantle [25]. The compressive stresses created by force *T* between the straight stem and proximal femur are much lower than in the curved stem. The straight stem undergoes auto cementing in the marrowy canal. To make the cement mantle more even in thickness, the Charnley curved stems are furnished with interlockers on the stem end.

The first designs of the straight stems had a collar to prop against the trim of the femur. M. Muller, however, refused from it later since he considered it difficult enough to resect the bone so as to reach an ideal fit with the collar. What is more, the stems with such collars are all the same subsiding in the femur's canal because of the lysis on its edges [26].

W. Link has suggested a so-called anatomic curved shape of the stem (Fig. 3.6) to lower stress concentration on the walls of the marrowy canal and ensure even thickness of the mantle. Porous coated short anatomic femoral stem (DePuy design) can diminish stress shielding of femur proximal area [27].

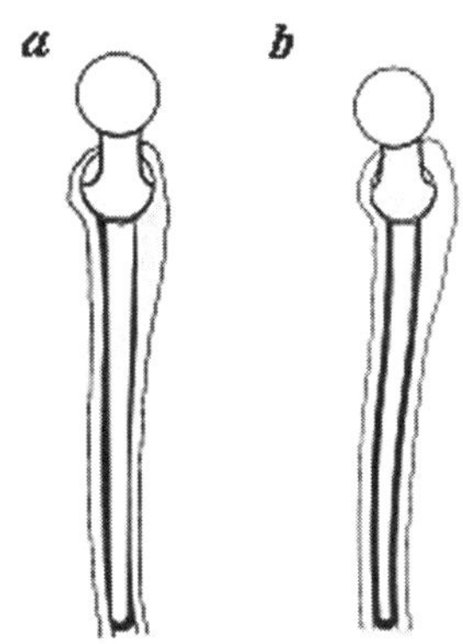

Fig. 3.6. Saggital section of femoral bones with implanted stems: *a* – "Straight stem", *b* – "Link SP" of anatomic form (cement mantle is blackened)

Endoprosthesis stems of the hip are commonly made of metals. Those of polymer composites are at present at the stage of testing and have not been yet adopted at the market unanimously.

The proximal stem constricts into a neck at an angle to the femur varying from 115° till 145°. Its value depends on the anatomic peculiarities of the patient and structure of the pelvic part. The neck ends in a standard cone.

Thanks to favourable outcomes of endoprosthetic operations with above-described stems the process of creating new ones has if not ceased lately then slowed down noticeably.

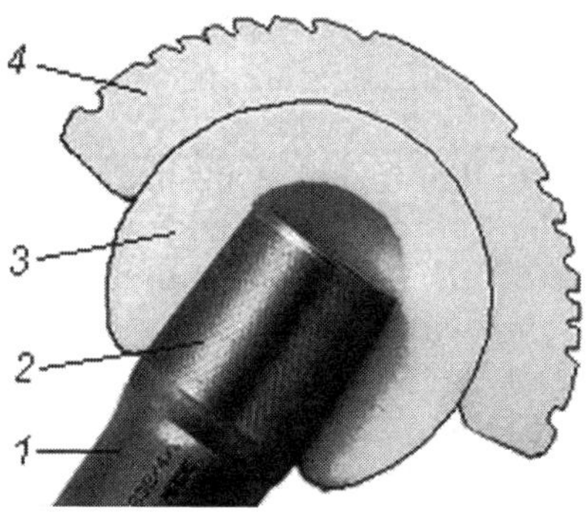

Fig. 3.7. Cross-section of the total hip joint endoprosthesis: 1 – stem neck, 2 – conical neck end, 3 – ceramic spherical head, 4 – UHMWPE liner

The ball head in the early designs of the hip endoprosthesis was made conjointly with the stem. The modern structures have the head fitted on a tapered neck (Fig. 3.7), which makes possible to use ceramic or metallic heads. The aperture on the head mates the taper on the neck. A standard cone of the modular endoprostheses is a machined Morse taper with a vertex angle 5°43'30''. In mechanical engineering such couplings are machined using mills, reamers, drills, and so on, and their taper shank is installed in a tapered cup of the work spindle. The high friction torque arising in this case keeps the tool from turning in the spindle during machining. Morse tapes ensure quick and handy coupling of the head and stem, accuracy of centring and reliability of the movable joints of endoprostheses. The commercial stems have acquired Morse's taper of 12 and 14 mm diameter. The head fits tightly the neck lest radial shifts of the head should initiate crevice corrosion or even breakage of the ceramic heads.

To adopt his concept of the low friction in artificial joints, J. Charnley recommended making the heads of the total hip endoprosthesis about 22 mm in diameter. The endoprostheses developed by M. Muller jointly with R. Mathys in the 1960s had 24 mm heads. The first Muller's straight head endoprosthesis (Setzholzprotese) was furnished with a 32 mm head to enlarge the span of the artificial joint motion and reduce the dislocation probability of the head. Up-to-date designs of the total hip endoprosthesis are manufactured with replaceable heads 28 and 32 mm. The initial roughness of the head is attached much importance to and has been specified in the standards [28].

To adjust the height of the head mounting on the stem according to the marrowy canal, the heads are made of three types: *s*, *m* and *l* depending on the depth of seating on the neck.

The cup of the hip endoprosthesis represents an acetabular prosthesis of the pelvic bone and functions in pair with the ball head. The friction part can be made of UHMWPE, metal alloys or ceramics.

In Fig. 3.8, one can see UHMWPE cups fixed by a bone cement (*a*, *b*), screws (*c*) or press-fit (*d*). Annular grooves and slots on the cups (c) along the meridians are facing the bone to allow the cement and bone in. One of the lower grooves contains a thin metallic ring for the X-ray evaluation of implantation correctness. The S-shaped cup (*b*) with flanges is able to maintain high pressure of the cement at fixing and its penetration into the osseous tissue.

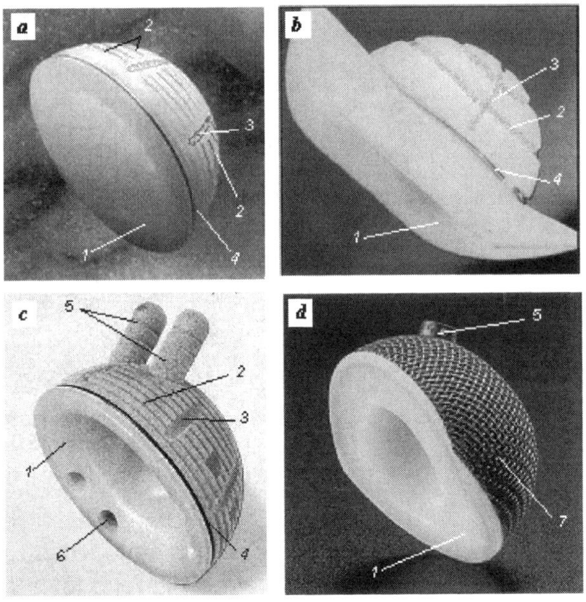

Fig. 3.8. UHMWPE cups of hip endoprostheses: *a* – Muller's, *b* - "Elite^{TM} Plus", *c* – Mathys', d – Morscher's. 1 – UHMWPE shell, 2 – annular grooves, 3 – slots, 4 – metal ring, 5 – support peg, 6 - screw hole, 7 –mesh coat

The cup illustrated in Fig. 3.8, *c* is fixed in the acetabulum by supporting pegs and special cancellous screws. To improve secondary fixation, the surface contacting the bone is covered with either titanium or hydroxiapatite coating. The life-table method predicted a 10-year survival rate of 97.9% with revision of a hydroxyapatite Mathys cup for any cause as the end point [29]. A special mesh coating on the external surface (*d*) presupposes cementless intraoperative fixation of Morscher cups by press-fitting. The pole of this cup is found lower that of the mating it sphere and the largest diameter exceeds that of the acetabulum by 1.5 mm. The mesh coating helps to fix the cup in the bone bed over the annular area when the cup is introduced into the acetabulum [30].

The pelvic component is most often composed of a few members, namely a metallic cup mechanically fastened in the acetabulum, and a friction counterpart immovably fixed in the cup in the form of the UHMWPE, metal or ceramic liner. The liner should be firmly fixed to provide reliability of the endoprosthesis. The main pelvic components of most common for today hip endoprostheses are shown in Fig. 3.9.

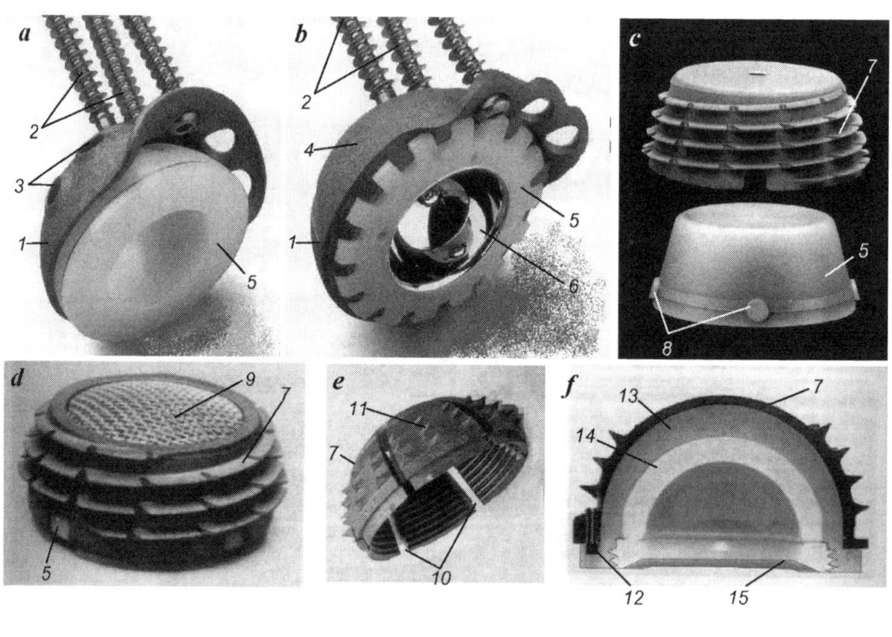

Fig. 3.9. Designs of pelvic components of hip endoprosthesis: *a* and *b* – Muller's with cemented and press-fit liners, *c* –Zweymuller's, *d* – Weill's, *e* – Spotorno, *f* – Link's. 1 – fixing ring, 2 – screw, 3 – screw hole, 4 – rough coating, 5 – UHMWPE liner, 6 – metallic friction part as a hollow sphere, 7 – threaded metal cup, 8 – retainer, 9 – mesh coating, 10 – cuts, 11 – tenon, 12 – screw, 13 – PE lining, 14 – ceramic liner, 15 – threaded ring

Pelvic components *a* and *b* of Muller's endoprostheses contain a metallic fixing ring screwed to the pelvic bone. The external surface of the rings has a rough coating for biological fixation in the acetabulum. Implanted porous-coated hemispherical press-fit cups of analogous structure (Stryker, Howmedica, USA) were classified as stable on radiography after mid-term observations (91.5 months) [31]. Bony in-growth on the smooth-surfaced hydroxyapatite-coated cementless ABG cup (Anatomique Benoist Girad, Howmedica, UK) is minimal. Macro-structure surface enhancing bony in-growth is necessary for long-term survival of these cups [32]. The application statistics of multihole-type Plasmacup-M (BiCONTACT total hip system) for displastic hips proves excellent results at 41 months follow-up [33]. The UHMWPE liner with grooves over its outer surface is fixed on the ring by the bone cement (*a*). A sandwich type structure is presented in Fig. 3.9, *b,* in which the UHMWPE liner is press-fit in the ring and incorporates, besides, a metallic or ceramic friction part.

Zweymuller's endoprostheses (*c*) have a threaded metal cup screwed immovably into the acetabulum. A tapered UHMWPE liner is inserted into the conical cavity of the cup. The retainers on the lower part of the liner enter the slots on the cup edge and hinder the liner from turning. A spherical cavity in the liner forms a friction pair with the ball head. Aseptic loosening of threaded cups is an unsolved problem. It is proposed in [34] to use a specific biomechanical procedure of assessment of long-term cup stability. Cups were screwed into high-precision excavations in foam blocks made of polymetacrylimide. The torque data were registered against the screw-in angle. These quantified results correspond to long-term stability of threaded cups.

The pelvic component of Weill's endoprostheses (*d*) consists of a tightened metal cup with screw ribs and a friction part in the form of a liner having a spherical cavity. A distinguishing feature of the design is a specific mesh coating made on the spherical surface of the cup to initiate grafting of the bone and osseointegration with the implant.

The Spotorno cup (*e*) is made as a hemispherical shell provided with cuts along the meridians. The sharpened tenons on the external surface of the shell are arranged in three annular rows. The cup is installed in the prepared for implantation acetabulum by fastening forceps. Upon removal of the forceps, the cup opens but partially. Using an expansion taper the cup is imparted a regular shape. Simultaneously, the tenons instil into the bone and fasten the cup. After this, the UHMWPE liner is screwed in the cup. The secondary fixation takes place upon ingrowth into the porous surface.

A cross-section of the pelvic component of the Link® endoprosthesis with a ceramic liner is presented in Fig. 3.9, *f*. It has a threaded over the external surface metal cup with a screwed to it polyethylene lining. A ceramic liner as a hollow hemisphere is mounted in the lining using a threaded ring. The advantages of this design are the next:

- reduced stiffness of the artificial joint in the ceramic/ceramic pair, relieved load peaks on the pelvic bone thanks to the elastoplastic lining (shock-absorbing);
- alleviated assembly/disassembly of the ceramic liner in contact with elastic PE.

In 1974 Burch initiated the construction of a custom-made acetabular implant to bridge a large posterior acetabular wall defect following a pelvic fracture. This implant remains in the use for more than 25 years. In 1976 E. Schneider embraced the idea and the Burch-Schneider anti-protrusio cage (BS) was established. From 1998 the BS was made of rough-blasted titanium (Fig. 3.10).

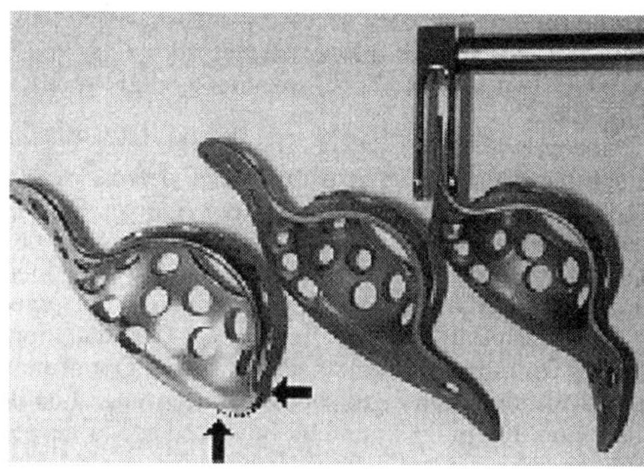

Fig. 3.10. Trial cage and anti-protrusio cage. The trial cage is without a distal flange.

The total hip replacement for secondary osteoarthritis is developmental hip displasia remains a difficult and challenging procedure. The reinforcement ring (RR) with the hook of R. Ganz in combination with autologous graft augmentation has been designed to address this issue [35]. The RR (Centre-Pulse, Switzerland) is made of pure titanium with an external frit-blasted porous surface.

Modifications of the hip endoprosthesis are numerous (about a thousand), though the majority of them differ just slightly from the basic variant. Of specific interest for the present study are the original designs of Isoelastic along with bipolar hip endoprostheses.

The first of them **"Isoelastic" prosthesis** developed by Prof. Mathys has a stem in the form of a polymer-coated metallic rod. In general, the idea of the elastic stem operating conjointly with the bone of a cementless hip endoprosthesis is extensively applied in practice today. Nevertheless, the clinical tests of the metallic elastic stems have proved once again that the metallurgy of today is unfortunately unable to give out a needed material. All available elastic structures of stems made as a bundle of rods, dissected or profile stems were found to undergo early breakage [36].

The first "Isoelastic" stems were cast of polyacetal reinforced by a metal cord. Polyacetal - a thermoplastic polymer $[-CH(CH)_3-O-]_n$ of a stereoregular crystallinity is identical in its physico-mechanical characteristics to a diaphyseal bone tissue. It does not, however, like all other engineering plastics, integrate with it biologically. This was the main reason of restrained attitude towards the "Isoelastic" stem. The problem was partly solved by the polyacetal of PEEK 450 G grade. It can be covered by a titanium coat known to be perfectly biocompatibile and inclined to osseointegration.

Figure 3.11 illustrates one of the hip joint components of the "Isoelastic" brand in which the stem is made as a metal-polymer rod with a polyacetal coating. The metal head is fixed on the polymer surface of the neck using a polymer liner fit in the head. The liner is fastened on the neck via a plastic ring inserted into the helical grooves on the neck and liner. The described type of fixation adds certain elasticity to the endoprosthesis.

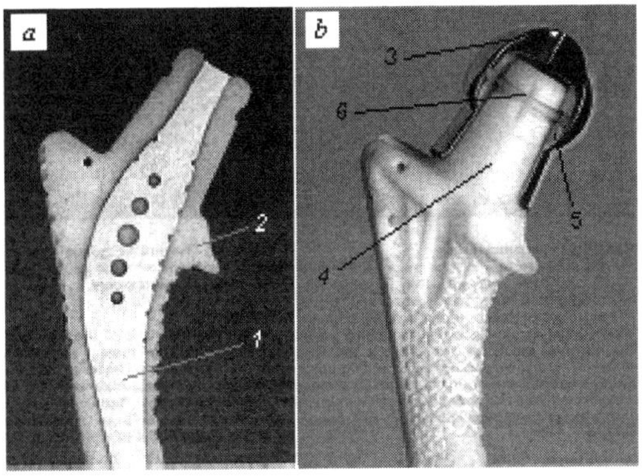

Fig. 3.11. Stem (*a*) and head (*b*) designs of RM endoprosthesis: 1 – metallic rod, 2 – polyacetal facing, 3 – metal head, 4 – neck, 5 – polymer liner, 6 – locking ring

The described Isoelastic stem was intended for young patients (30-40 years) for the cementless fixation and low osseous density to allow for the bone mass increment with recovery of the full load on the foot [37]. Long-term tests have not justified these anticipations due to aseptic instability of the stems. Nevertheless, the principle of isoelasticity in the hip replacement is important but requires improvements in the material and the design of the femoral components [38].

The bipolar endoprostheses of the hip are intended to replace femoral heads for articulation in pair with the natural acetabular cartilage. The cartilage is known to loose proteoglycanes at the early stages of friction against metals, which can be followed by the damage of the friction surface and degenerative changes in the cartilage (chondrolysis) [39]. Thanks to its original design the bipolar endoprosthesis ensures a mild friction regime over the cartilage (Fig. 3.12).

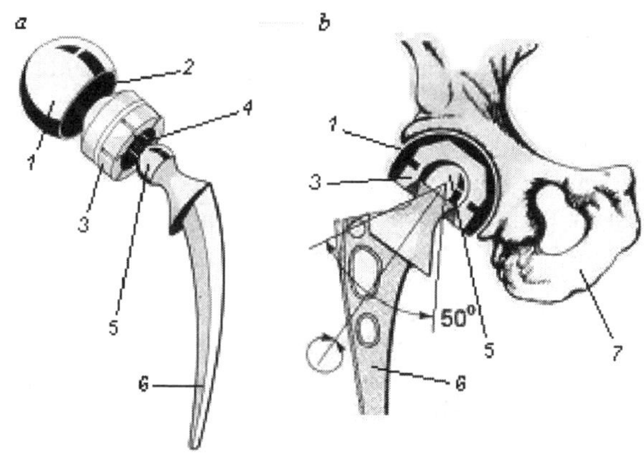

Fig. 3.12. Lima's (Italy) design (*a*) of bipolar endoprosthesis and its functioning scheme (*b*) in the acetabulum: 1 – external head, 2 – cylindrical cavity, 3 – cylindrical liner, 4 – spherical cavity, 5 – ball head of femoral component, 6 – stem, 7 – femoral bone

The outer ball head (40-58 mm in diameter) contacting the acetabulum is cast of a metal alloy. Its cylindrical cavity 2 contains a rotatable about the cylinder axis UHMWPE liner. The liner has, in its turn, a spherical cavity 4 incorporating a spherical femoral head (22.2 mm). The locus of the ball head turning angle inside the PE liner of the implanted endoprosthesis (*b*) equals to the volume of the cone with the apex in the head centre. The vertex angle is about 30° when the head is seated on the taper and reaches 50° in case the head is made conjointly with the stem. The head rotates inside the spherical cavity of the liner, and the liner in the external head limits reliably shifting of the endoprosthesis in the acetabulum and wear of the cartilage.

The original structures of the bipolar endoprosthesis show a coincidence between the centres of the external and stem heads (Fig. 3.13, *a*). Or otherwise, they would have a negative eccentricity in case the inner head centre is lower the external one (*b*).

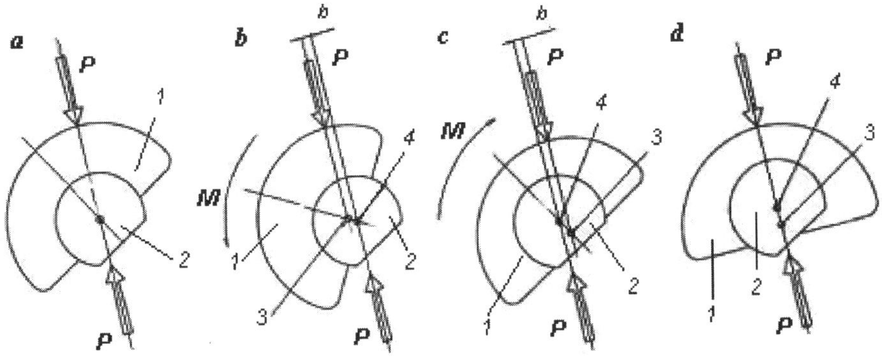

Fig. 3.13. Distribution of forces and torque in bipolar endoprostheses at alignment (*a*) of the external ball head centre with stem's one at negative (*b*) and positive (*c, d*) eccentricities: 1 and 2 – external and internal heads, 3 and 4 – rotation centres of external and internal heads, P – forces arising at loading of implanted endoprostheses

In the first case (*a*), the external head remains in the position it was mounted during the operation and shows no bias towards self-alignment under the varying load of the artificial joint. In the second case (*b*), loading of the endoprosthesis by load *P* brings about torque $M = P{\cdot}b$, where *b* is the centre-to-centre distance along the normal to force *P* action direction. This torque turns the external head in the acetabulum downwards, which disturbs stability of the endoprosthesis.

This is why the present-day endoprostheses originally acquire the positive eccentricity of their ball heads. It means that the internal head centre is found above that of the external one. This results (*c*) in a self-adjusting torque $M = P{\cdot}b$ that turns the external head in the acetabulum so that the centres of rotation of both heads align along the loading axis (*d*) [40]. Thus, the endoprosthesis acquires stability and sets in equilibrium independently of the original installation.

The bipolar endoprostheses "Lima" with a ceramic-ceramic friction pair contain an outer metallic head fit in a thin-walled UHMWPE cup in the cavity, inside which there is a ceramic liner. The liner, in its turn, is made with a spherical cavity articulating with the internal ceramic head mounted on the endoprosthesis stem. A locking ring threaded into the metal head is used to fix the internal head in the spherical cavity of the liner. The UHMWPE head does not in fact participate in friction but lowers much stiffness of the artificial joint and serves as a shock-absorbing means.

It is believed [39] that friction in the unloaded artificial joints with the described bipolar endoprosthesis is roughly similar in intensity in the cartilage-external head and liner-internal head pairs. As the load intensifies, friction over the cartilage drops to zero and all movements in the artificial joint are exercised inside the endoprosthesis. After 22 months of observations of the implanted cemented Leinbach bipolar prostheses they made a conclusion [41] that the inner motion of the bipolar head decreased over time. However the patients treated with primary bipolar hip prostheses had not acetabular destruction after 6 year [42]. Bipolar hip endoprostheses should be considered only in patients with a short life expectancy and unfavourable general conditions.

Above discussed designs of endoprostheses are on the main meeting orthopaedic needs of endoprosthetics of the femoral joints. This does not mean that the processe of their modification perfection has ceased. On the contrary, still new original structures are being created, especially for the revision operations and endoprosthetics of the hip affected by oncological diseases. These designs will be reviewed in sections 3.8 and 3.9.

3.3 THE KNEE

Orthopaedists commonly use four main types of knee endoprostheses for the total or partial replacement: sledge, hinge, rotational and combined ones. All of them are intended for replacement of degenerated or worn out friction surfaces in the articulating femur and tibia.

The sledge endoprostheses allow for mutual rolling sliding of the femoral and tibia components in the sagittal plane. A design of "St. George" endoprosthesis and a scheme of its usage in partial reconstruction of the knee are shown in Fig. 3.14 [43].

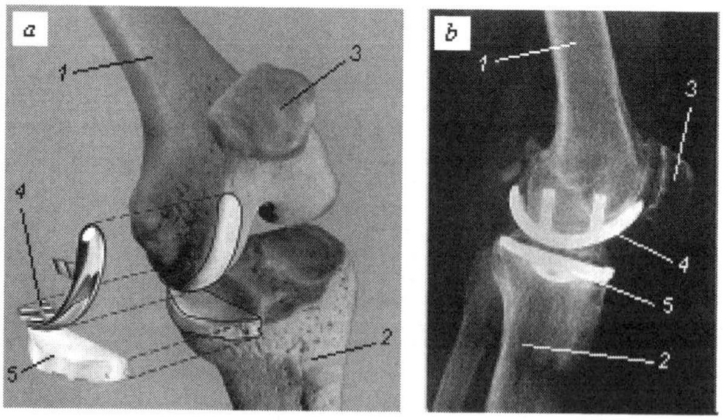

Fig. 3.14. Installation scheme (*a*) and a radiograph (*b*) of implanted sledge endoprosthesis "St. George". 1 and 2 – femur and tibia bones, 3 – patella, 4 and 5 – femoral and tibial components of endoprosthesis

In 1981 this endoprosthesis was modified and renamed into the "ENDO-Model". It includes a femoral (CoCrMo) and tibial (Ti) components (Fig. 3.15). The tibia contains an UHMWPE liner. The femoral component is attached to a prepared end of the femoral bone using shins of specific configuration. Its contact surface is similar to the anatomic shape of one of the condyles, and a corresponding oval depression is made on the liner. Three dimension-types of the "ENDO–Model" endoprostheses with four thickness-type liners are manufactured at W. Link GmbH & Co.

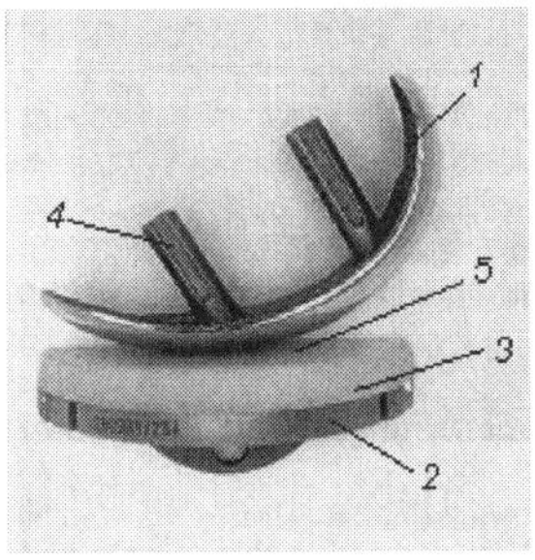

Fig. 3.15. "ENDO-Model" endoprosthesis: 1 and 2 – femoral and tibial components, 3 – UHMWPE liner, 4 – rod, 5 – recession

The main members of the Search® endoprosthesis produced by Aesculap Company for the total knee replacement are illustrated in Fig. 3.16 [45]. The femoral component (a) is made as a bracket, and is mounted on the prepared femur end in a manner depending on whether the crucial ligament and patella remain after the operation or not. The bracket is made of a CoCrMo alloy of 4–5 dimension-types for both legs. It can be fixed by bone cement or mechanically via supporting pegs (not shown) and biointegration.

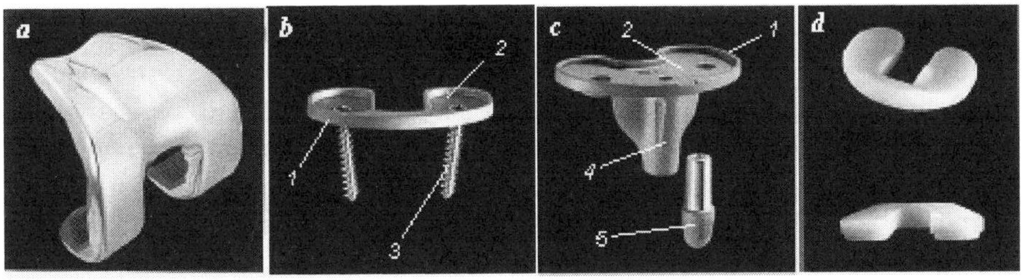

Fig. 3.16. Components of sledge endoprosthesis Search®: *a* – femoral, *b* and *c* – tibial components [45]. 1 – plate, 2 – seating, 3 – screws, 4 – stem, 5 – liner

The tibial component is made as a plate (4 to 8 types) repeating the configuration of the tibial plateau, and having a special socket. The plate is fixed on the trimmed tibial plateau by cement or screws (*b*), or is fastened by a profile stem of a regulated length (*c*), for which aim the endoprosthesis has the liners of four dimension types fit on a taper.

The friction member made of UHMWPE (*d*) is fit tightly (with a click) into the socket of the tibial component. The friction parts are manufactured of four thickness-types.

The Miller-Galante prostheses having been used since the mid 1980s are of analogous design. Uncemented fixation is achieved on the femur by clamping with secondary stability resulting from osseointegration into a multi-layered titanium fibre-mesh. The uncemented implant is fixed to the tibia with titanium screws and pegs. Metal-backed patella is fixed with fibre-mesh pegs. The Miller-Galante I system is characterized by a flat tibial UHMWPE only. To improve patellar centring, the Miller-Galante II system is designed with the patella bed more deeply recessed into the femur, and the femoral curvature more evenly curved. The tibial pegs are reinforced, and a 10° displacement of the tibial fixation crew allows variable positioning [46].

The patella clunk syndrome is a syndrome of patellofemoral dysfunction, consisting of a painful catching, grinding, or jumping of the patella on knee extension, and is a well recognaized complication after total knee arthroplasty [47]. Newer NexGen Legacy total knee prosthesis decreases the incidence of patella clunk syndrome. Its structure has the more posterior intercondylar box and femoral cam of the femoral component [48].

A requisite condition of implanting the sledge prosthesis is the presence of knee ligaments and the neighbouring muscles in a satisfactory state [43]. Most suitable indication for this kind of endoprosthetics is the medial gonarthrosis [49].

The hinge endoprostheses are intended for flexion-extension of the knee in the sagittal plane. Since the hinge unit connects the femoral and tibial components, all mechanical loads and friction of the artificial limb are concentrated here.

A hinge endoprosthesis design is shown in Fig. 3.17. This endoprosthesis consists of two main components with the stems fixed in the marrowy canals of the femur and tibia. The edges of the stems are furnished with the star-shaped PE tips to centre the stems in canals. The tibial component is provided with hinge 4 having an aperture for shaft 5. The hinge fits a slot in the bearing block 6 mounted in the femoral component. The block consists of two side plates containing UHMWPE plain bearings 7. In the final stage of mounting hinge 4 is inserted into unit 6 in the gap between bearings 7. The hinge mechanism is locked via a cross hole drilled in the medial condyle by the metal shaft 5.

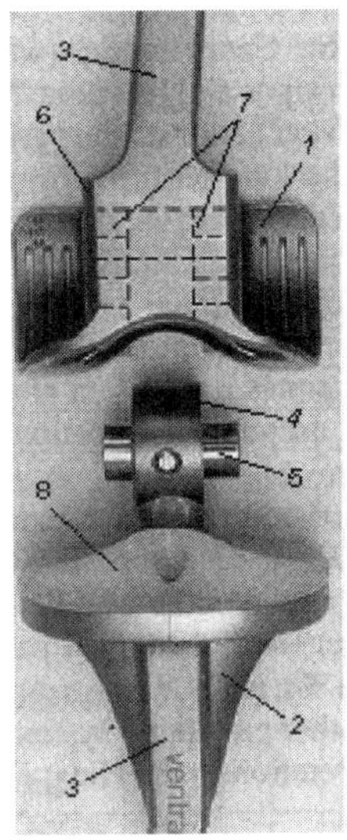

Fig. 3.17. Hinge endoprosthesis "ENDO-Model" [44]: 1 and 2 – femoral and tibial components, 3 – stem, 4 – hinge, 5 – shaft, 6 – block, 7 – plain bearing, 8 – plateau

All displacements and friction in such artificial limb are localized in the hinge mechanism. As a result, the contact between the femur and PE plateau 8 found on the tibial component turn to be practically unloaded.

The hinge endoprostheses are preferable at later stages of rheumatoid arthritis accompanied by the muscle atrophy and ligament failure [49].

The rotary endoprostheses incorporate two main friction joints: i) similar to the sledge-type one, aimed at transmission of the loads connected with the body mass on the tibia, and ii) the bearing unit in the tibial component is intended to fix the prosthesis components and confine their mutual shifts.

A prerequisite factor preceding the development of the rotary endoprostheses appeared two serious biomechanical complications revealed at analyzing a 5-year endoprosthetics of the knee using the hinge prosthesis. They are: loosening of the stems and fracture of the femur [50]. Certain modifications of the hinge endoprostheses have lead to the development of the rotary "ENDO-Model" one (Fig. 3.18).

The femoral 1 and tibial 2 components are fixed on stems 3 by the bone cement in marrowy canals of respective bones. Guides 4 are threaded into stem's ends to align their position in the canals. The tibial component in the form of plate 2 has on its rear part pin 5 with a spherical face. There is also an UHMWPE liner 6 on the plate with two oval cups 7 simulating the upper articulating surface of the tibia. The femoral component is made as a bearing unit 8 locked on a stem, and having thrust plates 9 from the left and right sides of the unit bent so as to accommodate anatomies of the condyles. The bearing unit is made with a through hole into which an UHMWPE tubular bearing 10 is mounted. The bearing is furnished with sleeve 11, which forms a T-joint and a metallic shaft 12. The clearance between the thrust plates 9 is correlated with the diameter of sleeve 11, while that of sleeve's hole – with peg 5.

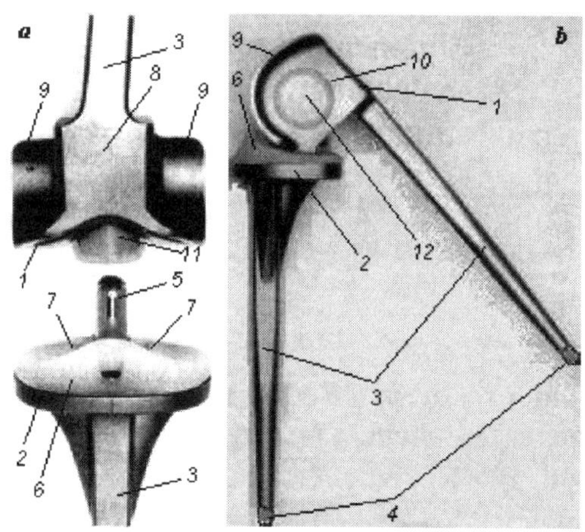

Fig. 3.18. Rotary endoprosthesis "ENDO-Model": *a* – front view, *b* – side view (femoral component in extreme lower position). 1 and 2 – femoral and tibial components, 3 - stem, 4 – guide, 5 – pin, 6 – liner, 7 – socket, 8 – block, 9 –thrust plate, 10 – bearing, 11 – sleeve, 12 – shaft

The implanted endoprosthesis is fastened by peg 5 in the hole of sleeve 11 to interlock the articulating components. During flexion of the knee and elevation of the hip their components occupy a position shown in Fig. 3.18, *b*. During motion, thrust plates 9 slide over sockets 7 of the polymer liner 6, and sleeve 11 shifts following pin 5 in the clearance between the thrust plates. This actuates rotation of bearing 10 in the clearance between the bearing body and shaft 12. The rotation of the tibia in the artificial limb is exercised by turning pin 5 in sleeve 11, which is restricted by the angular displacements of thrust plates 9 in the oval cups of polymer liner 6.

The rotary endoprostheses are employed at insufficient ligament functions but still strong muscles of the hip and shin. They are commonly implanted with indications of either virus or valgus deformations (O or X) of the knee complicated by the contraction or lack of the ligament system [49].

The combined endoprostheses unite the merits of above considered structures and contain additional elements aimed at elimination of their drawbacks. A typical representative is the combined GSB knee endoprosthesis, widely applicable in Europe.

The GSB prosthesis is made up of the first letters of its creators Gschwend N., Scheier A., and Bahler A. – orthopaedists of W. Schulthess clinics in Zurich. Figure 3.19 presents the photos of a GSB modification manufactured by SULZER Medica.

This endoprosthesis includes femoral 1 and tibial 2 components fixed with the help of stems and bone cement in the femoral and tibial bones. The femoral component is made as a hollow block 3 on which face a convex plate 4 is installed whose curvature accommodates the anatomic shape of the condyles and the patella surface of the femur. The plate has slot 5 communicating with space 6 of the block. The "ceiling" of the cup is shaped by a specific curvilinear surface of part 7. The tibial component includes a base 8 fixed on the stem and carrying a curvilinear flat latch 9. UHMWPE plane liners 10 are mounted on the base. The cementless patella endoprosthesis 11 has a slightly bulging contact surface that carries four sharp polyethylene pegs on the reverse side to fasten the bone, and a titanium mesh 12 for the osseointegration.

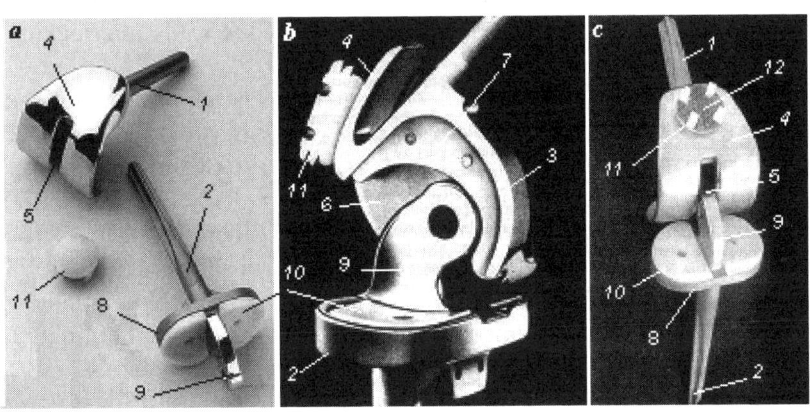

Fig. 3.19. Parts of GSB endoprosthesis (*a*), cross-section of the femoral component (*b*) and assembled view (*c*): 1 and 2 – femoral and tibial components, 3 – block, 4 – plate, 5 – slot, 6 –space , 7 – retainer part, 8 – base, 9 – latch, 10 – liner, 11 – patella, 12 – mesh

Upon fixing in the knee bones, latch 9 is mounted in space 6 of the femoral block. Thus assembled endoprosthesis operates as follows. Plate 4 props against liners 10 and transmits in this manner weight of the body onto the shin. At flexion in the knee the plate slides in the liners. The flexion is bounded by the shifts of latch 9 in space 6 of block 3. The latch contacts the retainer part

7 only in extreme positions of the shin. This prevents dislocation of the artificial limb in case of a poor functional state of the ligaments or muscles.

The GSB prostheses are recommended for in fact any disease of the knee that makes walking improbable and is incurable by the conservative therapy or osteotomy [49]. This endoprosthesis can be used for the correction of any axial deformations and instability of the knee ligaments. Endoprosthetics of the patella still remains disputable because of insufficient information on the wear rate of the patella cartilage against a metal plate of the tibia component, and UHMWPE liners against this plate.

The developers of GSB consider that the primary prosthetics of the knee gives rise to numerous problems even more complex than those at replacement of the hip. Therefore, it is necessary to estimate the probability of revision operations already in the course of the primary one, i.e., as J. Charnley called it, “the second defence line”. Note that GSB gives such an opportunity [51].

The endoprostheses of the knee undergo modifications mainly within the frames of the aforementioned designs. Along with already described structures we must name a sledge one [52]. It is made entirely of ceramics and is furnished with the stiffening ribs. It was attempted to improve the bearing unit of the rotary endoprosthesis [53] by assigning an optimum trajectory to the bearing using two pairs of cross pull rods. In the combined endoprosthesis [54] this problem is solved by the figured grooves in the block walls interlocking with projections on the block cam.

A great number of combined knee endoprostheses has been elaborated lately. For instance, Biomet Inc. (USA) manufactures “AGC HPPS” endoprostheses whose femoral component is similar to that of the “Search” one (Fig. 3.16) and the tibia is analogous to GSB’s (Fig. 3.19). The only difference is in the latch made conjointly with the UHMWPE liner. The “Dual Articular” (Biomet Merck Ltd, UK) endoprosthesis has analogous fixture unit. “AGS V2” also produced by this company is in its essence a sledge type of endoprosthesis and contains a plateau in the medial part of the polymer liner that limits lateral shifting of the contacting femoral component.

The knee is a most loaded human joint. So, in choosing the type of endoprosthesis the body weight is strongly important. Tis critical feature is reflected in a special orthopaedic term “body mass index” (BMI).

The BMI is defined as a body weight (kg) divided by the square of the height (meters) [55]. Measurements over 30 kg/m^2 correlate with moderate obesity; over 40 kg/m^2 with severe (morbid) obesity; whereas scores less than 30 kg/m^2 are classified as non-obese. The effect of BMI has been studied [56] as a predictor of outcome and survival in total knee replacement. It is proved that the revision and survivorship rates over the first decade are comparable to non-obese controls.

Recently, navigation systems have been developed to improve the accuracy of component alignment in total knee arthroplasty, and the first follow-up results of computer-assisted knee arthroplasty are promising [57].

As a matter of fact, an orthopaedist should be guided in the numerous family of knee endoprostheses by the patient's peculiarities as well as the technical and economic considerations, be familiar with designs and implantation techniques, and have a profound experience to give preference to an optimal specimen [58].

3.4 THE FOOT AND ANKLE

The endoprostheses of this type are implanted rather rarely and only in clinics developing certain narrow directions in orthopaedics. As an alternative to endoprosthetics the specialists may propose arthrodesis of the ankle and orthroplastics of the metatarsus and phalanges. As the experience of the orthopaedists in many countries proved, the alternative operations give a stable proximate and remote functional result.

The endoprostheses of the ankle are made as a pair of sliding plates, which replaces the surfaces of the talus and tibia.

A typical structure of this endoprosthesis is illustrated in Fig. 3.20.

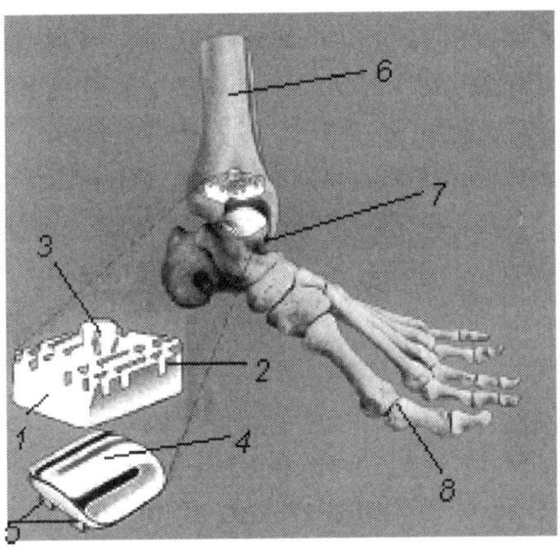

Fig. 3.20. Assembly scheme and parts of ankle endoprosthesis "Model St. George®", DBR: 1 – plate, 2 – tooth, 3 – rounded support peg, 4 – counterbody, 5 – square support peg, 6 and 7 – tibia and talus, 8 – the first metatarsus-phalange joint

The tibial component is made as an UHMWPE plate. Its upper surface is formed by rectangular teeth with a peg for cemented fixation of the distal tibia. The plate is movably joined with a CoCrMo counterbody in the form of a rectangle with the rounded off corners. The lower surface of the counterbody is furnished with square support pegs to fix the talus. The described endoprostheses are manufactured by a German company W. Link GmbH & Co.

Depending on the friction surface configuration of the plate and counterbody, we discriminate between three types of endoprostheses of the ankle: non-congruent, congruent and anatomic [59].

Non-congruent endoprostheses (to which "St. George" belongs, Fig. 3.20) are characterized by a minimum linkage between components, thanks to which they are able to articulate freely in all planes. Such endoprostheses have a reduced bearing capacity due to a small contact area between the plate and the counterbody. Among the drawbacks of these endoprostheses we can name low wear resistance of the polyethylene liner, early loosening and instability. The friction surface of the non-congruent endoprostheses is of a block type (Fig. 1.2, *a*) either concave-convex or convex-convex.

The congruent endoprostheses incorporate the components able to revolve only about the common axis of revolution. They are characterized by high extent of linkage. This presupposes that the artificial limb functions as a hinge and all its displacements occur only in one plane. As a result, this endoprosthesis is more stable in contrast to the non-congruent one, and its wear resistance is higher because of more uniform distribution of loads and deformations over the friction surface. The friction pair of the congruent endoprostheses can be made as a sphere-socket joint or be of a cylindrical, conical, spherical or barrel shape.

The anatomic endoprostheses simulate a two-condyle joint between the foot and shin. They are characterized by a high stability since their axes of revolution coincide with those of a healthy ankle. The success in the usage of named endoprostheses is rigidly related to the recovery of the anatomic centres of revolution and the degree of perfection of the components [60].

The endoprostheses of the ankle are commonly made as the polymer-metal, metal-metal, or ceramic-ceramic friction pairs.

The endoprostheses of the metatarsus-phalange joint represent a pair of rods contacting a spherical surface and fastened in the metatarsus bone and the main finger phalange.

Figure 3.21 illustrates an endoprosthesis of a joint [61] whose components are made of alumina corundum porous ceramics. A ball-shaped end of the metatarsus component is movably conjugated with a spherical plane of the artificial phalange thus simulating the anatomic form of the joint. The structure of this endoprosthesis is brought very close to the functional characteristics of the first metatarsus-phalange joint (see Fig. 3.20) by preserving natural resting of the metatarsus bone from the plantar side on the semasoid. When the components are press-fit, the osseous tissue grows into the ceramic pores and does not form a connective capsule, which elevates reliability of fixation.

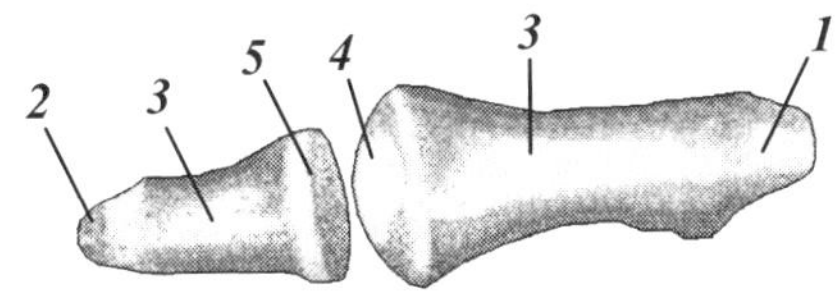

Fig. 3.21. Endoprosthesis of the first metatarsus-phalange joint: 1 and 2 – metatarsus and phalange components, 3 – rod, 4 – ball head, 5 – spherical cavity.

3.5 THE SHOULDER

There are three main types of the shoulder endoprostheses: unconstrained, constrained and semi-constrained [62].

The unconstrained endoprostheses were developed by Neer in 1973 to replace shoulder joints having undamaged joint lip. Figure 3.22 presents a modern variant of this endoprosthesis "Model St. George" produced by W. Link GmbH & Co.

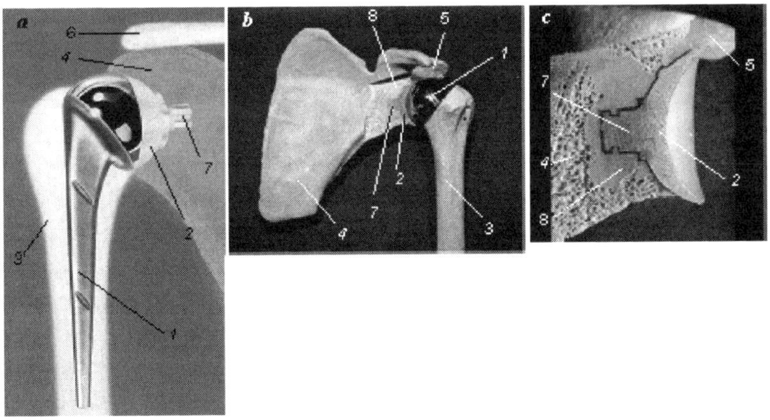

Fig. 3.22. Installation scheme (*a*) of unconstrained endoprosthesis "Model St. George", cross cuts of the humeral joint with implanted endoprosthesis (*b*), and articular cavity of the scapula with a cemented cup (*c*): 1 – stem with spherical head, 2 – cup, 3 – shoulder bone, 4 – scapula, 5 – rostral outgrowth, 6 – clavicle, 7 – support peg, 8 – cement mantle

Its UHMWPE liner is furnished with a support peg and is fixed in the articular cavity of the scapula by bone cement. The metal stem (CoCrMo) fixed in the shoulder bone has a spherical head 39 mm in diameter, which is close to anatomic dimensions. The described design makes possible the replacement with a minimal resection of the bones, and preserves bonded soft tissues, especially the joint capsule. The unconstrained endoprostheses restore a maximum possible volume of motion in the shoulder.

A new shoulder endoprosthesis "Articula" is produced by MATHYS Medical Ltd. (Fig. 3.23). It consists of a titanium (TiAlV4 alloy, ISO 5832-3) stem, a connector made of the same material, and a ceramic (Al_2O_3, ISO 6472-2) or metallic (CoCrMo alloy, ISO 5832-4) head. To make fixation of the stem stronger, the cement bed is covered by a ceramic (Al_2O_3) coating. This endoprosthesis has adjustable length by changing depth of the connector fitting into the stem, and retroversion or deflection of the shoulder head backwards. The "Articula" has been recommended for implantation to elderly people in cases of extensive osteoporosis, comminuted fracture of the shoulder head and posttraumatic necrosis.

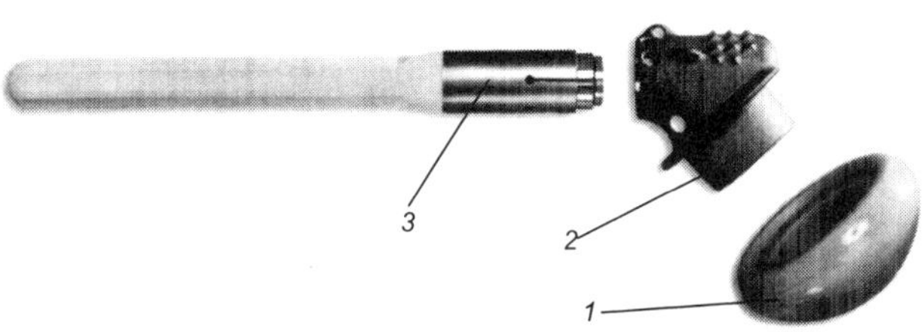

Fig. 3.23. Endoprosthesis "Articula" of shoulder joint: 1 – head, 2 – connector, 3 – stem

The cementless unconstrained endoprosthesis "Nottingham" is manufactured by Biomet Merck Ltd. (UK). Its polyethylene liner with a spherical shell is fixed on a metal substrate, which is screwed to the articular cavity. Metal parts of the endoprosthesis are porous-coated for osseointegration.

The semi-constrained endoprostheses differ from the unconstrained ones by their polyethylene cup shape, which is more complex and has a collar imitating a beak-shaped outgrowth of the scapula and the joint cartilage lip.

P. Grammont has introduced a reversed "Delta" prosthesis incorporating the principles of a large glenoid ball (the glenosphere) with hydroxyapatite-coated uncemented fixation [63]. The "Delta III" total shoulder prosthesis (De Puy, France) is a modular prosthesis available in either anatomic or reversed configuration (Fig. 3.24).

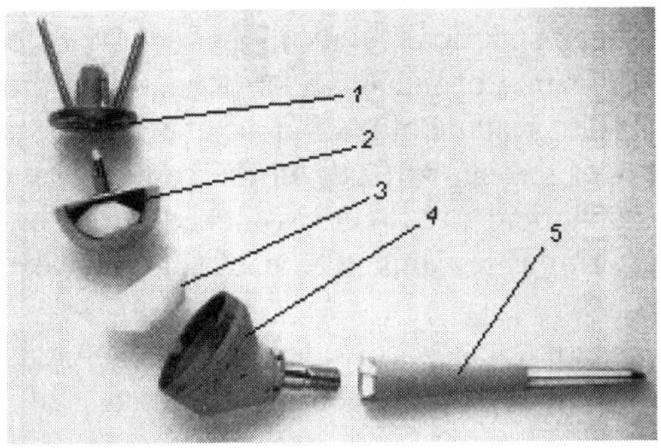

Fig. 3.24. "Delta III" total shoulder system: 1 – glenoid baseplate, 2 – glenosphere, 3 – UHMWPE liner, 4 – humeral component, 5 - stem

The stainless steel glenoid baseplate and humeral stem are fixing to the bone in a press-fit manner, enhanced by hydroxyapatite coating. Modularity of both components allows soft tissues balancing and adjustment of degree of constraint. This endoprosthesis is recommended for the patients with rheumatoid arthritis of the glenohumeral complicated by rotator cuff dysfunction [64].

The constrained endoprostheses include a shoulder component and a cup forming together a non-detachable movable unit. Their stability is independent of the functional state of the neighbouring muscles and ligaments.

Figure 3.25, presents a structure of the constrained endoprosthesis.

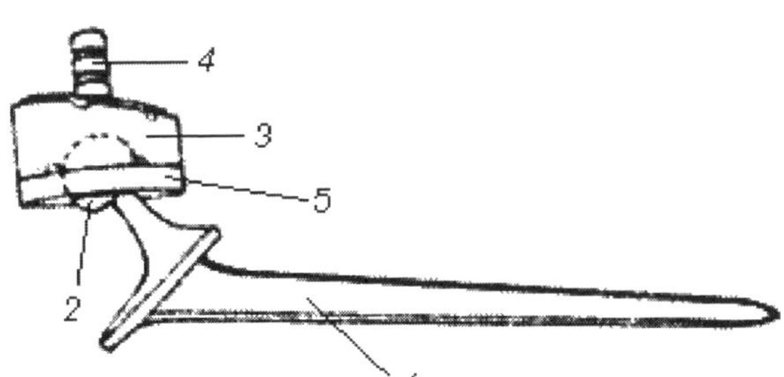

Fig. 3.25. Shoulder endoprostheses "M. Reese": 1 – stem, 2 – spherical head, 3 – cup, 4 – support peg, 5 – locking ring

It consists of a metal stem having a spherical head 22 mm in diameter. The head is movably joined to a deep spherical cavity in an UHMWPE cup. The stem is cemented in the marrowy canal of the shoulder bone, while the cup is fixed in the articular cavity of the scapula. Fixation of the cup is provided by the mechanical engagement of the support peg in the cement mantle. The head is brought into the cup's spherical cavity and is locked by a ring fastened in the cup. The diameter of the ring is less than the head's one to make the articulating joint as an entire unit.

The described design evokes considerable mechanical stresses at the bone-implant interfaces that may speed-up loosening of the prosthesis components.

The constrained endoprostheses are indicated for the replacement in cases of the expressed painful syndrome in unstable humeral joint with damaged joint lip. In this case, the unconstrained endoprostheses characterized by a relatively low risk and large service life are inapplicable. Arthrodesis may be an alternative if there is a necessity to implant the constrained endoprosthesis [65].

3.6 THE ELBOW

The achievements in endoprosthetics of the elbow are not so expressive as in, e.g. the total hip or knee replacement. This is, first of all, because of less quantity of operations on the elbow joint. The main designs and biomechanics of the elbow endoprostheses are the next.

The implants for the partial replacement of the elbow and the early hinge endoprostheses used in the 1940-1970s [66] are not discussed here since they present only a historical value. In replacements of the elbow four types of endoprostheses are used today [67]: unconstrained, semi-constrained, and a modern group of the hinge artificial limbs – free-hinged and semiblocked ones.

The unconstrained endoprostheses can preserve stability after implantation thanks to the congruent friction surfaces of the endoprosthesis and a stabilizing effect of the neighbouring muscles and ligaments of the medial collateral ligament on the first place.

The operation principle of these endoprostheses is illustrated in Fig. 3.26. The endoprosthesis "Lowe" (*a*) contains a Co-Cr shoulder member, which is press-fit in the distal shoulder bone followed by osseointegration. The elbow component made of UHMWPE is fixed by bone cement and support pegs to the proximal elbow bone. The artificial elbow joint articulates via sliding of the semi-cylindrical surface of the shoulder component over a hollow in the polymer elbow counterpart. This articulation is limited by the strength of ligaments and muscles of the elbow joint.

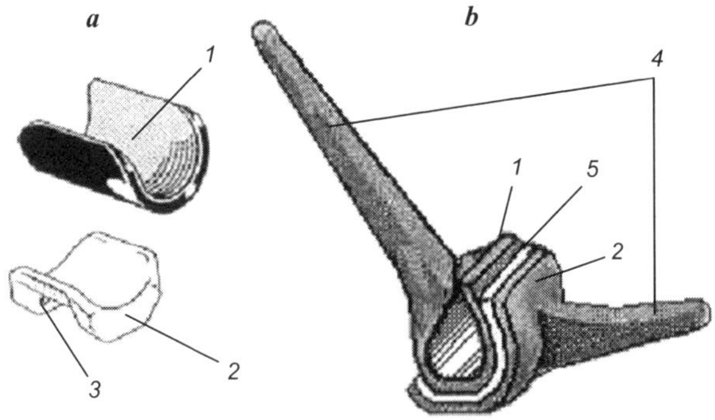

Fig. 3.26. Unconstrained endoprostheses "Lowe" (*a*) and "Kudo" type-4 (*b*): 1 and 2 – shoulder and elbow components, 3 – support peg, 4 – stem, 5 – polymer liner

The metallic endoprosthesis "Kudo" (*b*) differs from the "Lowe" (*a*) by the stems fixed in the marrowy canals of the shoulder and elbow bones and an UHMWPE liner mounted in the elbow cavity to ensure low friction and damp out impact loads. Due to the high incidence (70%) of proximal subsidence of the humeral component, the use of "Kudo" type-1 and type-2 prostheses was no longer recommended in 1985. In 1983 the type-3 prosthesis was developed. This prosthesis had an added humeral intramedullary stem to secure fixation. With the "Kudo" type-4 prosthesis the complications were fatigue breakage of the humeral stem, metallosis and a high rate of polyethylene wear. Results after an intermediate follow-up of "Kudo" type-5 prosthesis showed fair to excellent clinical scores, no revisions and no radiolucency [68]. Biomet Merck, Ltd. (G.B.) manufactures "Kudo" left and right in two modifications, i.e. cementless (elongated porous-coated stems) and cemented (truncated with less cross-section) ones.

The semiconstrained endoprostheses promote stability of the elbow joint with the help of retained muscles and ligaments. At the same time, they contain some structural elements that may restrict articulation.

Some authors believe that with progressive joint destruction and lack of ligamentous stability, a more constraint type of elbow prosthesis is indicated [69]. However, considering the constrained prostheses' high loosening rate, a semiconstrained prosthesis can be still indicated, even in elbows with severe destruction [70].

An example of the semiconstrained endoprostheses can be a shoulder-condyle one of "Ewald" brand shown in Fig. 3.27.

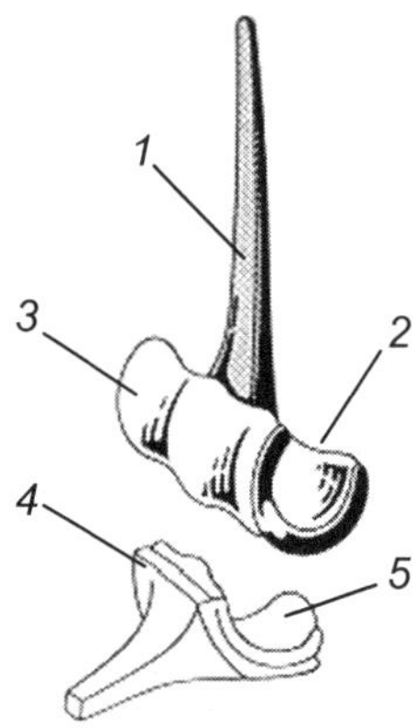

Fig. 3.27. Semiconstrained endoprosthesis "Ewald" of the elbow: 1 – stem, 2 – hollow cylinder, 3 – artificial condyle, 4 – elbow component, 5 – cavity.

Its metal shoulder component is made as a semi-cylinder fixed on a stem analogously to unconstrained Kudo's one. Its peculiarity is that the cylinder generatrix repeats the anatomic shape of the condyles of the distal shoulder bone. The metal condyles are arranged in the assembled endoprosthesis on both sides of the curvilinear cavity placed along the symmetry axis of the elbow component.

Such a design averts linear displacements of the components along the revolution axis of the endoprosthesis. The UHMWPE elbow component is cemented in the elbow canal. Late "Ewald" modifications propose the metallic elbow components incorporating an UHMWPE insert.

Another design of the elbow joint is "Souter" having analogous a two-condyle configuration of the friction surface.

The free-hinge endoprostheses are similar in structural principles to the hinge one but ensure independent rotation of the elbow bone about its longitudinal axis.

Let's consider a structure of extensively employed at present free-hinge endoprostheses on the example of the "Model St. George" design produced by W. Link GmbH & Co. (Fig. 3.28).

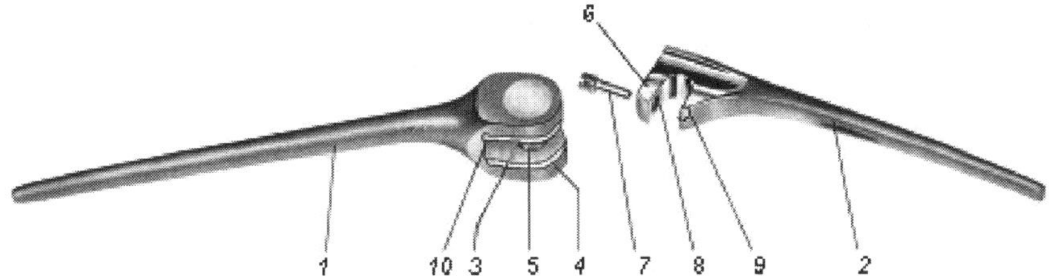

Fig. 3.28. Free-hinge endoprosthesis "Model St. George" of the elbow: 1 and 2 –humeral and elbow components, 3 – bearing, 4 – cheek, 5 – shaft, 6 –clevis, 7 – screw, 8 – threaded hole, 9 – groove, 10 - stopper

This endoprosthesis includes a Co-Cr shoulder 1 and elbow 2 components fixed on the stems. Its UHMWPE bearings 3 are mounted in cheeks 4 of the shoulder component and have a metal shaft 5. The hook-type clevis 6 of the elbow member is put on the shaft and locked by screw 7. The screw is fixed in groove 9 upon inserting into whole 8. The extreme position of the joint at extension is reached when clevis 6 props against the plastic stopper 10 mounted on the plateau between cheeks 4. Owing to a free fit of the clevis on shaft 5 and the clearance between the clevis and the cheeks, the elbow may turn about its longitudinal axis.

The merits of this endoprosthesis are its structural simplicity, low friction, and postoperative stability. The minimal resection of the bone at implantation helps to preserve the soft tissues and envisages resection arthroplastics in case of the endoprosthesis fracture [71].

The elbow endoprostheses "Dee-2", "Schlein", "Coonrad", and "Pritchard-Walker-2" are very close in their design to above described artificial elbow.

An original custom design makes provision for the elbow articulation about three axes [66]. This endoprosthesis is similar to Kudo's limb, and differs in the following: i) the cylindrical neck with a spherical head is fixed on the concave friction surface of the elbow component; ii) at assembly the head is press-fit with a click into the groove on a curvilinear plate of the shoulder component.

Semiblocked endoprostheses of the elbow are analogous in design to the rotary knee ones.

A typical semiblocked endoprosthesis GSB is presented in Fig. 3.29. Both shoulder 1 and elbow 2 components are installed on the stems. The shoulder component is made as a fork 3, whose cheeks have holes 4. A pair of curved plates 5 of a specific anatomic configuration are attached to the fork end.

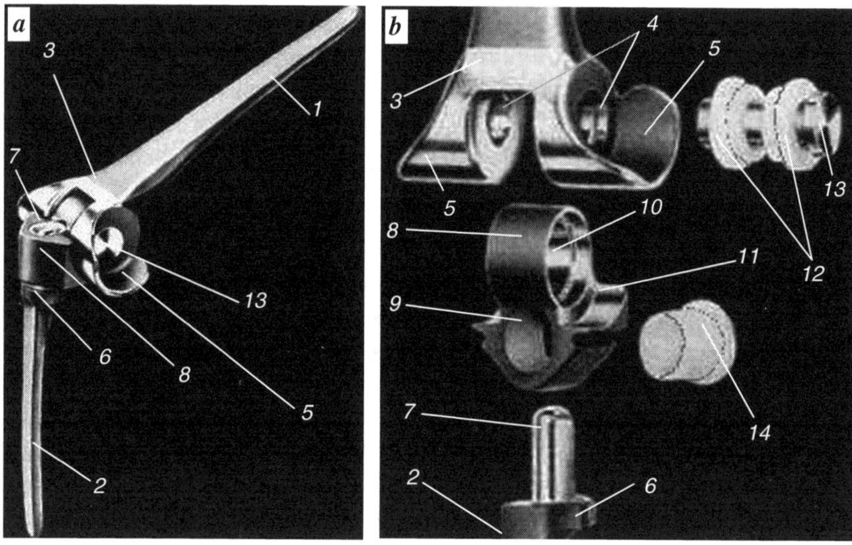

Fig. 3.29. Assembled (a) and non-assembled (b) elbow endoprosthesis GSB-3: 1 and 2 – shoulder and elbow components, 3 – fork, 4, 9 and 10 – holes, 5 – plate, 6 – plateau, 7 – pin, 8 – block, 11 – bearing surface, 12 and 14 – bearings, 13 – shaft

Pin 7 with a spherical butt is fixed to the plateau 6 of the elbow component. Block 8 has holes 9 and 10 normal to each other, and the tangs. Bearing surfaces 11 mate the shape of plates 5. Bearing 12 is inserted in whole 4 of the fork, while the tang of block 8 with whole 10 is installed between its cheeks. The block is then locked in the fork by shaft 13 inserted in the bearings and hole 10. A tubular bearing 14 is inserted into whole 9 of the block. The bearings are made of UHMWPE.

The endoprosthesis is assembled in steps using bone cement for fixing the shoulder component (with assembled bearing unit 8) and elbow. Then pin 7 is inserted in the hole of bearing 14.

Thus assembled endoprosthesis operates as follows. At flexion and extension plates 5 slides over the bearing surfaces 11 of block 8. The extreme position of the joint at extension is reached when block 8 props against fork 3. The elbow bone is able to revolve about its axis thanks to the rotation of pin 7 in bearing 14. Stability of the joint during revolving is conditioned by the functional state of surrounding muscles and ligaments.

One more variant of semiblocked endoprostheses [67] is "Coonrad". It is a hinge endoprosthesis with a metal shoulder and elbow components linked via a shaft installed on UHMWPE bearings. Thanks to wide enough gaps between the conjugated metal components, "Coonrad-2" and "Coonrad-3" modifications presuppose revolving of the elbow bone till 7° about its axis and varus/valgus deflections.

It is to be noted in conclusion that the elbow endoprostheses fixed by such stems undergo less loosening than the ones without the stems. The probability of loosening grows as the bearings and antifrictional liners wear out and violate surface congruence of the artificial limbs. Semiconstrained endoprostheses are considered the best means of solving post-traumatic problems of the elbow joint. Together with the unconstrained endoprostheses they are successfully applied at rehabilitation of the elbow affected by the rheumatoid arthritis [66].

3.7 THE WRIST AND FINGERS

The wrist (carpal) bones are linking the forearm and metacarpus to allow various movements of the most important and complex part of the upper extremity – the wrist. The wrist consists of numerous joints and this was the main reason of a great number of endoprostheses developed to simulate motions of this joint. The peak of their development fell on the 1970s. In the chapters to follow we are going to present the structures of the radiocarpal and carpal endoprostheses adopted in practice today.

The radiocarpal endoprostheses are intended to replace separate bones or the wrist joint as a whole.

Partial endoprosthetics of the wrist was exercised first of all because of unsatisfactory outcomes of the traditional anaplasty and bone replacement. When removed the affected bones may often shift the remaining ones and violate their congruence leading in some cases to the deformation arthrosis.

The endoprostheses of the wrist (scaphoid, semilunar and trapeziform) are commonly made of silicone elastomers. Due to their elasticity, the prostheses inflict less traumas to the cartilage and conjugated bones, and do not results in any degenerative processes. Design and replacement schemes of the scaphoid are shown in Fig. 3.30. This endoprosthesis repeats the scaphoidal anatomic shape and is sutured by the tapes extending from the endoprosthesis to the wrist capsule. In Moscow (Russia), such endoprostheses are produced by the Central Institute of Traumatology and Orthopedy named after N.N. Priorov (CITO).

Endoprosthetics of the scaphoid is considered as an alternative at comminuted fractures, non-union joints or the aseptic necrosis [72].

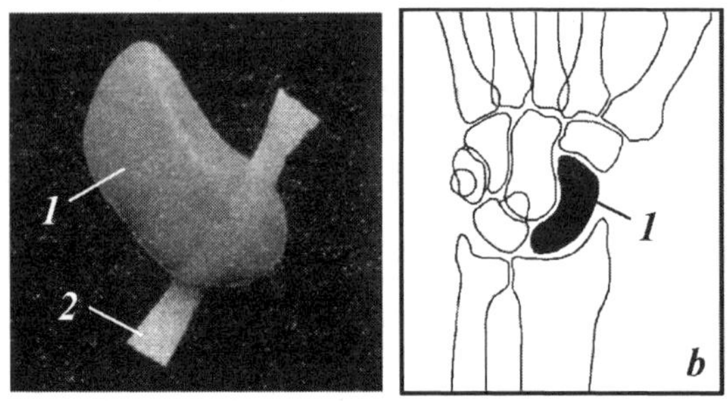

Fig. 3.30. Movshovich-Voskresensky's endoprosthesis of scaphoid (*a*) and replacement scheme (*b*) of the radiocarpal joint: 1 – endoprosthesis, 2 - tape

The total radiocarpal endoprostheses have been subject to modifications for more than 30 years (see section 1.2). The Meuli endoprosthesis belongs today to a most successful design thanks to a great freedom of motion. Some photos of the latest modification "MWP-3" 31 (Meuli Wrist Prosthesis) are shown in Fig. 3.31.

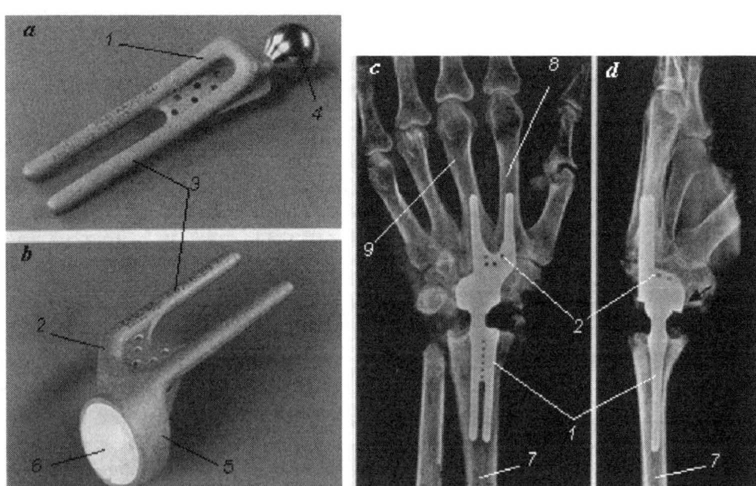

Fig. 3.31. Radial (*a*) and carpal (*b*) components of "MWP-3" endoprosthesis; radiographs of radiocarpal joint with implanted endoprosthesis: *c* – frontal view, *d* – side view. 1 and 2 – radial and metacarpus components, 3 – stem, 4 – head, 5 – cup, 6 – liner, 7 –radial bone, 8 and 9 – second and third metacarpus bones

The endoprosthesis consists of the radial and carpal components. The former is fixed in the spoke bone while the latter is attached to the metacarpus via a pair of stems and cement. The radial bone has a metallic ball head while the metacarpus has a cup furnished with an UHMWPE liner having a spherical cavity. Upon fixation in the bones, the head is mounted in the cavity and the positions of both head and cup are optimized relative to the stems to make the centres of rotation of the joint and endoprosthesis coincide [73].

The total endoprosthetics of the radiocarpal joint is recommended for the patients with strong cortical layers for surgical correction of deformations or biomechanical defects of the wrist. For strongly deformed or unstable joints the optional operations can be arthrodesis, synovectomy (excision of the synovial shell of the bone) or resection of the elbow bone head.

The endoprostheses of the fingers are subdivided into the polymer and metal ones.

Indications for endoprosthetics of the fingers are late stages of the rheumatoid polyarthritis with stable contractions, dislocations, heavy destructive changes or the anchylosis. Endoprosthetics helps in this case to remove deformations and restore mobility of the joints with sufficient stability of the fingers [72].

Silicone endoprostheses of the fingers (Fig. 3.32) are made as a rod with pointed from both ends stems that are divided by a semicylindrical element.

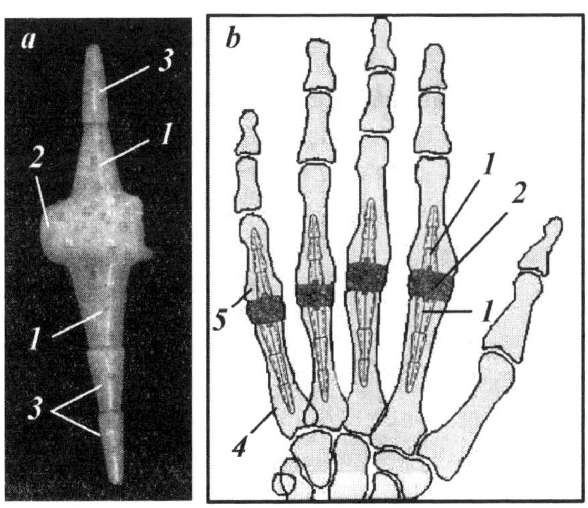

Fig. 3.32. Movshovich-Grishin's endoprosthesis of a finger joint (*a*) and its installation scheme (*b*): 1 – stem, 2 – semicylindrical element, 3 – cone, 4 – metacarpus bone, 5 – proximal phalange

Diameter of the cylinder exceeds that of the stems. To improve fixation in the marrowy canals the stems are made as inserted in each other cones whose larger diameters are closer to the semicylinder. Such primary fixation of the stems assists in rehabilitation of motions.

The metal finger endoprosthesis looks like a pair of rods conjugated via a spherical head with a recess from both ends that forms a movable contact (the endoprosthesis developed by K.M. Sivash) or via a hinge. In the postoperative period the metal wrist undergoes loosening with time, shifts into the soft tissue and may provoke erosion of the bones. There was an opinion [74] that the metal endoprostheses should be implanted in specialized clinics in which the staffs has acquired rich experience in operations on the wrist till the time their usage will become similarly stable as the implantation of the silicone substitutes.

One of the latest models of the wrist endoprostheses "RM Isoelastic Joints" is in its essence an attempt to unite the merits of the elastic and metal endoprostheses. This endoprosthesis is formed by a pair of hinged polyacetal stems reinforced by the stainless steel rods. The stems have a ribbed surface for better fixation in the bones. The conjugated surfaces of the hinge have the rests to restrict the articulation.

3.8 TUMOUR ENDOPROSTHESES

There are few fields in oncology where such dramatic improvement of treatment results can be observed as in primary bone tumours, especially in osteosarcoma and Ewing's sarcoma. Bone sarcomas are rare: approximately, the overall incidence for osteosarcoma is 1.5-2.5 million per year, for chondrosarcoma is 1.5-2.5 million per year and for Ewing's sarcoma 0.5-1 million per year. Because of the small number of new cases, centres have been developed with participation of orthopaedic surgeons, oncologists, radiologists, pathologists and radiotherapists for the interdisciplinary treatment of bone tumours. Besides the well-known centres and registries in Bologna, Birmingham, Prague, Munster, Paris and London, one of most respected bone tumour registry and oncological units is at the University of Vienna [75].

For the establishment of new treatment protocols, however, not even these centres have a sufficient number of patients. International teams have been formed, like the Cooperative Osteosarcoma Study Group (COSS) in the German-speaking area, or the Scandinavian Sarcoma Group (SSG), The European Osteosarcoma Intergroup (EOI), the Children's Oncology Group (COG) in the USA that together cover a population of approximately 400 million people. The advantage of such co-operation is that a sufficient number of patients for a prospective randomized study can be collected in a short time; therefore, results are really evidence based.

In the 1980s the majority of patients with bone tumour underwent either amputation or exarticulation of extremities. In rare occasions when the limb affected by the tumour escaped amputation, the joint was subjected to arthrodesis. The chances to rescue the limbs affected by the oncological disease have raised owing to the modern methods of determining anatomic boundaries of the tumour (first of all, computer tomography and nuclear magnetic resonance), preoperative therapy and special tumour endoprostheses. Such patients have to bear complex enough operations with subsequent tumour resection and reconstruction of the limb.

Tumour endoprostheses are intended for substitution of the affected bones and their joints. Such endoprostheses are either custom made because of unpredictable spreading of the tumour in the tissues or consist of the modules regulated to a needed size of each component. The modular endoprostheses have reached the perfection of today's custom made elements. In this connection, there arose an opinion [76] that in treating oncological diseases the qualitative resection of the tumour, and effective postoperative chemical and radiotherapy are more important than the type of endoprosthesis.

The endoprostheses of the hip used today for reconstruction of the femur and mating bones after the bone tumour resection are mainly cementless isoelastic ones. In Fig. 3.33 one can see modifications of such endoprostheses produced by Mathys Ltd. Bettlach.

The isoelastic hip endoprosthesis (*a*) is reinforced in its proximal part and contains a detachable distal part of the stem, a collar and a replaceable head. The heads are made 28 and 32 mm in diameter with 28 and 40 mm neck lengths. When the cup is not installed into the acetabulum, the head diameter is enlarged up to 40-56 mm. The endoprosthesis is equipped with special fixtures for buttock muscles.

Figures 3.33, *b–f* show the examples of using endoprosthetic components at various resection types of the femur. Upon the removal of its proximal part (*b*), the distal stem of the endoprosthesis is fixed in the marrowy canal of the remained bone using screws, and then the cup is mounted in the acetabulum in which the head is placed and the buttock muscle is fastened by a fixture.

Figure 3.33, *c* presents an operation scheme with the resected medial part of the femur. The endoprosthesis includes a femur insert 12 with its thin part (of *d* diameter) fixed in the canal of the proximal femur, while the thicker part is fixed in the distal fragment. The screws easily threaded in the polyacetal lining are used for fixation.

In the case the distal femur is resected (*d*), the endoprosthesis is to function as a tumour endoprosthesis of the knee. In this case, the femur insert is fixed like in the previous case in the femur's fragment. The tibial component 13 is fixed by a stem ($D > d$) and screws in the marrowy canal of the shin bone. The femur insert and tibial components form a hinge endoprosthesis of the knee interlocked to shaft 14.

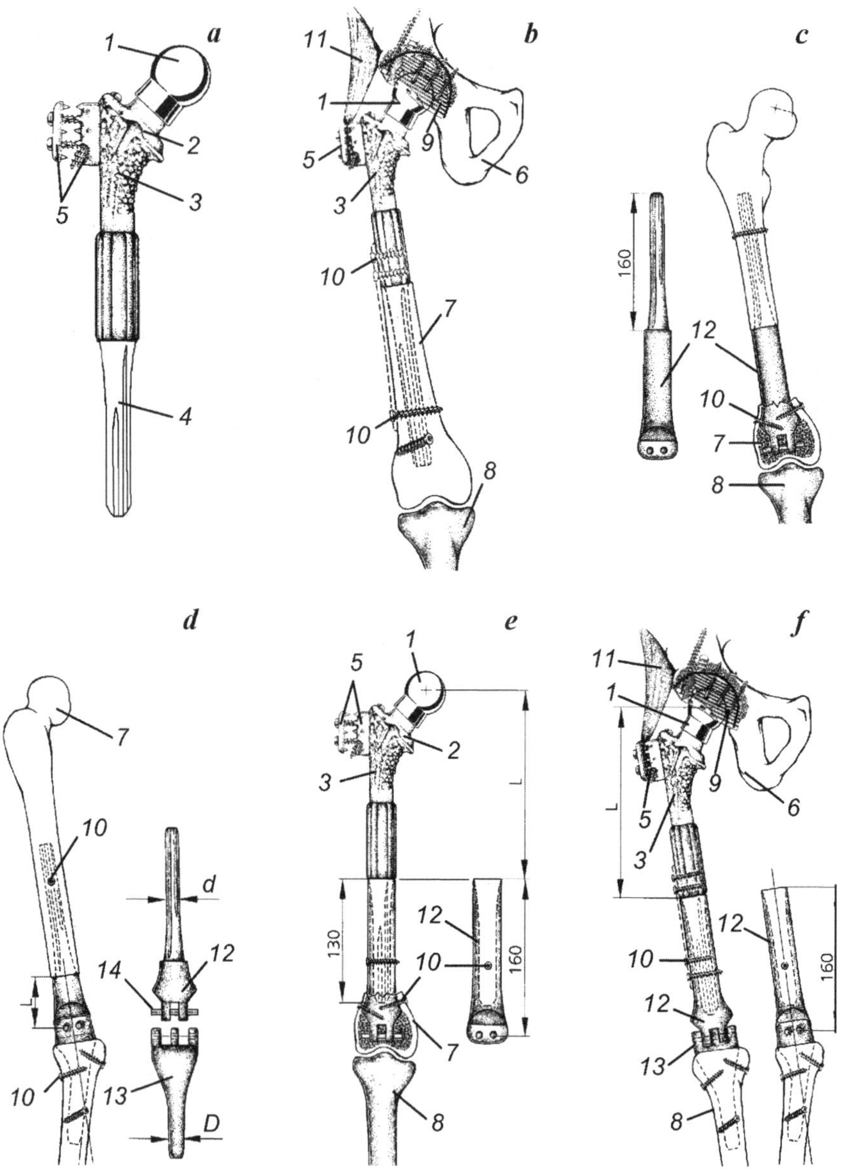

Fig. 3.33. A schematic view (*a*) and implantation schemes (*b – f*) of Mathys' tumour femur endoprostheses: 1 – head, 2 – collar, 3 and 4 – proximal and distal (replaceable) parts of the stem, 5 – fixture, 6, 7 and 8 – pelvic, femoral and shin bones, 9 – cup, 10 – screw, 11 – buttock muscle, 12 – femoral insert, 13 – tibial component, 14 – shaft

When it is necessary to resect almost the whole femoral bone except for its distal part, the endoprosthesis is to serve as a subtotal substitute of the femur (*e*). The replaceable stem 4 is removed from the proximal femoral part of the endoprosthesis and a thin part of the femoral insert 12 is mounted in the freed canal, and fastened by the screws. The distal femur is fastened by screws on the lower end of the insert. As a result of the operation, the endoprosthesis reconstructs the knee joint and its head is reset into the acetabulum or a cup.

The implantation scheme upon a most severe variant of resection with almost total removal of the femur is shown in Fig. 3.33, *f*. The body of the endoprosthesis is assembled similarly to the previous case. Cup 9 is fastened in the pelvic bone and tibia 13 of the endoprosthesis - in the shin bone. Then the knee prosthesis is locked by shaft 14, head 1 is reset in the cup and the buttock muscle is fixed in fixture 5.

During the preoperative stage, the radiographs define the optimal femur endoprosthesis length *L*, stem diameters *D* and *d* of the femur insert 12 and the tibial component 13 based on a case study.

A custom-made tumour endoprosthesis made of ceramics [77] is illustrated in Fig. 3.34. It is intended to reconstruct the resected pelvic bone with the acetabulum. This endoprosthesis includes pelvic and femoral components. The pelvic one simulates the bone without the ischium tuber and the obturator foramen. It contacts the opposite pelvic bone via a step having orifices for locating pins and its upper plane wing is fixed to the sacrum by the screws.

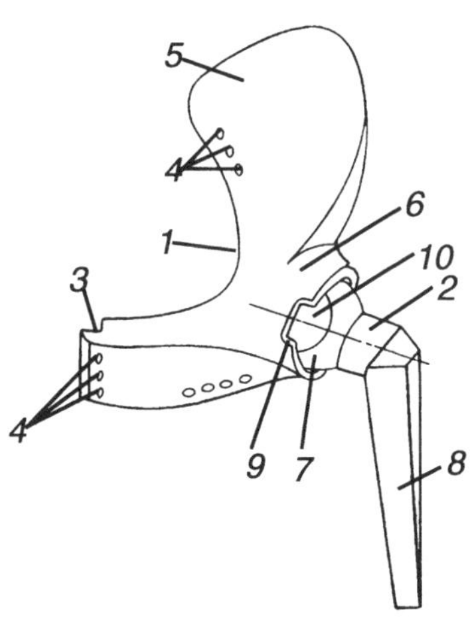

Fig. 3.34. Pelvic endoprosthesis with present hip joint: 1 and 2 – pelvic and femur components, 3 – collar, 4 – orifice, 5 – wing, 6 – acetabulum, 7 – head, 8 – stem, 9 – slot, 10 – plane plateau

The artificial acetabulum with a spherical cavity is mating the femoral head of the endoprosthesis fixed on a stem. The rectangular slots in the acetabular walls and the spherical head tapered from both sides alleviate introduction of the head into the acetabulum during assembly of the implanted endoprosthesis.

A saddle-shaped prosthesis of the Link America, Inc. Co. is widely applicable as a tumour endoprosthesis, which is implanted upon resection of the pelvis with the acetabulum. It consists of a saddle support resting on an elongated stem and has fixtures for the buttock muscles. The endoprosthesis is assembled as follows. The stem is fit in the femoral canal, then a plateau is formed in the pelvic for the saddle, and the buttock muscles are fixed on the stem upon mounting the saddle. This endoprosthesis is more often used in Europe than in the USA [76].

Among the advantages of endoprosthetics during which the femur and neighbouring joints are resected is a satisfactory volume of motion of the hip immediately after healing of the operative wound. The patients begin to walk already 1-2 months after the operation on their own or with an assistance and little lameness. In contrast, after the alloplastics the operated limb can not bear the patient's weight during 6 and sometimes 9 months after the operation until the engraftment of the bone fragments. Life of the tumour endoprosthesis is inversely proportional to the physical activity of the patient. It is believed that elderly patients with a limited activity may hope for a satisfactory functioning of the endoprosthesis during remaining life, while the young people have to choose whether to confine their activity or undergo a revision operation 10 or 15 years later.

Reconstruction of proximal femoral tumours with a modular megaprosthesis is a good procedure, but hip instability remains a major problem. The authors of [78] believe that either the tendon detaches or forms a thin fibrous tissue, which leads to the atrophy of the absuctors and loss of function. Other factors to consider, however, are correct anteversion of the femoral neck, the tension of tissue around the prosthetic hip after completion of the procedure by checking the telescoping effect, and sufficient reaming of the acetabular cup. The operation after-effect of may be dislocation in the early period (Fig. 3.35).

The principles of designing the tumour endoprostheses of the hip are common for all oncological endoprostheses. They are the following:

- modular or individual structure of endoprosthesis;
- elongated or split stems;
- the presence of both artificial couplings for the endoprosthesis with two joints;
- availability of fixtures for muscles;
- possibility of making a simplified rest for the endoprosthesis against a bone without forming a more refined artificial limb.

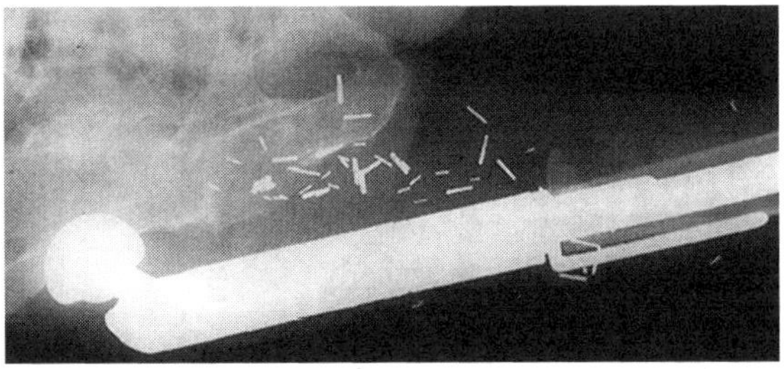

Fig. 3.35. Radiograph of a patient with recurrent dislocation after resection of the proximal femoral tumour and implantation with a modular megaprosthesis

3.9 THE REVISION ENDOPROSTHESES

The implanted endoprostheses may inevitably undergo breakdown or wear even without any infectious complications or engineering errors of the primary operation. This process starts, as a rule, from various defects in the osseous tissue. The revision endoprostheses are to allow for a satisfactory level of operation on the limbs despite the preoperative fractures.

The latest **revision endoprostheses of the hip** are made just as uncemented, so for the cement fixation.

The femoral components designed by H. Wagner are made of Ti-Al-Nb alloys (Protasul-100) and are noted for good biocompatibility. They are available in different length modifications (190-305 mm) and stem diameters (14-25 mm), Fig. 3.36 *a*. The straight stem does not simulate the natural curvature of the femur but looks like a cone with 2° vertex angle. It has eight longitudinal pointed at the end ribs (b) to ensure the primary press-fitting of the stem in the marrowy canal. The ribs dilate closer to the neck. Such stems are produced by Sulzer Ortopedics [79].

The advantage of the Wagner's stem is the simplicity of the design with no potential for mechanical breakdown other than at the head/neck junction. It also provides an excellent rotational stability. Wagner's stem is reserved for revision situations where fixation of conventional short-stem prosthesis is either not possible or fraught with an undue risk of periprosthetic fracture (Fig. 3.37). Careful preoperative planning and meticulous technique are prerequisites for successful outcome. The complications do not differ from difficult hip revision surgery using other implant systems, except for the subsidence, which can be well controlled by careful seating of the prosthesis. All these factors contribute to the substantial learning curve with the use of this implant [80].

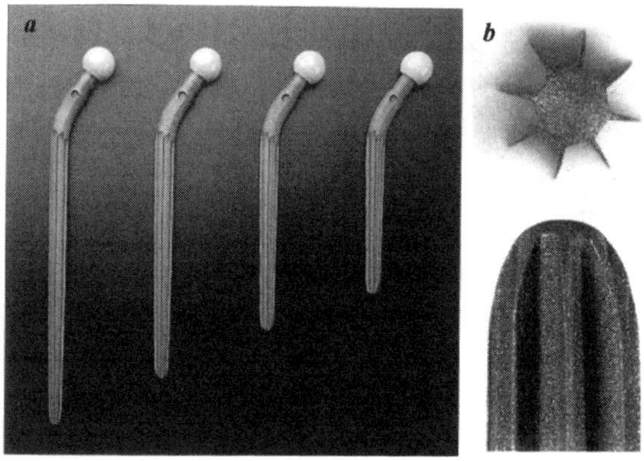

Fig. 3.36. Outward view of stems of Wagner's revision endoprostheses (*a*); frontal and side views (*b*) of stem's end

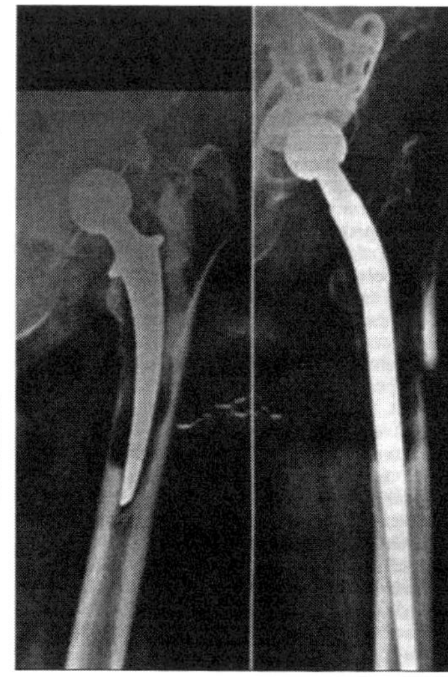

Fig. 3.37. Osteolytic destruction of the proximal femur, pre- and postoperatively (revision hip arthroplasty using the Wagner's stem)

It often turns that the only way to preserve serviceability of the limb is the use of the custom-made implants. Such endoprostheses are needed for the patients suffering from the malignant bone neoplasm, pseudoarthrosis and heavy fractures of the femur. These endoprostheses acquire rather lengthy stems (Fig. 3.38) having an additional anatomic bent to facilitate their fixing instead of the primary endoprosthesis.

The Mathys revision isoelastic endoprostheses have proved to be successful for these aims [81]. It is seen from Fig. 3.39, that the endoprosthesis is firmly screwed to the healthy bone and the polymer facing on the stem (*a*). In case the greater trochanter of the femur is damaged, the screw in the endoprosthesis facing (*b*) furnishes an additional stabilizing effect to the femoral component. The stability of the press-fit cup is also supported by the screws.

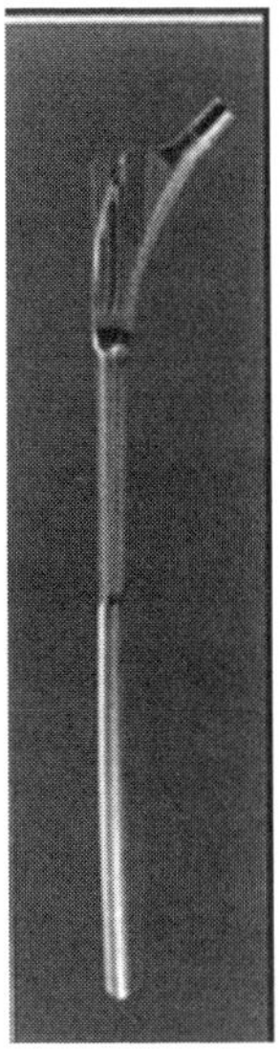

Fig. 3.38. A custom-made stem of hip endoprosthesis (*Sulzer Orthopedics*)

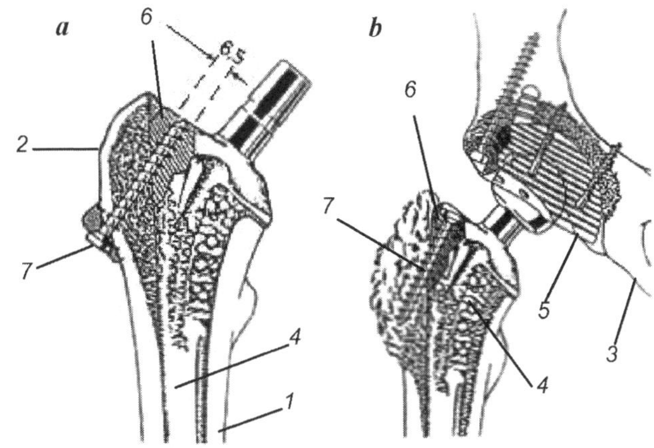

Fig. 3.39. Fixation of cementless femoral component of the Mathys revision endoprosthesis in the femur with intact (*a*) and damaged (*b*) trochanter: 1 – femur, 2 – trochanter, 3 – pelvic bone, 4 and 5 – femoral component and prosthesis cup, 6 – polymer facing, 7 – screw

A design of the revision Mathys cup shown in Fig. 3.40 is made of UHMWPE and is covered by a porous titanium coating. The cup is fit with two supporting pegs, tapped holes and the holes for balancing screws. The latters are used for the primary fixation of the revision cup and bear the main load on the femur. To fix UHMWPE liner in revision operations with Mathys' cups they often employ Bursh-Shneider's (Fig. 3.10) and Muller's metallic fastening rings (Fig. 3.9, *a*) and threaded Zweimuller's cups (Fig. 3.9, *c*). An approach to selecting pelvic components depending on defects of the acetabulum is proposed in [3].

The stems for the revision endoprosthetics of the hip produced by Biomet, Inc. (USA) are of the following modifications:

- cementless stem "Bi-Metric Revision Porous" of anatomic shape (right and left) of wrought titanium and a plasma-sprayed porous titanium coat;
- cement-fit stem "Bi-Metric Head/Neck Revision" is a straight anatomic stem of wrought Co-Cr alloy with a rough surface.

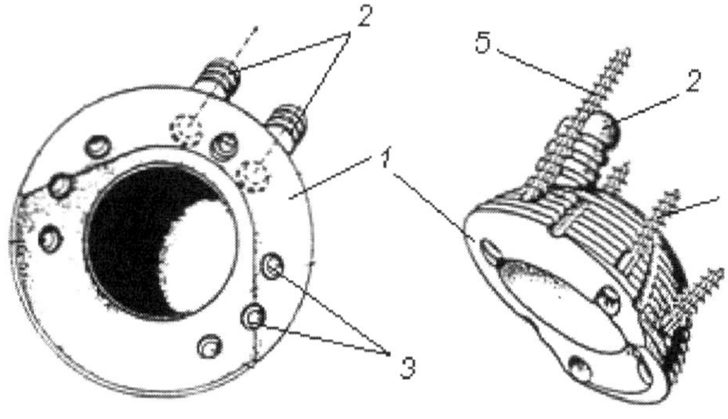

Fig. 3.40. Cementless RM cup for the Mathys isoelastic revision endoprosthesis: 1 – cup, 2 – supporting peg, 3 – holes, 4 – screw, 5 – balancing screw

The revision endoprostheses of the knee are in principle similar to those used at primary arthroplastics and range from the minimum constrained to the hinge and rotary ones.

Since the revision operations on the knee commonly subject the femur and shin-bones to the deep repeated resections, rehabilitation of the anatomic center of rotation of the knee is hampered drastically. Misalignment of the natural and artificial knee joints is a most frequent reason of instability of the revision endoprostheses [82]. This is why, it is important to retain the length of the hough and thus preserve the flexion-extension function without any lengthening of the linkage between the components.

The revision operations on the knee joint are usually accompanied by gross osseous losses, so they are often combined with the bone auto- or alloplastics. This requires specifically designed and custom-made endoprostheses. Still, the proper operation techniques and application of the "Natural Knee (TM) Revision System" (Intermedics Ortopedics, Inc.) gives perfect results as compared to other endoprostheses [82].

The revision endoprostheses of the shoulder are largely modifications of the basic models presented in section 3.5. The usage of special types of humeral endoprostheses "Neer-2" (unconstrained) and "Neer fixed-fulcrum" (fixed hinged endoprosthesis analogous to Michael Reese's ones, Fig. 3.25) instead of the "Neer" is described in [62]. A considerable central wear of the glenoid cavity after a primary unipolar endoprosthetics leads to a medial shifting of the endoprosthesis head. So, in revision operations, the heads of larger diameters are used, as well as osteoplasty and specific endoprostheses of the glenoid cavity.

The revision endoprostheses of the elbow are not mass-produced. The only indication for their use is in fact loosening of the primary prosthesis. When choosing the elbow prosthesis design for the revision operations, one should observe a rule that the unconstrained endoprostheses are substituted for the semiconstrained ones with an optional cement fixation of the components [66].

It is to be noted that arthrodesis and resection arthroplastics have not lost their actuality at revision operations on the majority of joints, especially on the ankle, carpal and finger joints.

◆ ◆ ◆

A number of companies are specializing at present in manufacture of numerous artificial limbs under different trade names. Many of them are just a reproduction of the basic models developed by the authorized endoprosthetic experts but are not sufficiently biocompatible, they make use of imperfect technologies and so are rather cheap. Sometimes, these endoprostheses violate the intellectual rights and patent laws. The essential features of the basic endoprostheses models were uphold at the expense of significant financial outlays of the designers and right owners. These basic models have to pass through a long-term clinical approbation that touched the health and fates of thousands of people. To rise the prestige of the doubling endoprostheses, they are patented. With this aim, they are imparted insignificant differences from the basic model, but the expedience of these differences has not been confirmed clinically. When choosing a needed type of endoprosthesis from a host of existing today structures it is recommended to head for the products manufactured by the companies having a multiyear experience in this field. The endoprostheses produced from cheap but low-quality materials can be a hotbed of future complications and are less durable than the basic models. The endoprostheses of joints are science-consuming and high-precision products designed for a long-term operation in the hostile biological media. Very few goods can stand on a par with the endoprostheses produced by the specialized manufacturers that commercialize only clinically verified durable specimens.

References

1. Shaposhnikov O.G. On some problems in joint endoprosthetics. *Bulletin of Traumatology and Orthopedy named after N.N. Priorov*, 1994, No. 4, p. 3–5.
2. Sherepo K.M. Diagnostics and curing tactics at aseptic instability and osteomyelities after the total hip endoprosthetics by Sivash's method.

Bulletin of Traumatology and Othopedy named after N.N. Priorov, 1994, No. 1, p. 7-11.

3. Nikolaev V.I. Aseptic instability of the acetabular component of endoprostheses: biophysical aspects, diagnostics, curing and prophylactics (clinical and experiemntal investigations). Ph.D. Thesis, Minsk, 2000.
4. Biomechanics. – Detroit: SAE Int., 2003, 170 p.
5. Optimizing structural design in composites by learning from bone. *Materials Today*, 2001, V. 4, No. 1, p. 16.
6. Gavryushenko N.S. Effect of various physico-mechanical factors on life of joint endoprosthesis and its functionality. *Bulletin of Traumatology and Othopedy named after N.N. Priorov*, 1994, No. 4, p. 30-34.
7. Toshchev V.D. The porous coating of endoprostheses as a factor of its stable fixing. *Bulletin of Traumatology and Othopedy named after N.N. Priorov*, 1994, No. 4, p. 34-38.
8. Sanchez-Sotelo J., Lewallen D.G., Harmsen W.S., et al. Comparison of wear and osteolysis in hip replacement using two different coatings of the femoral stem. *Int. Orthop.*, 2004, V. 28, p. 206-210.
9. Weber B.G. Total hip replacement: rotating versus fixed and metal versus ceramic head. *Proc. of 9th Hip Soc. Meeting*, St. Louis, Mosby, 1981, p. 264-275.
10. Weller S. Fefteen years of experience with the BioCONTACT hip endoprosthesis system – the past, the present, the future. What has been achieve? *Int. Orthop.*, 2003, V. 27 (Suppl. 1), S2 – S6.
11. Effenberger H., Witzel U., Lintner F., Rieger W. Stress analysis of threaded cup. *Int. Orthop.*, 2001, V. 25, p. 228-235.
12. Brusko A.T., and Omelchuk V.P. Experimental and theoretical substantiation of the mechanism of trophic effect of the function on structural organization of bone. Physiological recombination. *Bulletin of Traumatology and Orthopedy named after N.N. Priorov*, 1999, No. 1, p. 29–35.
13. Akulich Yu.V, Denisov A.S., Nyashin Yu.I., Podgayets R.M., and Akulich A.Yu. The influence of the scheme of loading variations on the recovering of the bone tissue elastic modulus. *Russian Journal of Biomechanics*, 1999, V. 3, No. 3, p. 63–72.
14. Belenkyi V.E., and Kuropatkin G.V. Which endoprosthesis is better? *Bulletin of Traumatology and Orthopedy named after N.N. Priorov*, 1995, No. 1–2, p. 47–51.
15. Nicholson J. Current trends in biomaterials. *Materials Today*, 1998, V. 1, No. 2, p. 6–8.

16. Structural biological materials: design and structure – properties relationships. Ed. by M. Elices. Amsterdam, Pergamon, 2000, 362 pp.
17. Smart materials and structures. Ed. by C. M. Friend. London, Chapmen and Hall, 1994, 354 pp.
18. Kestelman V.N., Pinchuk L.S., and Goldade V.A. Electrets in Engineering: Fundamentals and Applications. Boston, Kluwer Academic Publishers, 2000, 281 pp.
19. Malchau H., Herberts P., Ahnfelt L., and Johnell O. Prognosis of total hip replacement. University of Goteborg, Sweden, 1993, 10 p.
20. Lain H.-J., Pajamaki K.J., Moilanen T., and Lehto M.U.K. The femoral canal fill of two different cementless stem designs. The accuracy of radiographs compared to computer tomographic scanning. *Int. Orthop.*, 2001, V. 25, p. 209-213.
21. Walker P.S., and Culligan S.G. The effect of a lateral flare feature on uncemented hip stems. *Hip Int.*, 1999, V. 9, p. 71-80.
22. Pritchett J.W. Femoral bone loss following hip replacement. A comparative study. *Clin. Orthop.*, 1995, V. 314, p. 156-161.
23. Leali A., Fetto J., Insler H., and Elfenbein D. The effect of a lateral flare feature on implant stability. *Int. Orthop.*, 2002, V. 26, p. 166-169.
24. Mellor S.J., Ripley L.G., and Ricketts D.M. The femoral cement mantle in three total hip replacements. *Int. Orthop.*, 2004, V. 28, p. 40-43.
25. Ilizaliturri V.M., Bobadilla G., Espinosa R., et al. Plug migration and cement mantle assessment in total hip replacement. *Int. Orthop.*, 2004, V. 28, p. 11-15.
26. Fokin V.A. Twenty five years of the straight stem concept. *Margo Anterior*, 2002, No. 2, p. 1-3.
27. Niinimaki T., Junila J., and Jalovaara P. A proximal fixed anatomic femoral stem reduces stress shielding. *Int. Orthop.*, 2001, V. 25, p. 85-88.
28. ISO 7206/2. Implants for surgery – Partial and total substitutes of hip. Part 2. Metallic and plastic supporting surfaces. International Standard, 1987.
29. Salman Ali M., and Kumar A. Hydroxyapatite-coated RM cup in primary hip arthroplasty. *Int. Orthop.*, 2003, V. 27, p. 90-93.
30. Morscher E. Der press-fit cup. In: Endoprothetik. Ed. E.W. Morscher, Berlin, Springer, 1995, S. 208-218.
31. Spicer D.D.M., Schaper L.A., Pomeroy D.L., et al. Cementless cup fixation in total hip arthroplasty after 5-8 years. *Int. Orthop.*, 2001, V. 25, p. 286-289.

32. Chung Y.Y., Kim D.H., and Kim K.S. Bone in-growth on a smooth-surfaced hydroxyapatite-coated acetabular cup. *Int. Orthop.*, 2002, V. 26, p. 283-286.

33. Dohmae Y. Plasmacup – special design. Aspects of the dysplastic acetabulum. *Int. Orthop.*, 2003, V. 27 (Suppl. 1), S20–S23.

34. Klanke J., Partenheimer A., and Westermann K. Biomechanical qualities of threaded acetabular cups. *Int. Orthop.*, 2002, V. 26. p. 278-282.

35. Gill T.J., Siebenrock K., Oberholzer R., and Ganz R. Acetabular reconstruction in developmental dysplasia of the hip. Results of the acetabular reinforcement ring with hook. *J. Arthroplasty*, 1999, V. 14, p. 131-137.

36. Klyuchevsky V.V., Gilfanov S.I., Danilyak V.V., Kuropatkin G.V., and Fokin V.A. Isoelastic stems at a complex atypical endoprosthetics of the hip. *Margo Anterior*, 1999, No. 4, p. 1-4.

37. Huiskes R., Weinaus H., and van Rietbergen B. The relationship between stress shielding and bone resorption around total hip stems, and the effects of flexible materials. *Clin. Orthop.*, 1992, V. 274, p. 124–134.

38. Salaman Ali M., and Kumar A. Isoelastic femoral component in primary cementless total hip arthroplasty. *Int. Orthop.*, 2002, V. 26, p. 243-246.

39. Petty W. Bipolar hip arthroplasty. In: Total Joint Replacement. Ed. W. Petty, Phyladelphia, Saunders Co., 1991, p. 349–354.

40. Fantasia L., Cornacchia D., Agamennone M., and La Floresta P. A bipolar cup with ceramic insert for the hip degenerative pathology: early experiences. In: Performance of the Wear Couple BIOLOX forte in Hip Arthroplasty. *Proc. 2-nd Symp. on Ceramic Wear Couple.* Stuttgart, Ferdinand Enke Verlag, 1997, p. 34–37.

41. Rodop O., Kiral A., Kaplan H., and Akmaz I. Primary bipolar hemiprosthesis for instable intertrochanteric fracture. *Int. Orthop.*, 2002, V. 26, p. 233-237.

42. Hasan M.Y., Shinomiya F., Okada M., et al. Intracapsular hip fracture in patients with rheumatoid arthritis. *Int. Orthop.*, 2003, V. 27, p. 294-297.

43. Kunas R., Wiedmer V. Schlittenprothesen bei gonartrose, osteonekrose und posttraumatischer lasion – evaluation von 106 Arthroplastiken. *Aktuelle Rheumatologie*, 1995, V. 20, No. 4, S. 124–130.

44. Nieder E. Sledge prosthesis, rotating knee and hinge prosthesis Model St. Georg® and Endo-Model®. Differential therapy in primary knee arthroplasty. *Orthopedy*, 1991, V. 20, p. 170–180.

45. Total knee prosthesis. Aesculap, ISP S.A., 1994, 44 pp.

46. Rosenberg A.G., Barden R.M., and Galante J.O. Cemented and ingrowth fixation of the Miller-Galante prosthesis. Clinical and roentgenographic

comparison after 3 to 6 year follow-up studies. *Clin. Orthop.*, 1990, V. 260, p. 71-79.

47. Hozack W.J., Rothman R.H., Booth R.E., Balderston R.A. The patella clunk syndrome. A complication of posterior stabilized total knee arthroplasty. *Clin. Orthop.*, 1989, V. 241, p. 203-208.
48. Ip D., Wu W.C., and Tsang W.L. Comparison of two total knee prostheses on the incidence of patella clunk syndrome. *Int. Orthop.*, 2002, V. 26, p. 48-51.
49. Knee surgery. Complications, Pit falls, and Salvage. Ed. by M.M. Malek. Heidelberg, Springer-Verlag, 2001, 507 pp.
50. Engelbrecht E. The tibial rotating knee prosthesis "Endo" Model: Surgical technique. *J. Orthop. Surg. Techn.*, 1987, V. 3, No. 2, p. 83–98.
51. Gschwend N., Scheier H.J.G., and Munzinger U. GSB reoperation knee prosthesis. Baar: ALLO PRO, 1992, 23 pp.
52. Russia Patent 2121319, A 61 F 2/38. Knee endoprosthesis. S.L. Kabargin, G.I. Kuznetsov, L.P. Ivanova, et al., 1998.
53. Russia Patent 2123824, A 61 F 2/38. Knee endoprosthesis. V.I. Karptsov, N.V. Kornilov, A.F. Glinskikh, et al., 1998.
54. Russia Patent 2127096, A 61 F 2/38. Knee endoprosthesis. V.I. Karptsov, N.V. Kornilov, A.F. Glinskikh, et al., 1999.
55. Bray G.A. Overweight is risking fate. Definition, classification, prevalence, and risks. *Ann. NY Acad. Sci.*, 1987, V. 9, p. 14-29.
56. Spicer D.D.M., Pomeroy D.L., Badenhausen W.E., et al. Body mass index as a predictor of outcome in total knee replacement. *Int. Orthop.*, 2001, V. 25, p. 246-249.
57. Perlick L., Bathis M., Tingart M., et al. Minimally invasive unicompartmental knee replacement with a nonimage-based navigation system. *Int. Orthop.*, 2004, V. 28, p. 193-197.
58. Petty W. Prosthesis for total knee arthroplasty. In: Total Joint Replacement. Ed. by W. Petty, Phyladelphia, Saunders Co., 1991, p. 593–598.
59. Figgie H.E., Unger A.S., Inglis A.E., et al. Total ankle arthroplasty. In: Total Joint Replacement. Ed. by W. Petty, Phyladelphia, Saunders Co., 1991, p. 749–759.
60. Bouysset M. Bone and Joint Disorders of the Footh and Anakle. Heidelberg, Springer-Verlag, 1998, 357 pp.
61. Russia Patent 2152192, A 61 F 2/42. All-ceramic endoprosthesis of the first metatarsophalangal joint of the foot. V.M. Mashkov, E.L. Nesenyuk, I.E. Shakhmatenko, and N.I. Khomyak, 2000.

62. Turner J.A. Shoulder replacement arthroplasty. In: Total Joint Replacement. Ed. by W. Petty. Phyladelphia, Saunders Co., 1991, p. 601–658.

63. Grammont P.M., and Baulot E. Delta shoulder prosthesis for rotator cuff rupture. *Orthopedics*, 1993, V. 16, p. 65-68.

64. Woodruff M.J., Cohen A.P., and Bradley J.G. Arthroplasty of the shoulder in rheumatoid arthritis with rotator cuff dysfunction. *Int. Orthop.*, 2003, V. 27, p. 7-10.

65. Walch G., and Boileau P. Shoulder Arthroplasty. Heidelberg, Springer-Verlag, 1999, 444 pp.

66. Figgie M.P., Inglis A.E., and Figgie H.E. Total elbow arthroplasty. In: Total Joint Replacement. Ed. by W. Petty. Phyladelphia, Saunders Co., 1991, p. 659–706.

67. Gschwend N., Simmen B., Bloch H. Die ellbogenarthroplastik unter besonderer berucksichtigung der GSB-III-ellbogen-gelenkprothese. In: Endoprothetik. Ed. by E.W. Morscher, Berlin, Springer-Verlag, 1995, S. 409–423.

68. Kudo H., Iwano K., and Nishino J. Total elbow arthroplasty with use of a nonconstrained humeral component inserted without cement in patients who have rheumatoid arthritis. *J. Bone Joint Surg. (Am)*, 1999, V. 81, p. 1268-1280.

69. Wright T.W., Wong A.M., and Jaffe R. Functional outcome comparison of semiconstrained and unconstrained total elbow arthroplasties. *J. Shoulder Elbow Surg.*, 2000, V. 9, p. 524-531.

70. Reinhard R., van der Hoeven M., and de Vos M.J. Total elbow arthroplasty with the Kudo prosthesis. *Int. Orthop.*, 2003, V. 27, p. 370-372.

71. Intracondular total elbow joint endoprosthesis “Model St. Geprg”. Catalog. Hamburg, W. Link GmbH & Co., 1982, 19 pp.

72. Movshovich I.A., and Vilensky V.Ya. Polymers in traumatoloy and orthopedy. Moscow, Medicine, 1978, 320 pp.

73. Meuli H.C. Handgelenktotalprothese. In: Endoprothetik. Ed. by E.W. Morscher, Berlin: Springer-Verlag, 1995, S. 424–431.

74. Chidgey L.K., and Dell P.C. Arthroplasty in the hand. In: Total Joint Replacement. Ed. by W. Petty, Phyladelphia, Saunders Co., 1991, p. 725–745.

75. Szendroi M. Advances in orthopedic oncology. *Int. Orthop.*, 2002, V. 26, p. 195-196.

76. Springfield D.S. Joint reconstruction after tumour resection. In: Total Joint Replacement. Ed. by W. Petty, Phyladelphia, Saunders Co., 1991, p. 763–784.

77. Russia Patent 2117460, A 61 F 2/28, 2/32, 27/00. Endoprosthesis of the pelvic bone with the coxal joint. S.L. Kabargin, G.I. Kuznetsov, L.P. Ivanova, et al., 1998.
78. Ilyas I., Pant R., Kurar A., et al. Modular megaprosthesis for proximal femoral tumours. *Int. Orthop.*, 2002, V. 26, p. 170-173.
79. Wagner H. Uncemented self-locking revision stem for extensive bone loss. Baar, Sulzer Orthopedics, 1996, 20 pp.
80. Weber M., Hempfing A., Orler R., and Ganz R. Femoral revision using the Wagner stem: results at 2-9 years. *Int. Orthop.*, 2002, V. 26, p. 36-39.
81. Mathys R., Kaufmann S., and Decker S. Revision surgery of the hip using cementless, isoelastic prostheses and cups. *OP-Journal*, 1994, No. 3, p. 3–11.
82. Gustke K. Revision total knee replacement: technical aspects and experience with the Natural (TM) Revision Total Knee System. Berne, SULZER medica, 1993, No. 8, Lit. N 1893 e, 7 pp.

Chapter 4.
SOME CLINICAL ASPECTS OF ENDOPROSTHETICS

The chapter deals with main stages of joint endoprosthetic operations. The sequence of orthopaedist's actions and typical techniques for selecting fixation method, design and dimension of the endoprosthesis are discussed using hip joint as an example. Basic features and traditional methods for performing primary implantation and revisions are outlined. It is shown that usage of combined endoprostheses consisting of the components of hip joint endoprostheses of different design can compete with usage of special revision endoprostheses. The principles of composing the program for postoperative follow up of patients during several years are described. Dosed loading of an artificial joint aimed at approximating its functions to those of a natural joint is reported to be the main element in the nearest postoperative period. Basic postoperative complications and artificial joint failures as well as standard methods for their elimination are discussed.

In conclusion, the effectiveness of bone alloplasty in joint endoprosthetics and promises of artificial growing of transplants into synovial joints including gene engineering are evaluated.

4.1 PLANNING HIP JOINT REPLACEMENT OPERATIONS

Surgical intervention in a pathologically modified joint should be justified biomechanically. At joint endoprosthetics it is necessary to provide:

- matching of the rotation centres of natural and artificial joints;
- primary stability of endoprosthesis components;
- subsequent ingrowth of bone tissue into implants.

The final aim of endoprosthetics is maximal approximation of artificial joint functions to those of a healthy synovial joint.

Hip joint replacement is planned in the following way:

1) assessment of the bone tissue strength of proximal part of femur and acetabulum;
2) selection of fixation method and respective design of endoprosthesis;
3) determination of basic dimensions of endoprosthesis components;
4) planning of surgical intervention;
5) selection of drugs for postoperative complication prophylaxis.

Assessment of the strength of bone tissue of the proximal part of femur and acetabulum is needed to justify the selection of fixation method. Presently orthopaedists assess the strength of the bone tissue of the femur by standard technique including determination of Singh index and morphological cortical index.

Singh index, or index of the neck of the femur, allows determining the degree of osteoporosis. Figure 4.1 shows radiographic stages of osteoporosis of the proximal part of the femur.

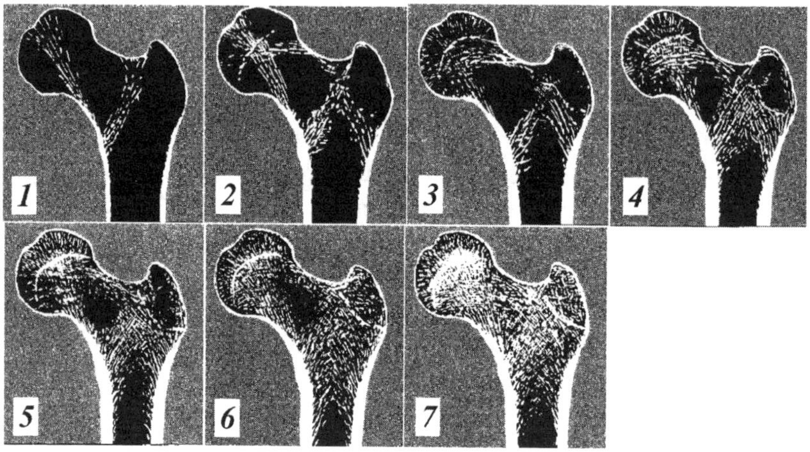

Fig. 4.1. Radiographs of the proximal part of the femur with different–pronounced osteoporosis. The figures in the pictures correspond to seven stages of osteoporosis by Singh

Singh has defined seven stages of osteoporosis [1]:

stage 7 - normal bone tissue structure, dense fine trabeculae (bone bars) are concentrated in the head and distributed uniformly in the trochanterian zone according to the directions of mechanical stresses, the intersecting arc-shaped bundles of trabeculae are not seen;

stage 6 – triangle osteoporosis zone is seen (Ward triangle) restricted by arc bundles of trabeculae of the greater trochanter and femoral head;

stage 5 – Ward triangle is seen clearly, the number of peripheral trabeculae is reduced;

stage 4 – peripheral trabeculae almost vanished in the greater trochanter region;

stage 3 – the number of trabeculae located in the in the form of arc in the subtrochanter zone is reduced;

stage 2 – the arc structure of trabeculae almost vanished;

stage 1– the arc-shaped trabecula bundle vanished in the subtrochanter zone, dense trabecula bundle is considerably rarefied in the head.

Stage 7 corresponds to normal bone structure, 6 and 5 – initial osteoporosis stage, 4 and 3 – pronounced osteoporosis, 2 and 1 - high degree of osteoporosis.

Morphological cortical index (MCI) is a criterion for assessing the shape of the femur and thickness of the cortical layer. It is a ratio of linear dimensions of the femoral bone which are measured in the radiograph (Fig.4 2):

$$MCI = CD / AB,$$

where *CD* is a distance between the external walls of the cortical layer in the front plane at the level of lesser trochanter measured by normal to the femur axis; *AB* is diameter of medullary canal of the femoral bone at a 7 sm –distance from the *CD* –line in the distal direction.

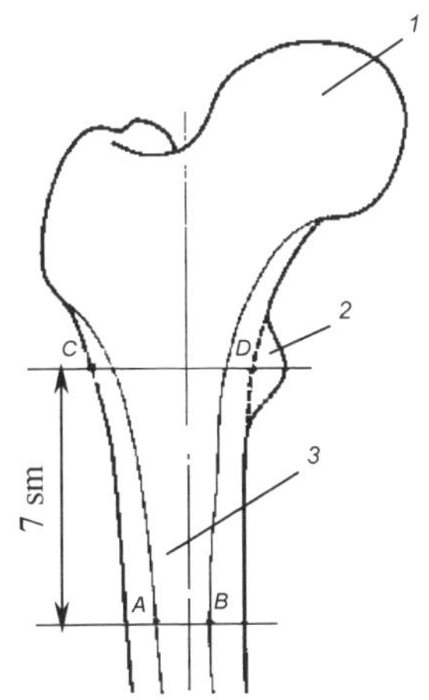

Fig. 4.2. Diagram of measurement of femoral bone parameters to determine MCI: *1* – head, *2* – lesser trochanter, *3* – medullary canal

Strength of the bone tissue around acetabulum and osteoporosis available are assessed by means of radiographs. The degree of osteoporosis in the zones of the pelvic bone (Fig. 1.4) proposed by J. Charnley [2] is of significance when selecting method for pelvic component fixation. The parameter is not as important as the strength of the femoral bone since the load to the pelvic component of endoprosthesis is distributed rather uniformly. Mechanical properties of the bone tissue in the femoral bone and acetabulum may be considered as being almost similar.

Fixation method and endoprosthesis design are chosen with account for the aforementioned parameters as well as age, pathological modifications of the joint and physical activity of the patient.

In work [3] four parameters are proposed to consider when choosing the fixation method of endoprosthesis stem (Table 4.1)

By the sum of points the method for stem fixation is determined: 0–4 – cementless fixation is preferred, 5 – cementless fixation is possible, 6 and more – cement fixation. Cementless endoprosthetics is advisable to apply to the patients before 50 years old since in revision operation it is less problematic to remove an "old" implant and to set a new one. In work [4] are presented the clinical and radiological results of uncemented total hip arthroplasties for patients over 65 years of age. A conclusion has been made that the uncemented arthroplasties can give a very satisfactory result in the elderly patients. They may be comparable to the results after cemented total hip arthroplasties.

Table 4.1.
Parameters and respective points when choosing stem fixation method

Sex	Points	Age, year	Points	Singh index	Points	Morphological cortical index	Points
Male Female	0 1	before 50 51–60 61–70 after 70	0 1 2 4	7 5–6 4–3 2–1	0 1 2 4	3.1 and more 3.0–2.7 2.6–2.3 2.2 and less	0 1 2 4

Unfortunately, an equally simplex and reasonably justified procedure for choosing the fixation method for acetabulum component has not been developed yet. In this case the main criteria are the osteoporosis degree in the pelvic bone zone and sphericity of the acetabulum. Orthopaedists – designers of hip joint endoprostheses have accumulated a large scope of statistical material concerning reliable fixation of acetabular component of their designs. Thus, J. Charnley preferred cement fixation of the polymeric liner. Its latest modification ("Ogee" cup, Fig. 3.8, *b*) allowed for enhancing long-term fixation strength owing to more uniform distribution of the cement in the bone-implant clearance. On M. Muller's opinion, the best long-term results can be achieved by using a reconstructive supporting ring (Fig. 3.9, *a*, *b*) fixed by screws. With a sufficient sphericity of the acetabulum K. Zweimuller and D. Vale recommend to use screwed-in cups of conical (fig.3.9, *c*) and cylindrical (fig.3.9, *d*) shape. A.I. Voronovich has proved advisability of using Zweimuller cups even with significant deficiencies of the acetabulum using a large scope of clinical material [5]. In E. Morsher's opinion, in this case *press-fit* fixation of the cup with netted coating (Fig. 3.8, *d*) is sufficient. More reliable mechanical fixation of the pelvic component is achieved when using R. Mathis cup provided with supporting pins (Fig. 3.8, *c*) or L. Spotorno widening cup (Fig. 3.9, *e*). The authors believe that the method of fixation of polymeric liner with a supporting ring allows achieving strong primary fixation of the pelvic component of endoprosthesis practically in any situation (osteoporosis, dysplasy, defects of the acetabulum, etc). The same point of view is argued in [6] where the treatment results are reported of patients with severe acetabular bone-stock deficiencies due to developmental dysplasy of the hip. It is shown that reinforcement ring (RR, Center-Pulse, Switzerland) with hook provides adequate in poor bone-stock settings and prevents bone-graft resorption. The recommendations are not final in connection with development of new designs of cups, methods for their fixation, and accumulation of long-term results in endoprosthetics.

The role of total hip arthroplasty for treatment of displaced intracapsular fractures is controversial. Some authors have strongly voiced against total arthroplasty in active elderly patients with no pre-existing acetabular disorders [7]. Other has shown that treatment with bipolar hip prosthesis may cause development of acetabular erosion, which might later require further revision. In work [8] the treatment of patients (average age – 66.4 years) with rheumatoid arthritis and femoral neck fracture was reviewed. First group of patients was treated with primary total hip arthroplasty. Second group of patients was treated with primary bipolar hip prosthesis and after 6.1 years there was no acetabular destruction. However, long-term results were inferior to patients treated with total arthroplasty. It was concluded that bipolar prostheses should be considered only in patients with a short life expectancy and unfavourable general condition.

To choose among the accessible variety of implants the following criteria are used: method of fixation, anatomical and pathological features of the acetabulum and femur, endoprosthesis cost, surgeon's experience in implanting these or those endoprosthesis designs, traditions of a clinic, and personal preferences of a surgeon.

Basic dimensions of endoprosthesis are determined with account for anatomical ratios of geometrical parameters of patient's femur and pelvis to recover the rotational centre and biomechanical functions of the hip joint.

All joint endoprostheses under usage in clinical practice have individual templates. They are two (pelvic and femoral) sets of images of the component of each size on transparent plastic sheets. Traditionally the procedure for determining dimensions consists in matching the radiograph of the hip joint and templates corresponding to femoral and pelvic components of endoprosthesis. In standard cases of primary endoprosthetics front photographs of both hip joints and that of the joint under operation are used in side projection made with a focus distance of 120 cm. They are matched by the templates made at 1.15 power compared to the true dimensions of endoprostheses. The principle forms the basis of all methods for determining endoprosthesis dimensions, both classical ones introduced by J. Charnley [9] and those adapted to the latest endoprosthesis designs [10, 11]. Abundance and intercorrelation of biomechanical criteria of the procedure make endoprosthetics the most precise and intellectual division of orthopaedic surgery.

Experts from "MATHYS Medical Russia, Ltd" suggest an alternative technique for choosing dimensions and parameters for setting hip joint endoprostheses. It allows for correcting length and biochemical defects of limbs. The scheme of the procedure is given in Fig. 4.3.

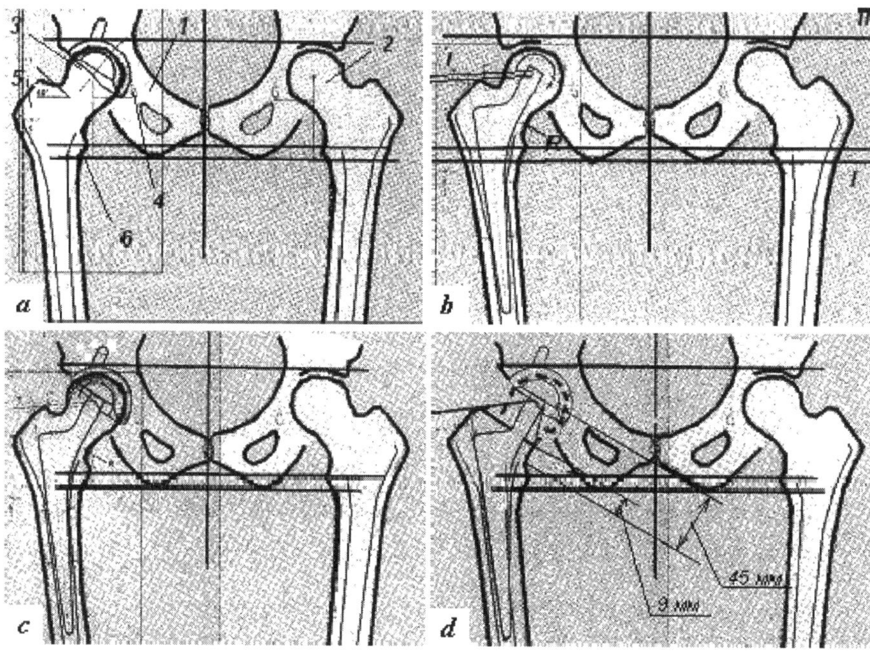

Fig. 4.3. Stages of selection of hip joint endoprosthesis: *a* – determination of the size and position of the cup; *b* – the same for the stem; *c* – setting level of the pelvis; *d* – final pattern. *1* – pelvic bone, *2* and *3* – healthy and operated joints, *4* – "tear" contours, *5* and *6* – greater and lesser trochantri

Four basic lines are applied to the radiograph, viz. vertical axis of pelvis and three horizontal lines: I – tangent to the contour of the tuber of the ischium, II – tangent to the upper edges of acetabula, III – line connecting lesser trochantri of femoral bones. With equal leg lengths lines I - III are parallel whereas their non-parallelism indicates the difference between leg lengths because of wrong disposition and pathological modification of femoral bone and acetabulum.

First the size and optimal disposition of the endoprosthesis cup is determined (Fig. 4.3, *a*). The rotational centre of the healthy joint is then found and symmetrically transferred to the damaged joint. The cup template is superposed to the radiograph in such a way that its rotational centre coincides with the transferred rotational centre of the joint. Then such a template size is chosen that the external cup contour passes through the upper edge of "teardrop" (cavity in the bone forming the true bottom of the acetabulum). After this the template is turned around the rotational centre in such a way that the cup axis was at 40–45° angle to the vertical axis of the pelvis. With the clearance between the radiographic front edge of the acetabulum and the external contour of the cup the conclusion is drawn on arthroplasty.

The tracing paper is superposed to the radiograph and the template so that its edge was parallel to the vertical axis of the pelvis. The contours of the pelvis and cup are superposed with a red pencil and then the paper is removed from the radiograph.

Dimensions and optimal position of the stem are determined by superposing the stem template to the radiograph (Fig. 4.3, *b*). The following criteria are used when choosing a stem:

- cross-sectional dimensions of the template should correspond to the width of the medullary canal of the femoral bone;
- depth of "penetration" of the stem into the canal should be such that one of the three T-lines on the template (T-lines are perpendicular to the stem axis and correspond to *s*, *m* and *l* , viz. short, medium and long necks of the femoral component of endoprosthesis) contacts the top of the greater trochanter.

The difference in the leg lengths is eliminated in the following way. The tracing paper with the pencil contour of the pelvis and cup is superposed to the radiograph without removing the chosen stem template (Fig. 4.3, *c*). If it is required to elongate the leg the tracing paper with the pencil contour is lifted on the axis of the pelvis above its X-ray picture by a required value. If the leg is to be shortened the contour of the pelvis on the tracing paper is lowered by a respective value below the X-ray picture of the pelvis. In accordance with the correction needed the size of the stem and cup is specified.

The procedure results in the tracing paper with the pencil contours of pelvis, femoral bone, and chosen stem and cup (Fig. 4.3, *d*). The following lines are drawn on the tracing paper: 1) from the top of the greater trochanter to the rotational centre of endoprosthesis and 2) the contour of the external edge of the stem is continued to intersection with the first line. The second line is a projection of the vertical plane of the resection of the greater trochanter and specifies the direction for rasping the plane. The distances are measured and recorded:

- between the upper edge of the stem and lesser trochanter (45 mm in Fig. 4.3, *d*);
- from endoprosthesis heel to T-line I (20 mm);

- from lesser trochanter to guide line *R* on the stem template (9 mm); *R* is projection of the plane of the neck of femoral bone resection.

The drawing in the tracing paper should correspond to the postoperative radiograph of the joint. The suggested model of endoprosthesis and dimensions of the cup and stem are recorded on the tracing paper.

On the basis of long-term observations of the patients with implanted different types of endoprostheses the experts from "MATIS Medical Russia" recommend setting standard stems to the following depth (depth of setting is the distance from the endoprosthesis heel to the top of greater trochanter) (Fig. 4.4):

- stems of Muller's endoprosthesis – about 10 mm;
- those of CLS Spotorno endoprosthesis – on average 20 mm;
- isoelastic/isotitanium endoprostheses – from 5 mm to 0.

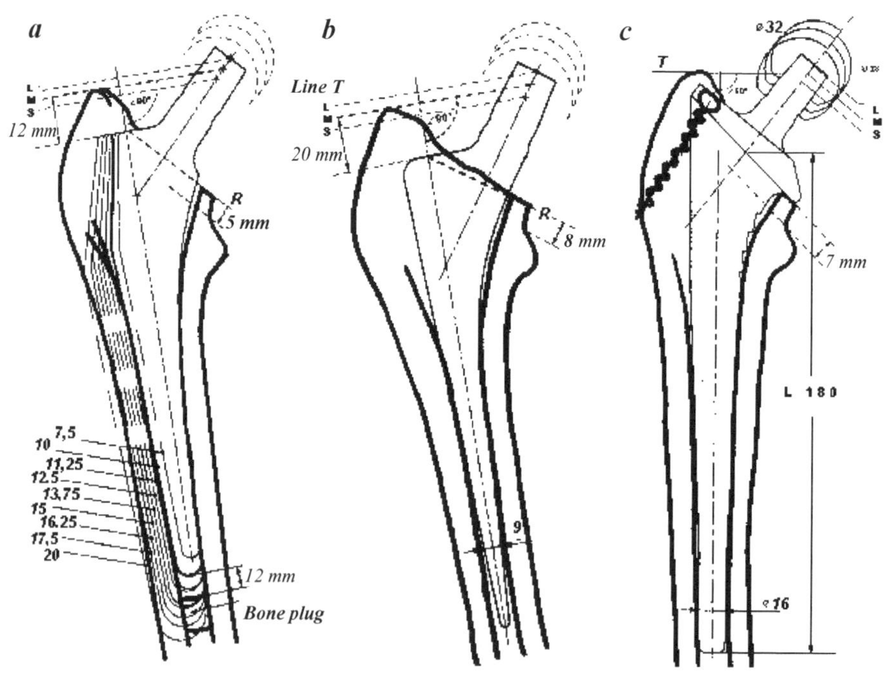

Fig. 4.4. Diagrams of optimal setting of the stems of endoprostheses: *a* – Muller's, *b* – Spotorno's, *c* – Isoelastic/Isotitanium

The plan of surgical interventions in endoprosthetics regulates the standard in modem orthopaedics procedure irrespective of the endoprosthesis fixation method. Below are described main stages of hip joint operation with usage of total cementless prosthesis "Lubinus Classic Plus" (WALDEMAR LINK GmbH). It is recommended for implanting in patients with critical non-inflammatory degeneration process: osteoarthrosis, avascular necrosis of the head of femoral bone, posttraumatic arthrosis, after-effects of epiphysiolysis of the head of the femoral bone, ankilosis, after-effects of the fracture of the acetabulum, etc. Its cementless implantation is not contradicted in patients under 40 years old.

The patient is prepared for operation by a standard scheme (see 1.3). Prior to or within operation antibiotics are introduced prophylactically.

The hip joint is metaphorically compared to a house and the access to the joint – to its doors. The anterolateral access ("front door") is located between the muscle straining the femoral fasciae from the front and the gluteus medial muscle. The posterolateral access ("back door") is between the medial gluteus muscle from the front and the upper edge of the large gluteus muscle from behind. "The front door" through the anterolateral approach provides the highest throughput into "the house" but is opened by osteotomy (section) of the greater trochanter, increases the operation period and blood loss, probability of non-union, etc. Most orthopaedists prefer posterolateral approach of Kokher-Langenbeck, which allows exposing the acetabulum and proximal part of the femur sufficiently. It is convenient since there are not large blood vessels along its line, the danger of intersection of the sciatic nerve is absent and the greater gluteus muscle is damaged less. Any standard surgical access is appropriate at "Lubinus" endoprostheisis implantation.

The advantages and disadvantages of the lateral and supine positions, respectively, in hip replacement surgery have been discussed among orthopaedic surgeons. One advantage mentioned for the lateral position is decreased bleeding during surgery. In work [12] the positions and their effect on blood loss is compared. All other factors (surgical approach, anti-thrombotic treatment, non-steroidal anti-inflammatory medication) were the same. An advantage is shown in using the lateral position in reducing postoperative blood loss, on average 200 ml.

Minimally invasive techniques for total hip arthroplasty have been introduced in the last several years and are becoming popular mainly in North America [13]. These techniques are designed to allow hip arthroplasty to be done through smaller incisions, potentially with less soft-tissue disruption (Fig. 4.5).

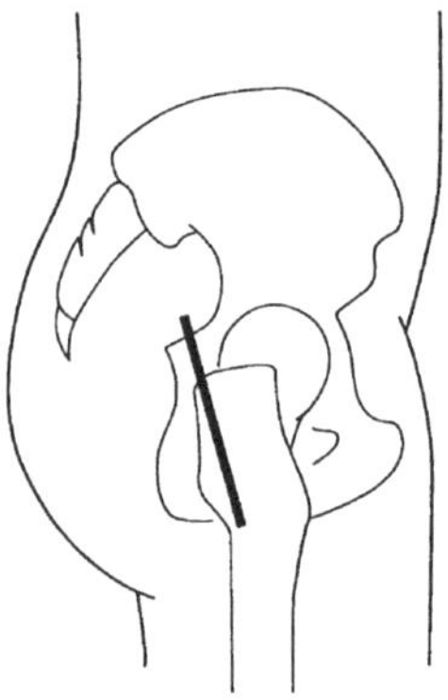

Fig. 4.5. Scheme of skin incision for the mini approach

The tip of the posterior aspect of the greater trochanter was marked. A straight skin incision spanning 3–4 cm proximally and 6–9 cm distally was made through the marked point. The incision was made slightly oblique from posterior-superior to anterior-inferior [14]. There are three main methods including a mini-incision anterior approach [15], a mini-incision posterior approach [16], and a two-incision technique [17]. In work [14] it is concluded that favourable short-term outcomes with less surgical invasion can be expected via a mini-incision posterior approach if the surgeons are already familiar with a conventional posterior approach.

The main stages of total hip joint replacement from posterolateral approach [18] are shown schematically in Fig. 4.6.

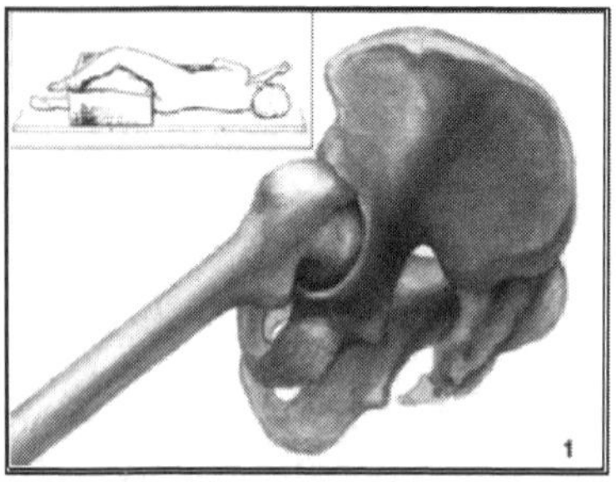

1. The patient is placed on his/hears side, in the dorsolateral position.

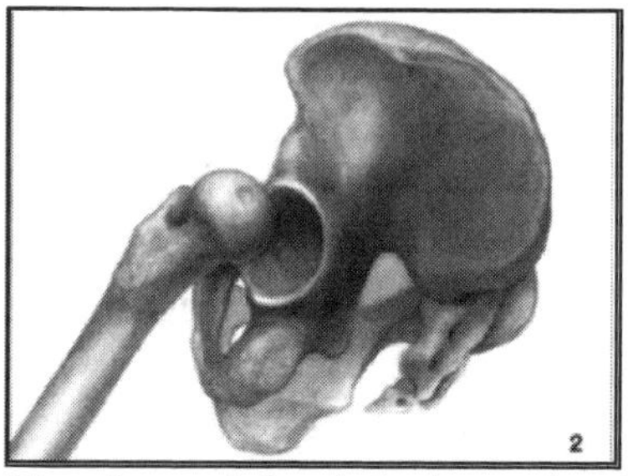

2. The femoral head is luxated using rotation and 90XXX flexion of the femur.

To be continued

Fig. 4.6. Scheme of main stages of hip joint replacement operation

Fig. 4.6. Continuation.

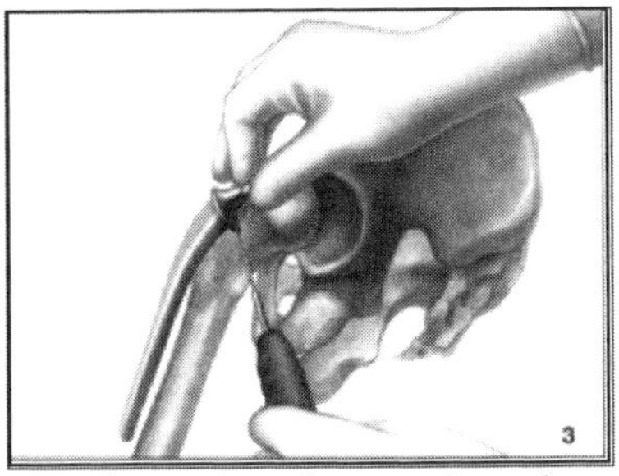

3. The femoral neck resection plane is determined. It will be approximately 90° to the neck of the femur. Sparing resection allows optimal seating of the prosthesis as additional reaming or resection remains possible.

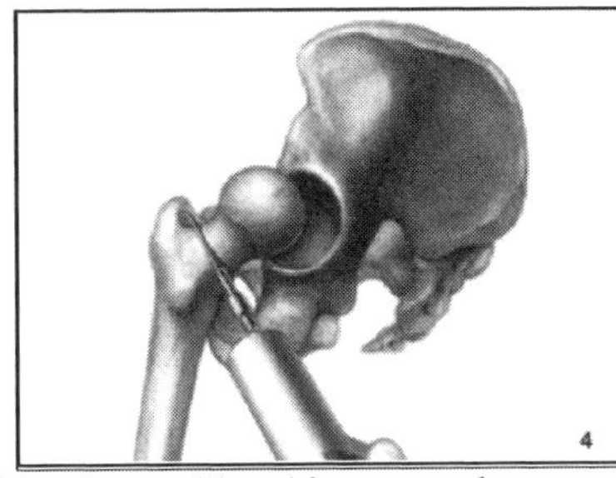

4. The femoral head is resected.

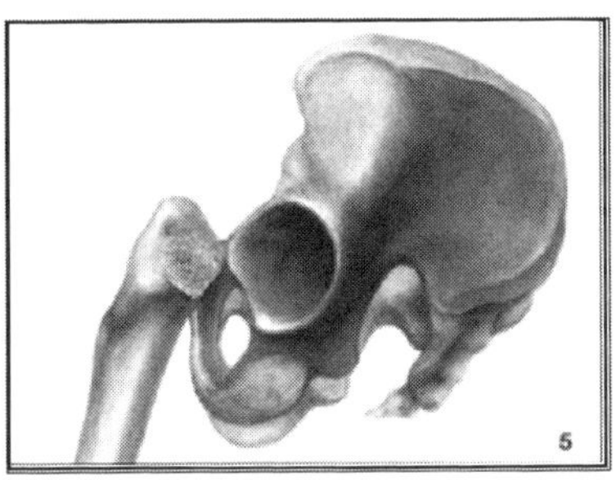

5. The acetabulum is exposed.

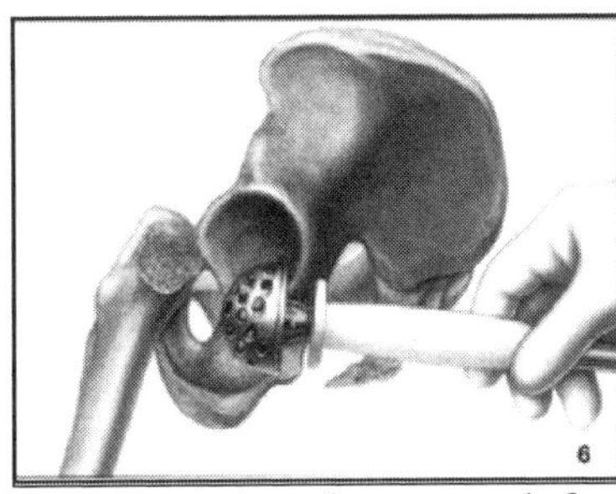

6. The acetabulum is prepared for an acetabular cup with a reamer. The reamer is used to expose cancellous bone for the acceptance of chosen cup. The diameter of the reamer is determined by comparison with the acetabulum.

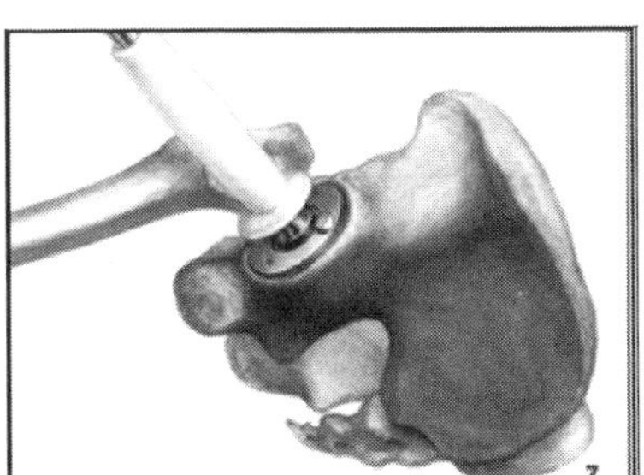

7. Reaming out the acetabulum. The outer cortical layer is removed only to the point where the cancellous bone is barely visible.

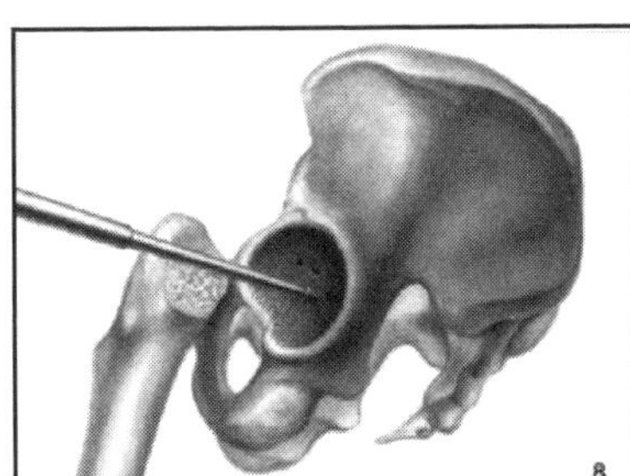

8. The anchoring holes for the bone cement are drilled primarily in the loading zone of the acetabulum using the penetrating drill.

To be continued

Fig. 4.6. Continuation.

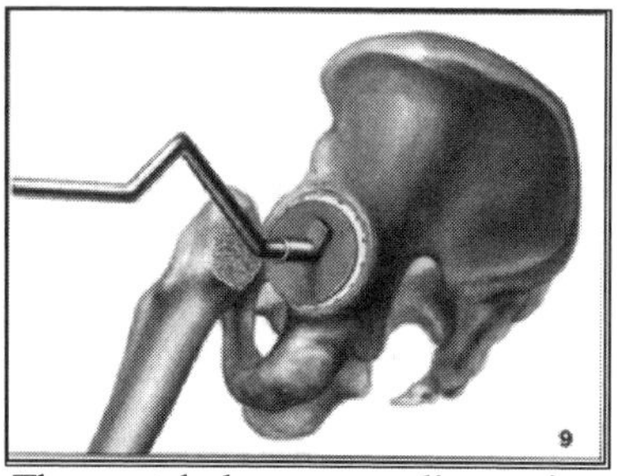

9. The acetabular cup applicator is used to introduce the acetabular component into the soft cement layer. When introducing the cup it is important to observe directional agreement (see Fig. 10 and 11).

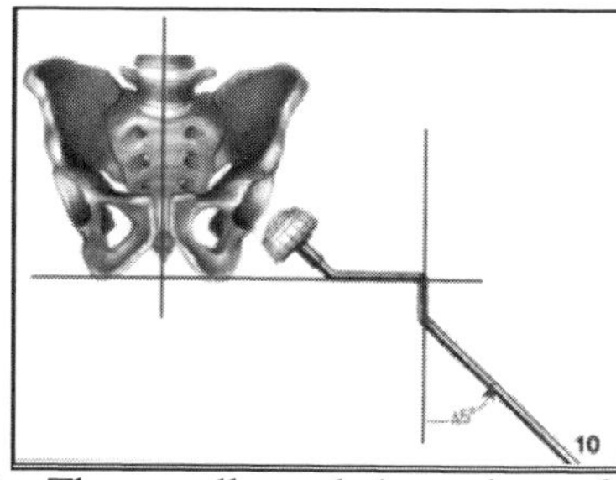

10. The small angled section of the application shaft must be aligned parallel to the vertical body axis (thus a 45° angle to the body axis is produced by the handle-near long applicator shaft section).

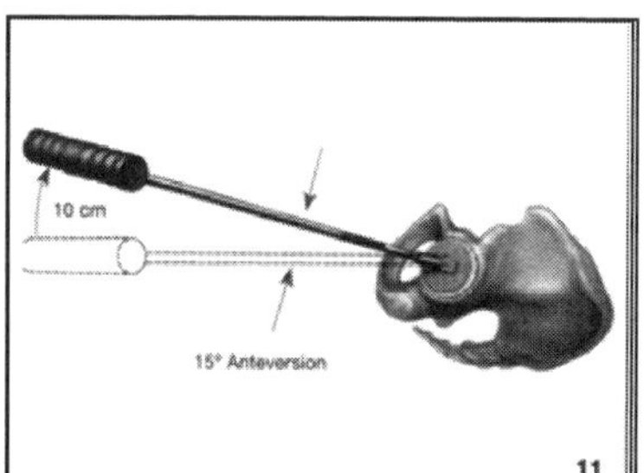

11. Simultaneously the application handle, parallel to the front plane of the patient's body, is lifted approximately 10 cm to adjust the anteversion of the acetabular cup to 15°.

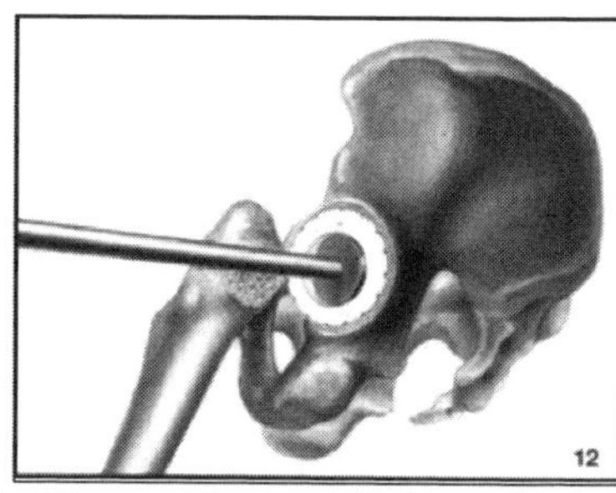

12. To avoid transmission of possible micromotions to implant site, which might to be produced by the surgeon during cement hardening, the ball-shaped acetabular cup pusher is consequently used to hold the acetabular cup until cement hardening has taken place.

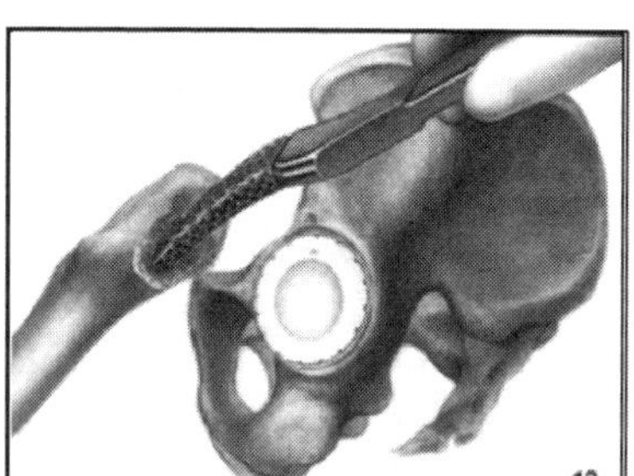

13. A rasp stem is driven into the centre of the femur using a detachable handle.

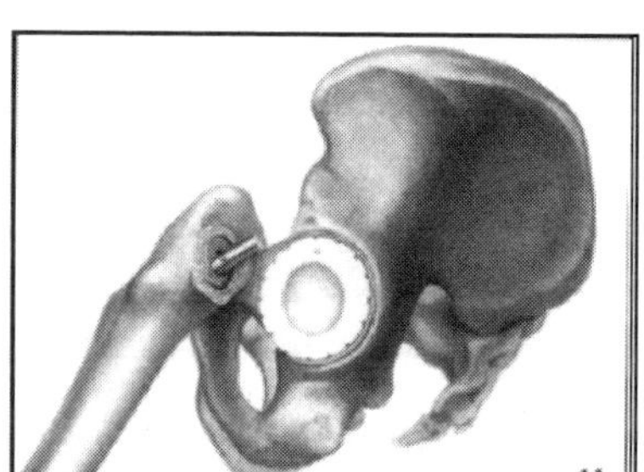

14. The rasp stem seated in the femur.

To be continued

Fig. 4.6. Continuation.

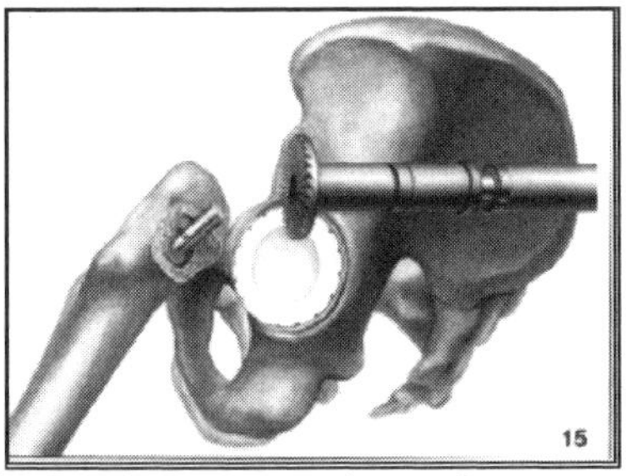

15. A calcar reamer is used to prepare the resection zone above the rasp stud.

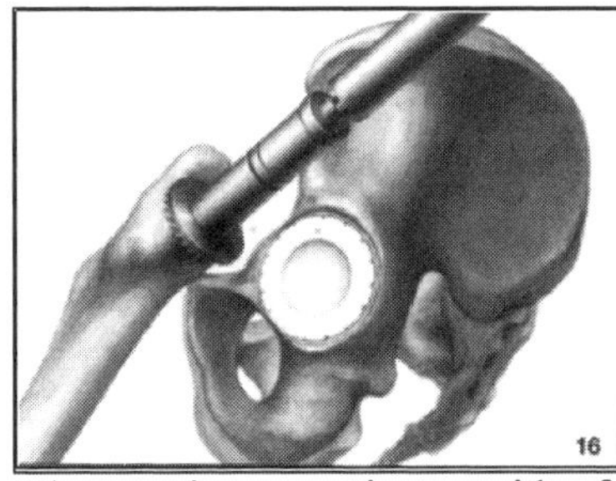

16. The reaming operation provides for a precise seating of the prosthesis collar on the end of the proximal femur.

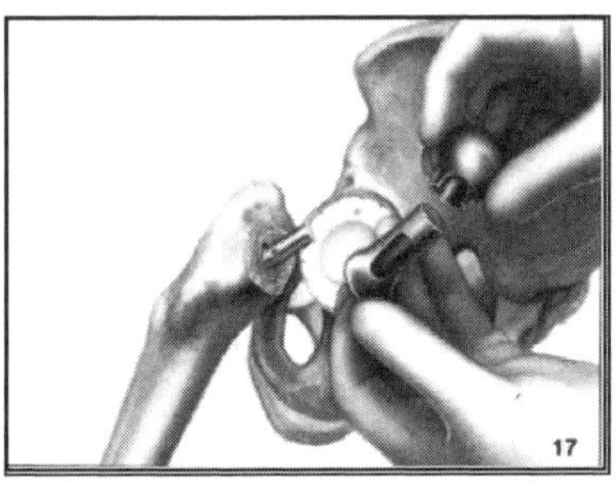

17. The rasp now serves as the trial prosthesis stem. A trial neck section is fitted on the rasp trunion. Trial heads are available in four neck lengths corresponding to standard type heads, and three neck lengths corresponding to A or B type heads.

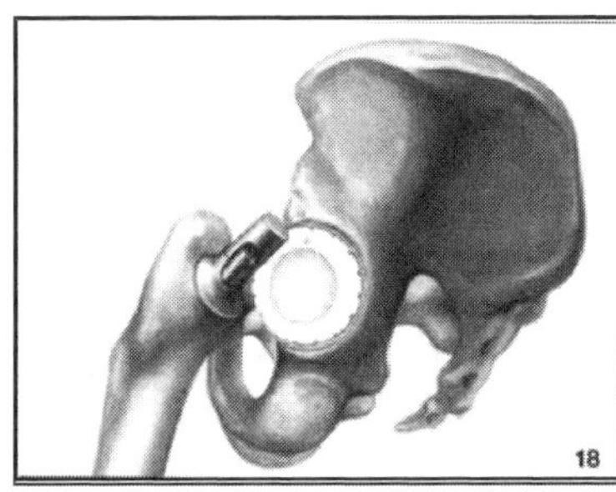

18. The trial neck section is fitted on the rasp stem with collar resting on the femoral calcar.

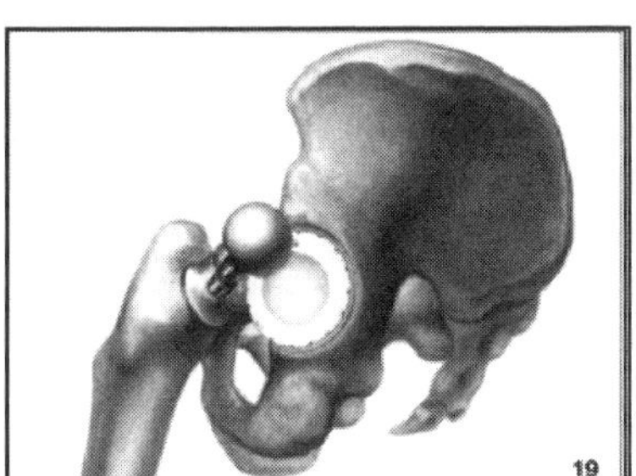

19. Rasp with trial neck section and trial head prior to trial reduction.

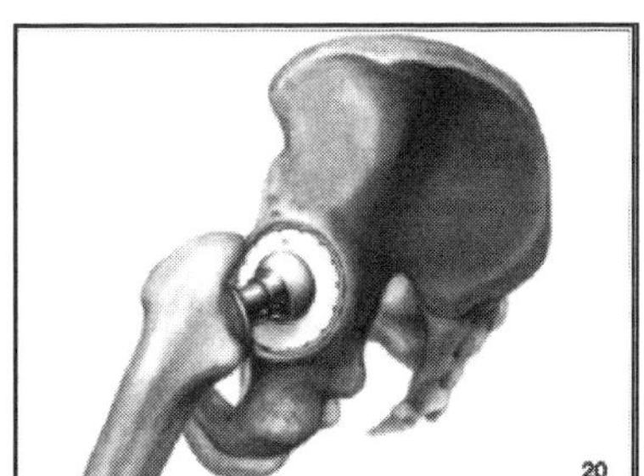

20. Trial reduction. If adjustments are necessary, the calcar can be reamed further using the calcar reamer, or a prosthesis head of different length can be used. Sometimes the use of an implant with different CCD angle becomes necessary. Consequently the respective trial head and trial neck are used.

To be continued

Fig. 4.6. Continuation.

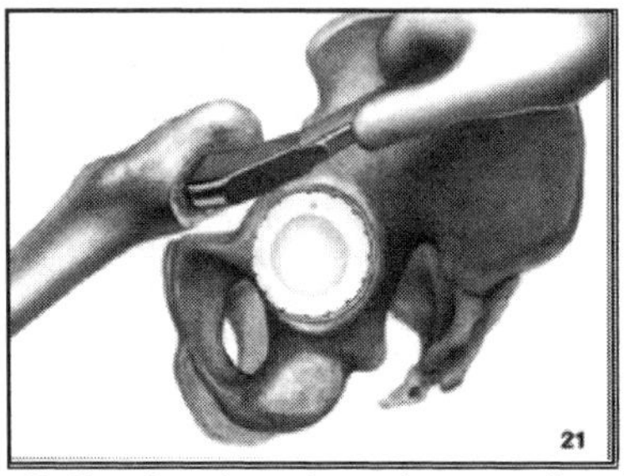

21. Following luxation, the rasp handle is re-attached to the femoral rasp and the stem is removed.

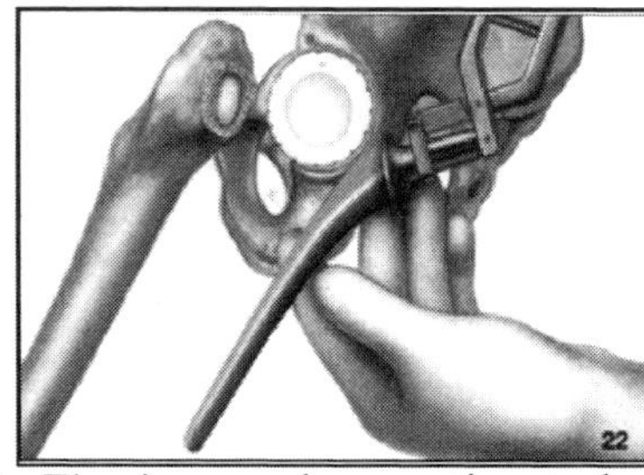

22. The improved cementing technique requires an introducing forceps to place the femoral component into the femur. The introducing forceps avoids damage to the taper and enables to introduce the femoral component securely into the cement bed.

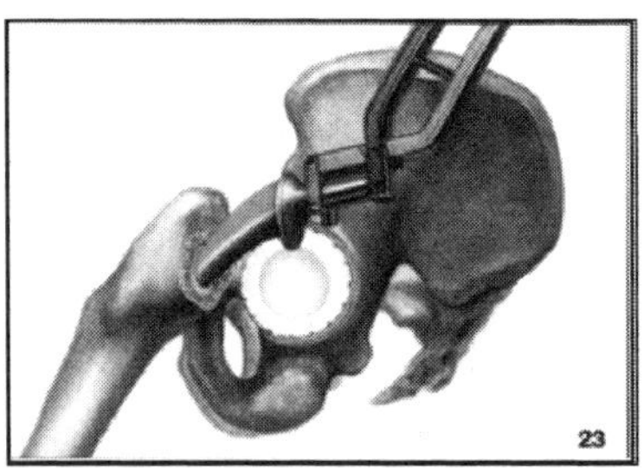

23. Introduction of the femoral component into the femur. A cancellous bone cylinder or polyethylene plug of appropriate size is placed in the femoral canal, down slightly beneath the anticipated seating of the femoral component's distal tip, prior to cement introduction. The introduction forceps is removed, as soon as the femoral component has been inserted far enough, so that the neck of the component protrudes a few millimetres only above the resection plane of the proximal femur.

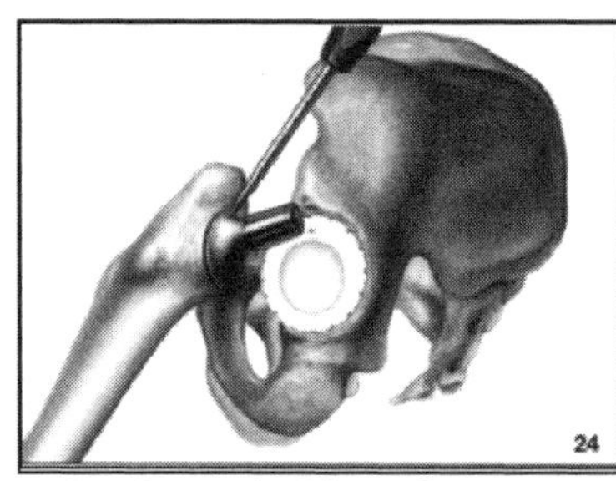

24. The femoral driver forms a needle joint together with the bore in the cranial neck of the prosthesis. Using the femoral driver, the femoral component is finally driven home, using light taps on the driver's handle. The needle joint between femoral driver and femoral component protects the implant site against damaging micromotions, which might inadvertently be produced and transmitted to the still soft cement bed by the hand of the surgeon in case of rigid coupling between driver and component.

To be continued

Fig. 4.6. Continuation.

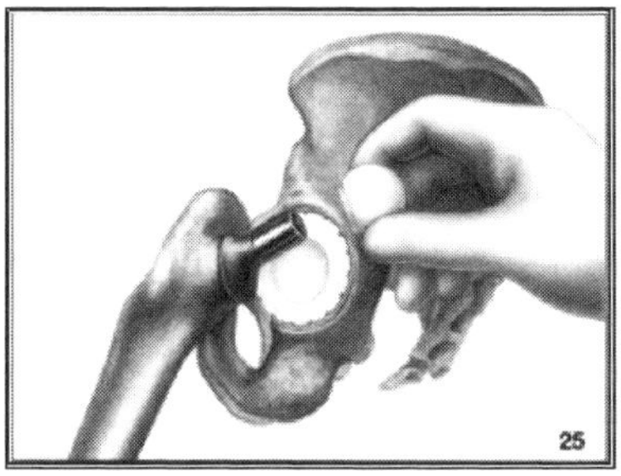

25. After cement hardening the final trial reduction is performed using the appropriate size of the white plastic trial heads.

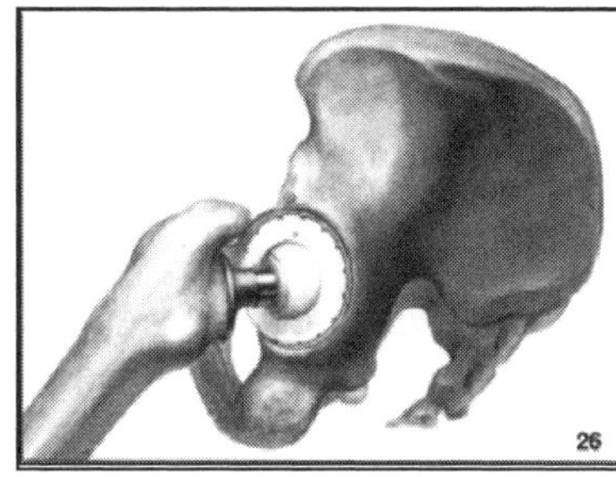

26. Luxation test following trial reduction.

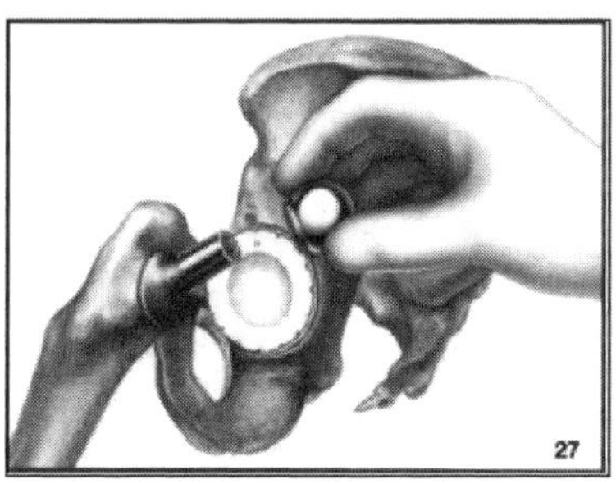

27. A prosthesis head is fitted to the prosthesis stem. It is crucial that all surfaces of taper and bore remain undamaged and free of contamination.

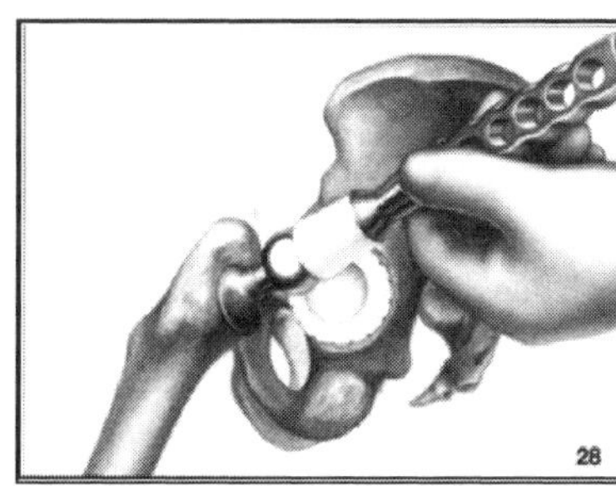

28. The prosthesis head is fixed to the stem with a light blow to the head driver. Again, it is crucial that the head as well as the plastic impact surface on the driver remain absolutely clean and undamaged. If necessary, the plastic impact section on the driver may be replaced.

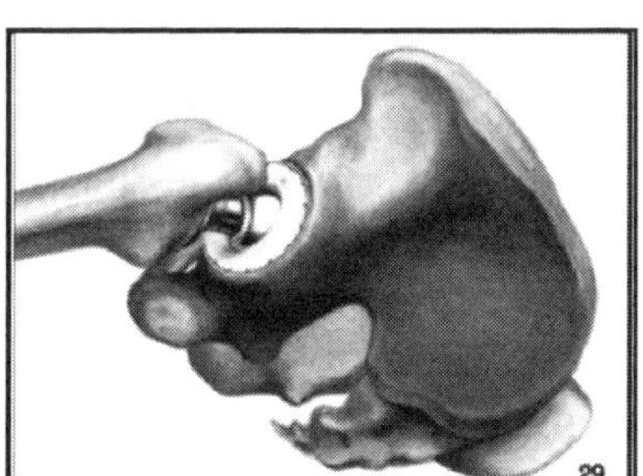

29. Completed implantation of the Lubinus Classic Plus with acetabular cup.

Fig. 4.6. Scheme of main stages of hip joint replacement operation

Fokin V.A. ("MATHYS Medical Russia") recommends treating the acetabulum in the following way. A pseudobottom of the acetabulum is exposed with a tooth-shaped curette making rotational motions directed medially downward (Fig. 4.7, *a*). When performing this unsafe and delicate procedure it is not recommended to use a chisel. Then (*b*) a central exostosis (osteocartilaginous growth) is taken away with a small-sized reamer 32–38 mm up to the true bottom of the acetabulum. The latter takes the form of a cavity consisting of two arcs of different diameters (*c*). After this the protrusion arising at the arc-arc interface is evened with the reamer. Further the cavity is treated with the reamers of sequentially rising diameters up to the subchondral layer. The last reamer is used manually observing appropriate inclination angles (external deviation from the longitudinal pelvis axis in the front plane is 40–45°) and anteversion (forward deviation from the longitudinal pelvis axis in the sagittal plane is 10–15°). As a result, the acetabulum must acquire a proper half-sphere shape.

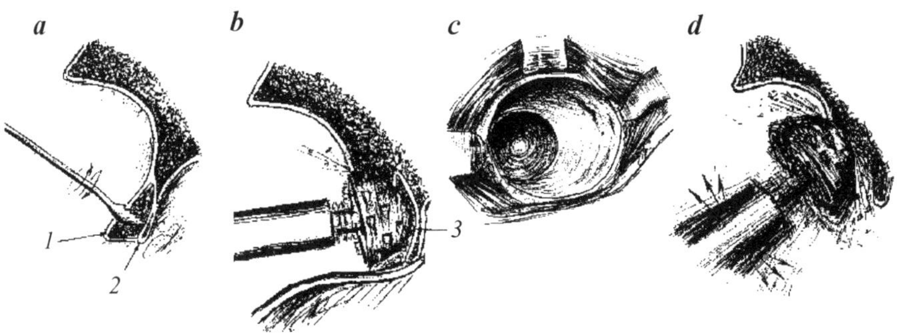

Fig. 4.7. Technique of treatment of the acetabulum: *a* – exposure of pseudobottom; *b* – taking away central exostosis; *c* – view of the acetabulum after stage *b*; *d* – protrusion reaming: *1* –central exostosis, *2* – «teardrop», *3* – true bottom

In total hip arthroplasty optimal placement of both cup and stem is essential for a satisfactory result [19]. Malpositioning of one or both components significantly increases the risk of failure or various complications like high rate of wear, early loosening or postoperative dislocation of the prosthesis may occur.

In order to improve surgical precision, computer aided techniques have been recently introduced in some specialities. Intraoperative surgical guidance can be provided with the help of preoperative planning and by semiautomatic devices (robots), which can perform operative steps under supervising by the surgeon. M. Tannast et al. [20] demonstrated that the cup position could be severely miscalculated without a standardized reference system. Using computer tomography (CT) based calculation, an exact reference plane can be defined. Such reference provides highly accurate information of the three-dimensional cup position [21]. In order to optimize implantation techniques in hip replacement ISS (Integrated Surgical Systems, Sacramento, USA) developed a computer-based system to work with a surgical robot [22]. In work [23] the quality of computer-assisted orthopaedic surgery in arthroplasty of the hip is evaluated. Femoral stems were implanted either manually by an experienced surgeon or by a robot in fresh human cadaveric femora. Implantation by robot showed higher precision in reconstructing the true anatomic situation as well as providing a better press fit.

Recently, navigation systems have been developed to improve the accuracy of component alignment in total knee arthroplasty and the first follow-up results of computer-assisted arthroplasty are promising [24]. OrthoPilot navigation system affords a possibility to place femoral and tibial components precisely with less axis deviation than with the conventional technique [25]. Using a CT-free navigation system and minimally invasive unicondylar arthroplasty with specific instruments leads to a superior accuracy in reconstruction of limb alignment [26].

Choice of drugs for prophylaxis of postoperative complications should meet the criteria cited in 1.3. First of all, they are prevention of purulent complications, thromboembolism, and heterotopic (associated with tissue displacement) bony formations (built-on).

Blood loss is replenished intra- and postoperatively. Because allogeneic transfusions carry risks of vital disease transmission, allergic reactions, and posttransfusion immunosuppression orthopaedic surgeons have investigated various blood management strategies in orthoplastic surgery to reduce patient exposure to allogeneic blood [27]. They include the use of pharmacological agents, particularly recombinant human erythropoetin (epoetin alfa). The simulation of red blood cell production by erythropoetin therapy is one means of treating this anaemia preoperatively [28]. Long-term experience of autoblood reinfusion of patients subjected to arthroplastic surgery has revealed a tendency that postoperative autologous blood reinfusion is alternative to homologous, banked blood transfusions [29].

Postoperative vacuum drainage of the operated joint prevents haematoma formation and, consequently, development of inflammatory complications. The drain is removed not later than in 24–48 hours to avoid infection through the place of its setting.

The dressing is changed next day after operation and beginning from the 4-th day film-forming means can be used instead of dressings. The sutures are removed in 12–14 days.

Unfortunately, within the frameworks of the chapter we could only outline general principles of planning and some details of operation procedure. Joint prosthetics even when using conventional implants is an individual procedure for each patient. Therefore in planning and especially when performing operation one ought to take into account much more factors than those mentioned above. Thus, when the damages of the locomotor system are combined with cranial-cerebral trauma the treatment assumes the patient's survival with further planning operations on bones and joints [30].

4.2 REVISION OPERATIONS

Presently primary joint endoprosthetics provides a high precision of revision operations. Unfortunately, despite good closest results a need rises in replacing an artificial joint or its component proportionally to the time passed. One cannot but agree with the opinion [31] that statistical data on the frequency of revision operations have become of concern of even zealous followers of endoprosthetics since the figures imply human lives.

The following data seem to be the most reliable and convincing among numerous statistical data on hip joint revision operations. From 20 to 60 % of endoprostheses reveal loosening in 5–10 years. One third of loosened endoprostheses turn out to fail within first 5 years [32]. From 30 to 40 % of femoral and 10–20 % of pelvic components of endoprostheses work not more than 10 years [33]. Therefore, the ratio of cases of primary and revision endoprosthetics in largest clinics worldwide is 4:1 and even 3:1 and in the nearest future for each two primary arthroplasty there will be one replacement of endoprosthesis or its component [34].

The reaction of the organism to the primarily implanted endoprosthesis is developed in the following way [35]:

1) a fibrous capsule arises in the contact of implant with living tissues;
2) under cyclic loading when walking the capsule is thickened and separates the implant from the bone; thus endoprosthesis wear products are formed;
3) bone tissue resorption occurs, aseptic inflammation develops aggravated by wear debris; the fibrous capsule forces back from the implant the surrounding tissues forming a protective buffer medium. At this stage the endoprosthesis is not already stable;
4) destruction and lysis of bone tissue occur, which is a prerequisite for revision endoprosthetics.

Main indications for revision operation is instability of endoprosthesis caused by aseptic loosening of its components and more rarely by early or late deep suppuration, recurring luxations, fatigue failure of endoprosthesis components, fractures of femoral and pelvic bones. Analysis of the data in Fig. 4.8 indicates that the frequency of revision operations is proportional to the number of primary endoprosthetic operations that was stabilized in European countries in the beginning of 1980-ies [36]. Most revision operations are caused by aseptic instability of endoprostheses, whereas the frequency of revisions caused by suppuration, engineering errors when performing operations, and some other causes remains almost the same. Revision joint endoprosthetics is a serious intervention having a particular philosophy, unconventional technologies, particular designs of endoprostheses and instrumentation. It is complicated by the necessity to resolve the following issues: 1) insufficiency of bone support for a new implant; 2) necessity to pack the defect of the bone available; 3) difficulties in removing the "old" endoprosthesis as well as the cement; 4) ageing of the patient within the span between primary and revision arthroplasty; 5) availability of cicatricial tissues remaining after primary operation; 6) increased probability of purulent complications.

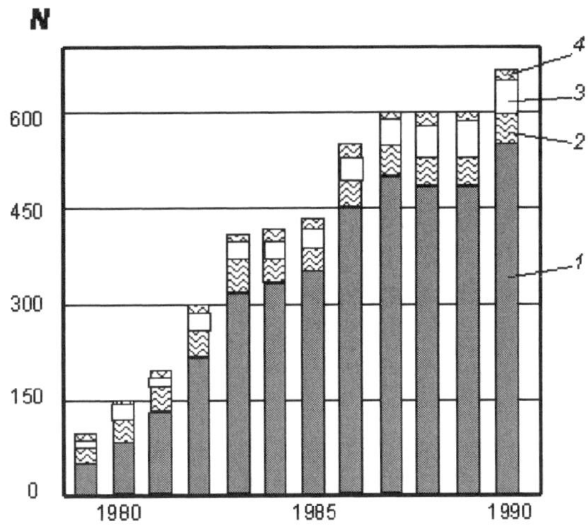

Fig. 4.8. Number of revision operations (*N*) on total hip joint replacement made in Sweden in 1979-1990–is versus: *1* – aseptic instability, *2* – suppurations, *3* – engineering errors when performing operations, *4* – other causes [36]

In work [34] the following main principles of revision endoprosthetics are formulated:

- non-invasive removal of an unstable component or the whole endoprosthesis with special tools;
- provision of stability of a new implant, which means proper operation planning and optimal for the given conditions choice of a revision endoprosthesis;
- making up lacking parts of the bony bed;
- prophylaxis of purulent complications.

The engineering features of joint revision operations are characterized by the following.

Revision operations tend to be performed under super-sterile conditions with using operation pressure suits for surgeons and prophylactic antimicrobic therapy. However in most cases the operation is performed in a conventional operating room with observing the simplest cautions including talks and unnecessary motion prohibited, prescribing prophylactic antibiotic therapy, etc. In work [37] (1) a strong correlation between surgical site infection and the probability of developing deep wound infection is confirmed and (2) it is indicated that hematoma formation and persistent postoperative drainage increase the risk of site infection. The authors hypothesize that postoperative monitoring of patients for hematoma and persistent drainage enables earlier intervention that may lower the risk of developing site infection and subsequently deep wound infection. In work [38] the experience is reported in a situation where there had been a high incidence of postoperative infection with methicillin-resistant *Staphylococcus aureus* after joint replacement.

Revision operation is followed by a more massive than the primary one blood loss, viz. from 550 to 3750 ml and on average 1750 ml. To compensate it the hemotransfusion is used, autohemotransfusion being preferred. For this purpose the blood is preliminarily taken or the blood collected during surgical intervention is reinfused [34].

When trying to retrieve the femoral component of the hip joint endoprosthesis without preliminary removal of the cement from the lateral stem part the greater trochanter can be fractured. The cement is removed with a special tool called an osteotome while frequently washing the medullary canal with physiological solution. Sometimes it is needed to “dip out” the cancellous bone in the region of the greater trochanter in order to direct the osteotome distally. If the cement is not removed totally it can impede the passage of reamers when preparing the bony bed for a new endoprosthesis. It causes the danger of cracking of the femoral bone or eccentric drilling of the canal. There are unconventional methods for removing the bone cement, e.g. by introducing fresh polymethylmethacrylate into the old cement mantle in which a threaded rod is immersed and after the fresh cement is seized the rod is screwed out from the canal. Ultrasonic tools are proposed to grind the cement mantle [39]. In a number of cases it is necessary to form a “window” in the tubular bone. In Fig.

4.9 the diagram is shown of the cement removal through an oval "window" in the cortical layer of the bone in order to facilitate the stem removal and prevent accidental damage of the femoral bone [40]. The absence of the acute angles in the "window" allows reducing stress concentration in the bone. The ends of the bony plate are tapered inward the window to provide its safe anatomic setting and primary fixation with cerclage sutures.

Similar operations of cement mantle removal are also performed when retrieving pelvic components of endoprostheses. Therefore, in revision of the latter in 26 % of cases the defect of the acetabulum bottom is revealed, in 36 % – its edges, and in 26 % – both [34].

Due to these features a general strategy of revision endoprosthetics is based on the evaluation of the deficiency of the bone support for a new implant and aimed at solving the problem of making up the lacking bone.

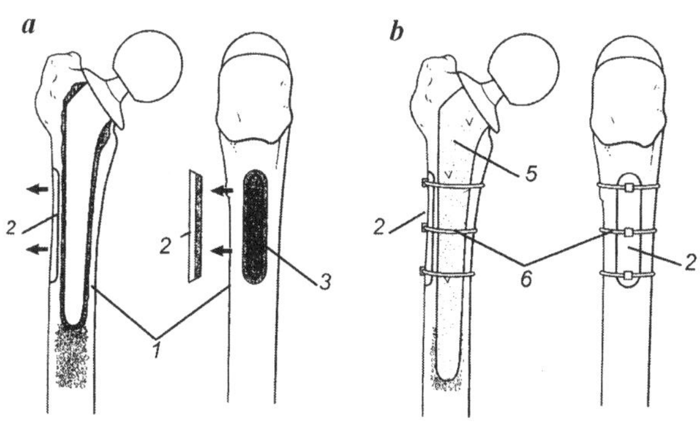

Fig. 4.9. Diagram of femoral bone revision operation: *a* – front view and side view of the bone with the femoral component of the endoprosthesis set in primary operation, *b* – the same after revision operation. *1* – femoral bone, *2* – bony plate, *3* – "window", *4* and *5* – primary and revision stems, *6* – cerclage suture

Planning of revision operation assumes a particular concern of the orthopaedist about the restoration of the anatomic centre of joint rotation. The solution of the problem is complicated by the bone mass losses in joint replacement and operation. To assess the losses their scientifically justified classification is needed.

Classification of the defects of hip joint bones can follow several criteria.

American Academy of Orthopaedic Surgeon (AAOS) recommend to use the classification proposed by D'Antonio in 1989. It is given in Tables 4.2 and 4.3 [40].

Table 4.2
Classification of acetabulum defects

Type	Defect
I	**Segmental** Peripheral (relative to acetabular lip) 1. Anterior 2. Posterior 3. Superior 4. Central
II	**Cavitary** 1. Anterior 2. Posterior 3. Superior 4. Medial (protrusion of medial wall)
III	**Combined** (segmental and cavitary) 1. Superior segmental – superior cavitary 2. Medial segmental – medial cavitary (absence of medial wall) 3. Posterior segmental – posterior cavitary
IV	**Divergence of pelvic bones**
V	**Consolidation**

Table 4.3
Classification of femoral defects

Type	Defect
I	**Segmental deficiences** 1. Proximal a) partial b) complete 2.Intercalary 3. Greater trochanter
II	**Cavitary deficiences** 1. Cancellous 2. Cortical 3. Ectasia – " bulgding" of femoral bone without failure of cortical layer
III	**Combined segmental and combined cavitary**
IV	**Malalginment** 1. Rotational 2. Angular
V	**Stenosis (hip narrowing)**
VI	**Discontinuity**

In the acetabulum the bone loss is believed to be cavitary if it is shaped as a cavity or hollow in the cancellous or cortical layer whereas from outside the cortical wall of the pelvic bone is not damaged. Bone losses in the anterior or posterior part of the acetabulum or in its medial wall are referred to the segmental defects. Cavitary defects always occur beneath the loosened cup. Radiographic displacement of the cup by 1 cm and more as well as osteolysis, i.e. bone tissue destruction with polyethylene wear products indicates the segmental defects of the acetabular lip or the medial wall. The defects are shown schematically in Fig. 4.10.

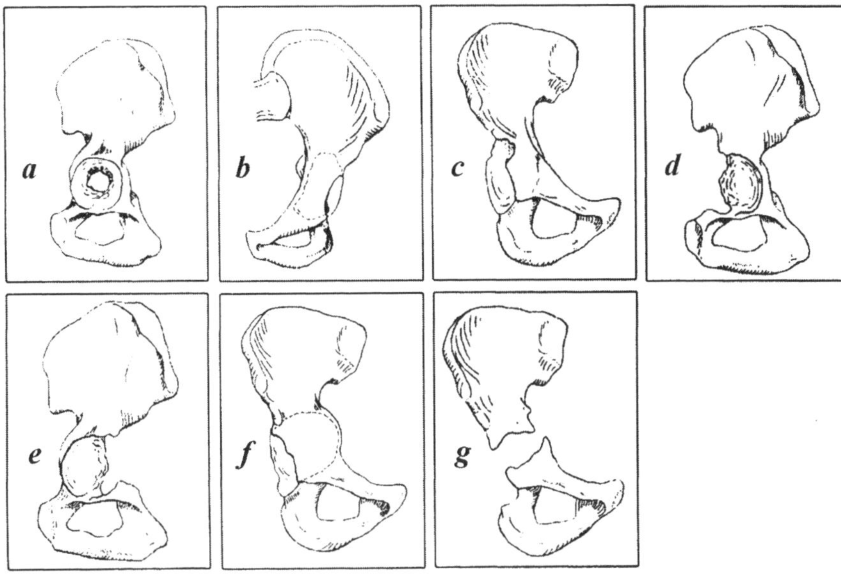

Fig. 4.10. Typical defects of the acetabulum: *a* – central segmental (side view); *b*– anterior-inferior cavitary (front view); *c* – superior segmental (front view); *d*– posterior segmental (side view); *e* – anterior segmental (side view); *f* – protrusion in the form of central cavitary defect; *g* – divergence of pelvic bones caused by anterior, posterior, and central defects

Figure 4.10, *a* shows the segmental defect (Table 4.2, type I), which is the bone loss in some part of the support half-sphere, e.g. in the medial wall. Type II of the cavitary defect (*b*) is the bone loss in the acetabular lip including the medial wall not ruined totally. Both segmental and cavitary defects can be peripheral or central, while peripheral ones can be superior, posterior, or anterior (*c*, *d*, *e*). Combined defects (type III) have some features. By way of example, such a defect can consist of the posterior and superior peripheral defects. Protrusion of the acetabulum (*f*) is a central cavitary defect, which can be complicated by the central segmental defect. Defects of types IV and viz. divergence (*g*) and symphisis of pelvic bones complete the classification.

Defects of the femoral bone illustrating Table 4.3 are shown in Fig. 4.11. Segmental defects of the femoral bone are spread to its cortical layer and classified as partial or total. Partial (*a*) ones can be located in the anterior, medial, or posterior parts of the femur, total (*b*) damage the cortical layer over the whole perimeter of the bone.

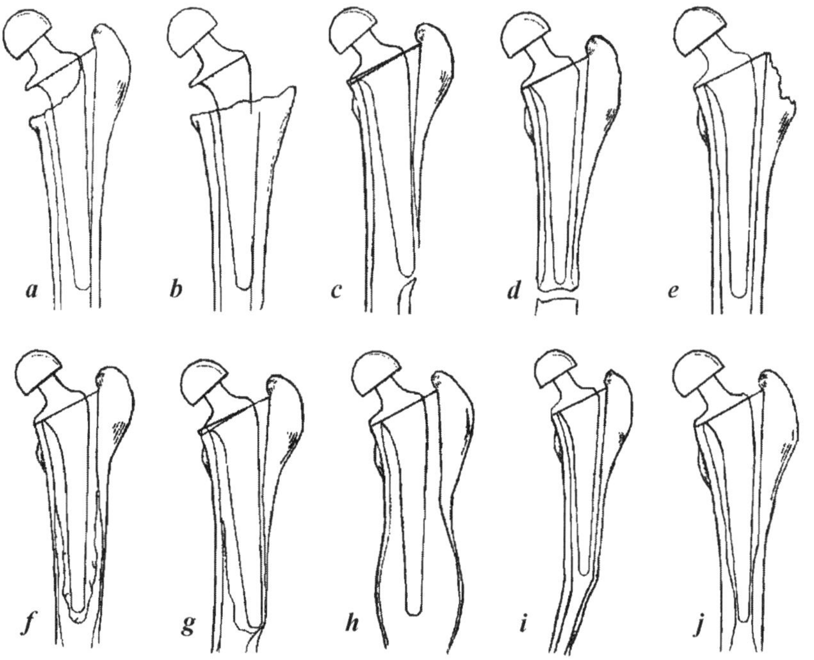

Fig. 4.11. Typical defects of the femoral bone [40]: *a* and *b*– partial and total segmental proximal; *c* – segmentary from penetration; *d* – fracture; *e* – segmental defect of the greater trochanter; *f*, *g*, and *i* - cavitary cancellous, cortical, and extasy; *j* – angular deformation; *κ* – stenosis

The segmental defect from penetration (*c*) arises at wall breaking (perforation) with the distal end of the endoprosthesis stem and is always partial. The total segmental defect from penetration is classified as hip fracture (*d*). Absence or deficiency of the greater trochanter (*e*) is a particular form of the segmental defect. Cavitary defects damage the cancellous (*d*) or cortical (*g*) layers of the femoral bone or can cause bone "inflation" (*i*), viz. extasy. Segmental and cavitary defects are often combined.

Deformation defects can be rotational, angular (*j*) or a combination of these deformations. Stenosis, viz. femur narrowing (*k*) completes the classification.

The level of the femur defects (Fig. 4.12) is their important characteristic. Level I is spread from the proximal part of the femur downward to the lesser trochanter. Level II has a length of 10 cm below the lesser trochanter. Level III corresponds to the distal part of the femoral bone below level II.

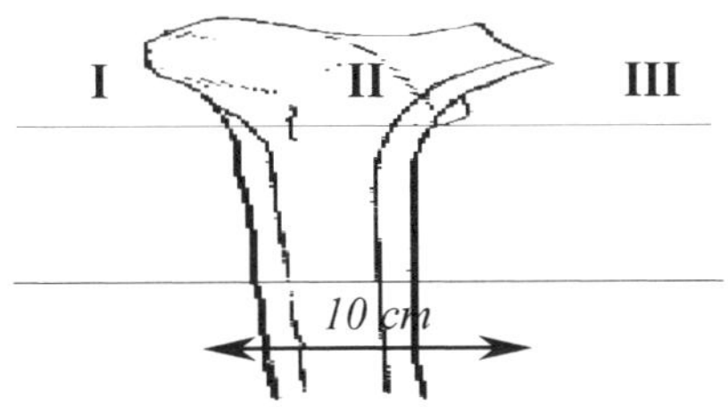

Fig. 4.12. Diagram of the level of femoral bone defects

When planning revision operations Austrian orthopaedist K. Perner prefers to use the Engelbrecht and Heinert classification that is given in Table 4.4 [41]:

Table 4.4.
Classification of hip bone defects by Engelbrecht and Heinert

Destruction degree	Characteristic
0	*No bone loss, the signs of endoprosthesis component displacement are absent*
1	*Insignificant bone tissue loss* Acetabulum: resorption zone is noticeable, clinics – pains when moving but without cup migration Femur: resorption zone in the proximal part of the femoral bone, clinics – pains when walking without changes in mutual dislocation of the endoprosthesis components. Fracture of the endoprosthesis stem is possible with its partial fixation being retained
2	*Moderate loss of femoral tissue* Acetabulum: increasing resorption zone, clear displacement of the cup to the widening lysis zone. Femur: the lysis zone is spread along the whole stem, bone resorption reaches the proximal part of the medullary canal, the endoprosthesis components can be mutually displaced

3	*Pronounced loss of bone tissue* Acetabulum: clear instability of the cup with noticeable displacement. Femur: ectasia of the medullary canal with increasing transversal dimensions of the bone, availability of bone defects in the proximal part with possible perforations
4	*Very large (up to total loss) deficiency of the bone tissue* Acetabulum: considerable displacement of the cup. Common defects, destruction of the hollow edge is possible. Femur: spread or total destruction of the proximal and medial third, pronounced lesion of the distal part (osteoporosis, thinning of the cortical layer), support loss

K. Perner believes that revision operations on hip joint is advisable to perform when there is no considerable loss of the bone mass, viz. at the 1–2 degree of destruction. It causes the necessity of regular examination of the patients with artificial joints since the pain being the main indication for operation frequently appears at 3–4 degree of the bone tissue destruction.

Many orthopaedists [42] when planning revision operations use the DGOT (Deutshe Gesellschaft fur Orthopadie und Traumatologie) classification of the femoral bone defects. Its diagram shown in Fig. 4.13 [43] does not need any comment.

The classification if often used in combination with Pak J. and Paprosky W. classification that regulates the destruction degree of the bony bed on the basis of radiographic analysis [44].

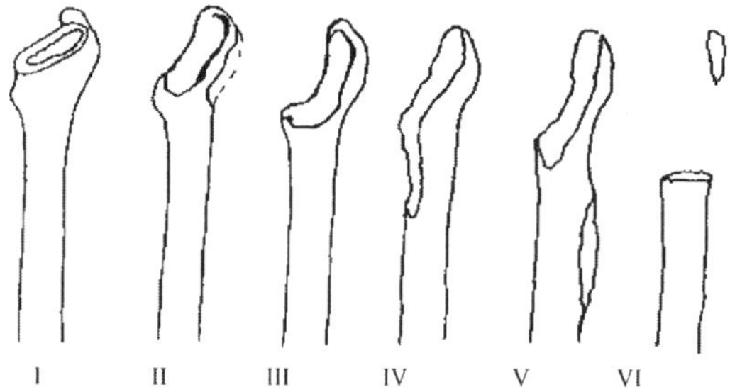

Fig. 4.13. Diagram of the defects of the proximal part of the femoral bone according to the DGOT classification: I – spongious (in cancellous bone) defect, II – trochanteric, III – osteophyte, IV – osteophyte and medial defects of diaphysis, V – osteophyte and lateral defects of diaphysis, VI – segmental defect of diaphysis

Defect size depends on numerous factors, viz. initial deformation of bones, endoprosthesis loosening degree and migration of its components, method of fixation, number of preceding operations on the joint, osteolysis degree, etc. The classification allows us terminologically precisely determine bone defects whose objective assessment is important when planning revision operations.

The acetabulum can be operated with using the following variants of the pelvic component and method of its fixation [45]:

- cemented UHMWPE liner or it in the metallic shell,
- cementless porous-coated cup,
- cementless threaded cup,
- bipolar endoprosthesis,
- special revision endoprostheses designed for implantation at large bone loss.

Each of theses components can be used in combination with the bone allo- or autoplasty reconstructing cavitary or segmental defects of the acetabulum by means of ground bone or bone blocks. Below there are given the examples of such reconstruction [40].

Cavitary defects can be filled with osseous cement or ground bone tissue followed by endoprosthesis cup setting (Fig. 4. 14). In these situations the usage of reconstructive supporting rings can be useful. The method of cup fixation is determined finally intraoperationally depending on the true size of the cavitary defect and osteoporosis degree.

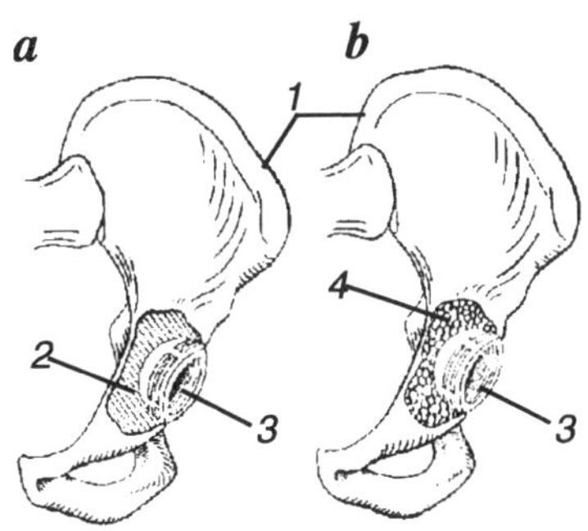

Fig. 4.14. Diagram of reconstruction of the acetabulum having cavitary defect with the osseous cement (*a*) and ground bone (*b*): *1* – pelvic bone, *2* – cement, *3* – cup, *4* – osseous-plastic material

Considerable difficulties occur when reconstructing the acetabulum having large cavitary and superior segmental defects. The diagram of the stages of this operation is given in Fig. 4.15.

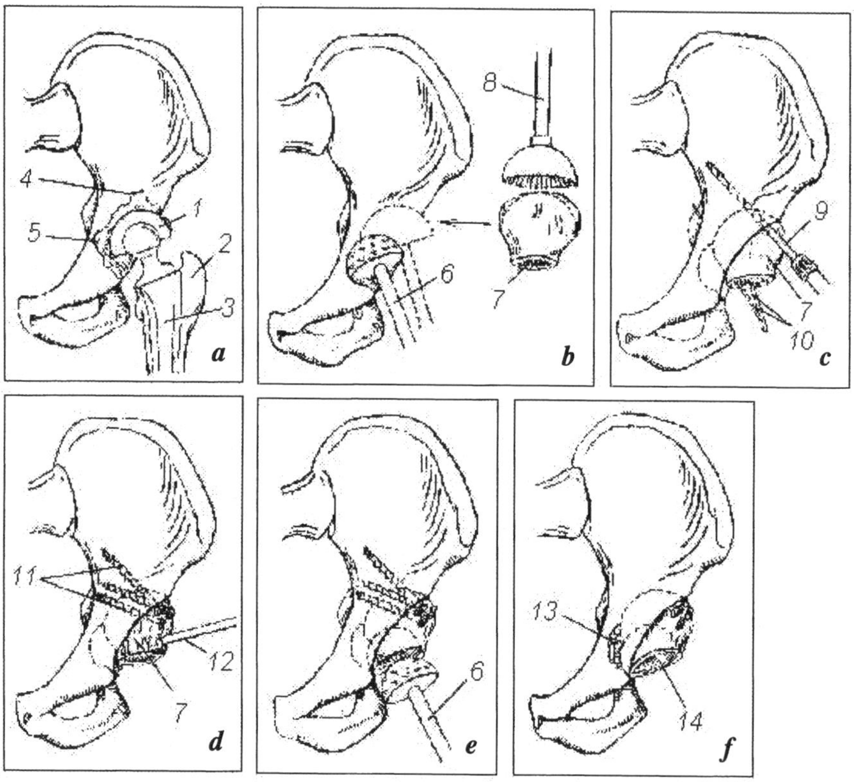

Fig. 4.15. Diagram of reconstruction of the acetabulum with extended cavitary and superior segmental defects: *a* – preoperation state of the artificial joint, *b* – smoothening of the defects by reaming and preparation of the transplant from the femoral bone head, *c* – temporary fixation of bone transplant, *d* – final fixation of transplant and its primary treatment, *e* – formation of cavity for endoprosthesis cup, *f* – reconstructed acetabulum. *1* – cup of primary endoprosthesis, *2* – femoral bone, *3* – femoral component of endoprosthesis, 4 and 5 – superior segmental and cavitary defects, *6* – reamer, *7* – bone transplant, *8* – hollow reamer, *9* – driller, *10* – spindles, *11* – cancellous screws, *12* – cylindrical reamer, *13* – ground bone, *14* – revision cup

The cup of the primary endoprosthesis is removed and the surface of the segmental defect in the acetabulum is smoothed with a reamer (*b*). The allotransplant head is treated with a hollow reamer of the same diameter. Such preparation provides congruency of the transplant surfaces and the cavity in the acetabulum. After fixing temporarily the transplant with spindles the holes for screws are drilled in it and the pelvic bone (*c*). The heads of cancellous screws are disposed in the upper part of the transplant so that they do not hinder its primary reaming with a high-velocity reamer (*d*). The cavity is cut to fit the cup of the revision endoprosthesis (*e*), filled the remaining cavitary defect with ground allobone and the revision cup is set (*f*).

One more example of using allobone to reconstruct extended anterior, superior, and posterior segmental and cavitary defects of the acetabulum is shown in Fig. 4.16. The surface of defects is smoothed with a reamer and the allotransplant is set in the cavity in the form of the proximal part of the femoral bone (*a*). The transplant is fixed to the pelvic bone with screws and then a spherical cavity is cut in it to set the cup of revision endoprosthesis. The odd parts of the transplant are removed before fixing it in the pelvic bone (*b*).

Before setting bipolar endoprosthesis the acetabulum is treated in the same way as when fixing the cup of total endoprosthesis.

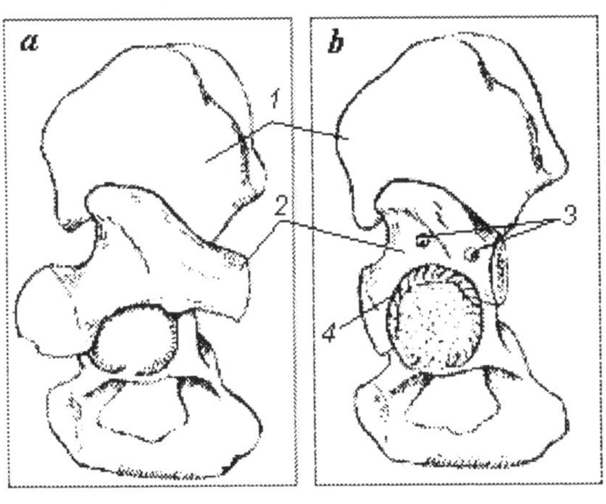

Fig. 4.16. Reconstruction of the acetabulum with extended defects: *a* – primary setting of the bone transplant, *b* – fixed transplant is treated with a reamer: *1* – pelvic bone, *2* – transplant, *3* – screw, *4* – cavity cut out in the transplant to fit the endoprosthesis cup

The cavity in the acetabulum is cleaned from the wear particles, smoothed by reaming and filled with the ground bony mass (Fig.4.17). A half-spherical hollow (a) congruent to the head of the bipolar endoprosthesis is formed by a special reamer with reversing motions. While making the cavity the transplant is consolidated.

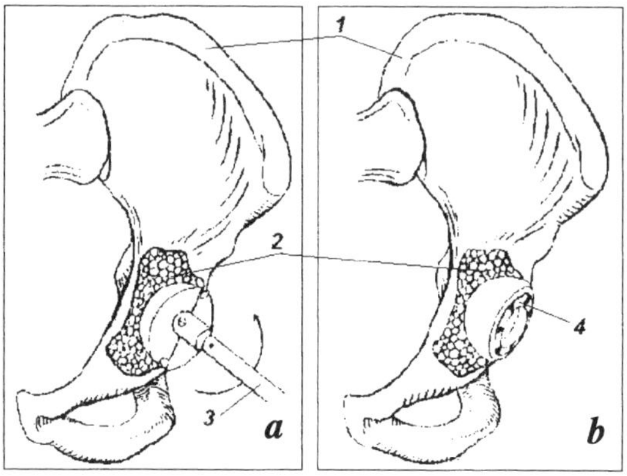

Fig. 4.17. Diagram of setting of bipolar endoprosthesis into the acetabulum having extended cavitary defect; *a* – formation of the cavity to fit bipolar head, *b* – setting of the head into the reconstructed acetabulum: *1* – pelvic bone, *2* – cavitary defect, *3* – reamer, *4* – endoprosthesis head

Significant differences in designs of primarily implanted endoprostheses, methods for their fixation, and individual features of surgical instrumentation impede the choice of optimal solutions in revision endoprosthetics. In work [46] the procedure of revision operations on hip joints with extended defects of the acetabulum is discussed. It uses bone allotransplants and several endoprosthesis designs differing by the method of fixation of the components. In work [41] the Zweimuller cup implantation is exemplified with using osseoplasty at acetabulum destruction degree 3-4 (by Engelbrecht, Table 4.4). Burch-Schneider and Muller rings as well as Weiling wide rings are used as an alternative. The application of Burch-Schneider anti-protrusio cages for revision cup arthropslasty has been highly estimated in [47].

The above data indicate the absence of consensus on many issues of revision enoprosthetics, viz. diagnosis of instability of primary endoprostheses, terms of revision operations, choice of revision endorpostheses, restoration of the insufficiency of bone support providing primary stable cup fixation and reconstruction of the anatomic centre of joint rotation.

In work [48] it is suggested to choose the cup design and technique for revision endoprosthetics by the criterion of the deficiency of the acetabulum, Analysis of the radiographs supported by intraoperative evaluation of true size of defects allowed distinguishing four degrees of destruction of the actetabulum.

I. Chondrolysis of the acetabulum after one-polar endoprosthetics. Radiographic signs: absence of the slit of hip joint and osteoporosis in 3 zones of the acetabulum. Typical pattern of chodrolysis development is shown in Fig. 4.18.

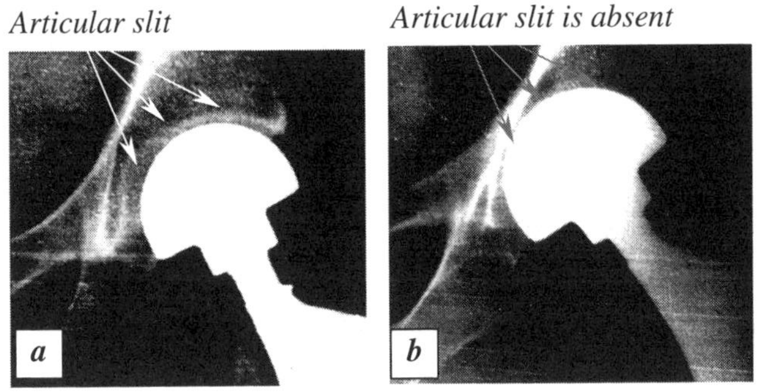

Fig. 4.18. Radiographs of hip joint after primary operation (*a*) and 1 year after operation (*b*) with signs of chondrolysis

II. Slight degree of the acetabulum defect. Radiographic signs (Fig. 4.19): after one-polar endoporsthetics (*a*) a defect 4–5 mm - sized is seen in one of the zones of pelvic bone by Charnley, after total arthroplasty (*b*) the width of the dividing line of pelvic component is 2–4 mm.

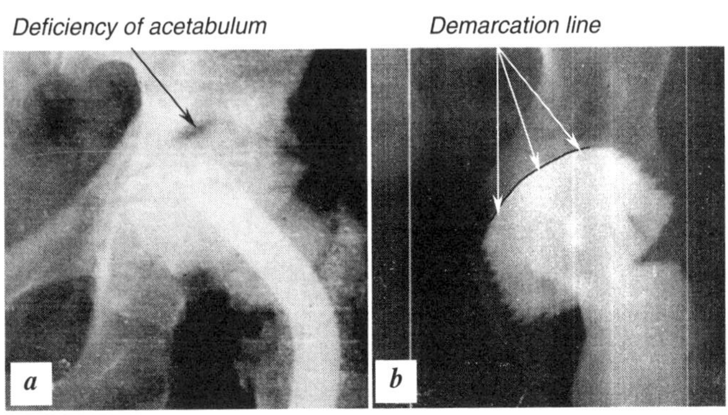

Fig. 4.19. Radiographs of hip joints with a slight degree of the acetabulum defect following: *a* – one- and *b* – bipolar endoprosthetics

III. Intermediate degree of the actetabulum defect. Radiographic signs: maximal size of the acetabulum on the front radiograph exceeds the diameter of the pelvic component of endoprosthesis by 5-5 mm, cavitary defects are insignificant (Fig.4.20).

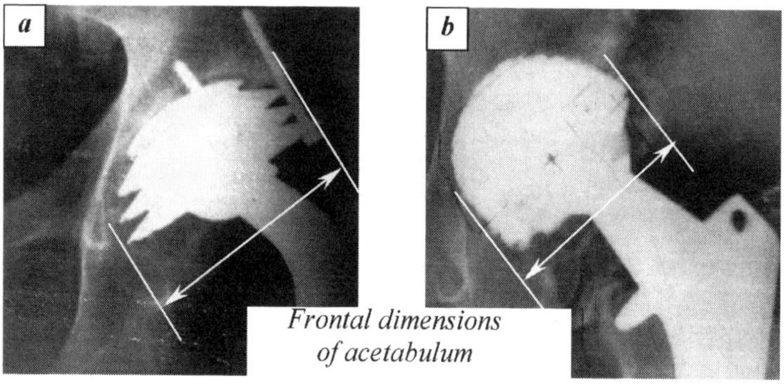

Fig. 4.20. Radiographs of hip joints with intermediate degree of the acetabulum defect. Primary endoprostheses: *a* – Sivash's, *b* – Gerchev's

IV. Severe degree of the acetabulum defect. Radiographic signs: significant widening of the acetabulum in all directions and cavitary defects up to its protrusion (Fig. 4. 21).

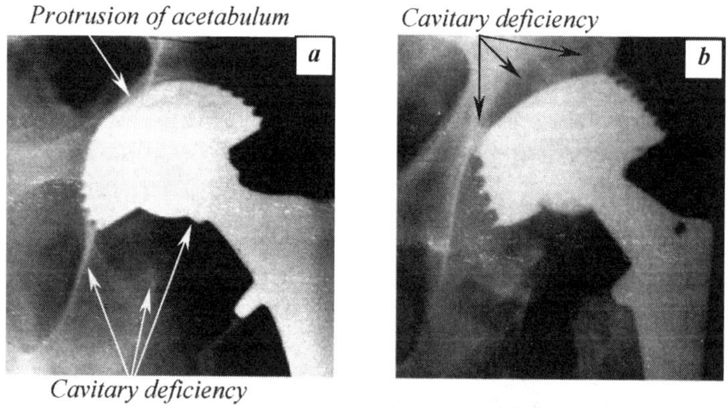

Fig. 4.21. Radiographs of hip joints with severe defect of the acetabulum after implantation of endoprostheses: *a* – «Foenix», *b*– Gerchev's

Histological examination of the joint capsules whose radiographs are given in Figs. 4.18 – 4.21 indicate the availability of conglomerates of colloidal polymeric and metallic wear debris incorporated in the shells of the connective tissue elements. Endoprosthesis wear debris is distributed practically over the whole thickness of the joint capsule. It is shown pathomorphologically that the cartilage areas with chondrolysis lesion are destructed chondrocytes. Wear debris is found in the acetabulum at a depth of 5–10 mm. Histological modifications of the surface layer indicated dystrophic process of osteocytes. Removal of the pathologically modified bone tissue results in increasing defect of the acetabulum and complication of revision operation.

The following recommendations can be given on planning revision operations on the acetabulum [46, 48]. After primary one-polar endoprosthetics in the acetabulum chondrolysis (type I defect) and slight defects (type II) commonly occur. In these cases the revision operation can be performed without osteoplasty using the Muller support ring and Mathys cup as a pelvic component. With defects of type II and III combining different kinds of osteoplasty is advisable with using Muller ring and Zweimuller cups. With defects of type III usage of Zweimuller cup allows avoiding osteoplasty. The cup has stable fixation even when 2/3 of its perimeter contact the pelvic bone. Revision operations on the acetabulum with defects of type IV assume obligatory osteoplasty in combination with Zweymuller cups or Burch-Schneider reconstructive rings.

In work [49] factors influencing the revision rate of Zweimuller acteabular cup has been analyzed. This may indicate that hard-hard paving with either metal-metal or ceramic-ceramic and/or cross-linked UHMWPE inserts should be implanted in patients under age 60. In concurrence with other studies, patient age and small cup size appear to be risk factors for early insert revision.

Revision operation on the femoral bone assumes the choice of cement or cementless fixation of the stem as the main issue. After retrieving the loosened primary stem the cancellous bone layer is often absent on the inner surface of the medullary canal. Availability of the layer is a prerequisite for reliable mechanical fixation of cement. In these cases the following variants of performing revision operations are possible [45]:

- long cemented stem allowing for overcoming the deficiency of the bony tissue in the proximal part of the femoral bone;
- cementless stem with the porous coating on its proximal part;
- the similar stem with the extended porous coating to provide a more reliable distal fixation;
- filling of the medullary canal with ground bony tissue and cemented stem of common length;
- individually made stem of a required length.

With significant bone losses in the proximal part alloplasty might be needed, which has become very popular in recent 20 years. The tubular bone is impacted with "grafting" osseous fragments and cerclage wire suture. Then the ground cancellous bone is pressed into the canal cavity and a cemented polished stem is set. The tapered form of the stem provides the cement flowing in the canal and consolidation of the ground bone. In work [50] the results are reported of impaction bone grafting for total hip revision (average patient age – 68 years, average duration from last surgery – 9 years). Morselized allograft bone was used for the graft, and the femoral component was a collarless polished tapered stem. The authors are optimistic that if complications in the perioperative period can be avoided the long-term results will be excellent.

In works [42, 45] the following variant of revision is suggested in case of fracture of the primary endoprosthesis and significant proximal losses of the femoral bone. The cement fixation of the stem is carried out in the proximal part of the femur restored by osteoplasty and *press-fit* fixation of the revision Wagner stem is performed in the distal part of the femoral bone. Such a method is used in patients with great losses of bone at which the use of long stems is impossible. It provides a reliable load transfer in the artificial joint and good connection of the allotransplant and host bone.

Advisability of applying allotransplants in the form of bone blocks to reconstruct the defects of the femoral bone can be discussed because of insufficient (75–79 %) probability of successful operations during a 4-year observation period [51]. This occurs due to insufficient incorporation and collapse of the bone graft. In work [52] a new reconstructive method was developed using a pedicled iliac graft. A surgical reconstruction of the acetabular roof was simulated in cadaver specimens. It was shown that the use of such a pedicled structural graft might provide good primary stability and allows local remodelling and incorporation under load.

Thus, main trends in revision endoprosthetics of the femoral component can be outlined in the following way. After removal of the loosened stem the state of the cortical and spongy layers of the proximal parts of the femur should be evaluated. In patients with sufficient spongy layer the revision stem fixed by third-generation cementing techniques provides reliable long-term fixation (distal restrictor or plug, cement pistol, decreasing cement porosity by vacuumizing and holding constant pressure of liquid cement till mantle hardening). For a sclerotic (hard) canal deprived of spongy layer the porous-coated stem is optimal, which is *press-fit* and fills the canal volume. The stem length is governed by the size of cavitary or segmental defects that should be properly compensated. In work [53] the results are given indicating that biological revision with long BiCONTACT stem with distal interlocking option is a concept to solve problems associated with high bone-loss situations.

Loosening of primary femoral components frequently results in great cavitary defects in the proximal part or in extasy of femur. In this case the stems with porous coating over the whole length are used, which are rigidly fixed in the distal part at least to 5...10 cm. In elderly patients with totally disturbed osteogenesis *press-fit* setting of even long stems can be non-rigid. In such cases the osteoplasty and cement fixation of stem is used. Osseotransplants in the form of blocks are used at great losses of bone when there is no alternative.

Fracture of the ceramic head is considered a rare complication in total hip replacement. According to information supplied by manufacturers their case represent only the second fracture of more than 7 500 ceramic heads which had been implanted with cement [54]. When this happens the replacement of damaged components leads to satisfactory results [55].

These principles remain also valid in revision operation on other joints. Thus, in work [56] it is said that the main task in knee joint revision is maximally keep the bone that is removed together with the primary endoprosthesis and take steps to reduce new resection of the bone to minimum. Despite some losses of distal and posterior parts of the femoral bone the use of revision Natural Knee (TM) endoprosthesis allows avoiding osteoplasty and gives an opportunity to apply cementless fixation of endoprosthesis. Revision stems of regulated length and thickness do not create a dangerous concentration of stresses in contact with the bone and not always need cement fixation. Errors in reconstruction of the anatomical axis of knee joint rotation are the main cause of instability of revision endoprostheses. In contrast to primary endoprosthetics it is much more difficult to find a natural location of the axis in revision operation since the ends of the bones forming the knee joint have been resected in primary operation. In this connection intraoperative determination of anatomic location of the axis of flexion – extension of the joint becomes of major importance. One more difficulty is to reconstruct the previous tension of lateral ligaments. It is believed that [40] untightened ligament is more preferable than loosening of the tightened ligament in a misbalanced artificial joint.

The infection rate after total knee arthroplasty ranged from 0.5 to 2 % [56] and is a cause of revision operations despite modern technology and rigorous prophylaxis. A variety of techniques and devices have been developed to improve the management of infected total knee arthroplasty. Antibiotic-loaded bone cement spacers in the two-stage re-implantation technique allow early joint and patient mobilization, a short hospital stay and potentially a reduced rate of re-infection [57]. Knee arthrodesis is the treatment of choice for chronic periprosthetic infections, when there is a persistent infection in a revised knee, when the infection is caused by highly resistant micro-organisms, when the extension apparatus is absent or when soft-tissue coverage is impossible [58]. External fixation, plate and screw osteosynthesis and intra-medullary nails are classical methods used when performing a knee arthrodesis. Surgical approaches with the Wichita fusion nail (Stryker Howmedica) have recently been introduced. It is a solid nail made of vitallium. It's composed of four basic components: a la bolt, a femoral component, a tibial component and transverse locking screws. The nail permits intra-operative joint compression while maintaining post-operative dynamisation [59].

Recommendations cited in the given paragraph do not provide unambiguous answers to all questions that can arise in planning of revision operations. However, they are based on clinical practice and reflect main engineering aspects governing revision results. In addition, they form ground to critically evaluate traditional designs of endoprostheses and principles of their fixation both in primary and revision joint endorposthetics.

4.3 POSTOPERATIVE PERIOD

Endoprosthetics of large joints refer to the surgical interventions of III–V categories of risk that significantly modify the organism systems. Response of the organism to this traumatic damage reveals itself in local and total inflammations.

Inflammatory reaction of the organism to joint endoprosthetics is non-specific (general). When developed, it favours the removal of necrotized tissues and cells as well as reconstruction of tissue structures. It has the following stages [34].

A. Bleeding occurs when soft and bone tissues as well as vessels of different size are cut. In the damaged parts of tissues the blood is accumulated and forms hematoma. Some blood imbibes (impregnates) the muscles transforming them into the parabiosis state (coexistence of two states of the muscle, viz. normal and imbibed). When the sum of blood pressures in hematoma and the tissues surrounding the wound balances the pressure in the blood channel the bleeding stops. At the same time mediators (intermediaries) of inflammation are released, viz. chemical substances whose molecules react with receptors of the cellular membrane and modify its permeability for certain ions and generate an electric signal, i.e. a potential of action. This causes the spasm of arterioles, viz. blood vessels that end the ramification of arteries. Dzeta-potential of erythrocytes varies resulting in clots and decreasing blood flow around the operation wound. Insoluble protein – fibrin – precipitated from the clots generates thrombi which “block” damaged tissues.

B. Local acidosis (shear in acid-alkali balance in the organism to hyperacidity), mediators of inflammation and pressure of intercellular liquid irritate nerve endings giving pain. Through endothelium (a layer of cells covering the inner surface of blood vessels) an active migration takes place of leucocytes (white blood cells) and macrophages (the connective tissue cells actively engulfing and digesting bacteria, cell remnants and other foreign or pathogenic particles). The stage of utilization of damaged tissues and hematoma begins. It includes two simultaneously running processes, viz. decomposition of nonviable structures and transfer of their components into the microvessel channel. Increasing number of ferments (biological catalysts) coming with the blood flow to the operation wound accelerates the utilization of nonviable tissues.

C. Inflammatory organism response is regulated medicamentally by analgetics and non-steroid anti-inflammatory drugs. In work [34] the results of application of hydrolytic ferments as therapeutic medicine having antiedematic and anti-inflammatory effect are reported.

Physical rehabilitation of patients after joint endoprosthetics starts next day after operation. After hip joint replacement the following exercises are recommended (see Table 4.5)

Table 4.5.
Exercises in bed

Day	Exercise
1	Flexion-extension in ankle joint. Ankle circles gluteal and quadriceps sets Alternating hip hiking Simulation of bicycling.
2 or 3	Additionally – abduction of raised hip and knee flexion If quadriceps is strong enough straight leg raising may be done

Patients, who cannot immediately do the exercises, commonly manage doing them some time later after strengthening the quadriceps. In order to decrease pains in the lumbar part of the backbone the patients are recommended to bend legs in knee and hip joints, sit down in bed, bend in the backbone and slow return to laying position. The exercises are taught several days before operation.

Most patients are able to get up 1–2 days after operation on hip joint. On the third day they are recommended to use crutches. The axial load to the operated leg should not exceed 1/3 of body's weight [60]. Normally patients begin walking with short steps with the operated leg resting on the toes. It should be taken into account, otherwise a gait appears that compensates insufficiency of the muscles that ensure the functioning of hip joint. To recover the normal gait patients are taught to make equal-length steps resting on the heel with the knee extended and the head not hung when walking. After this one can begin to be trained to walk on stairs. It is reported in work [61] that patients can walk with full weight bearing safely immediate after hydroxyapatite-coated non-cemented total hip replacement.

With no complication a patient leaves clinics on 10–12th day. Most surgeons recommend gradual increase in loading if the prosthesis has *press-fit* fixation. Even more sparing loading regime is prescribed in case of porous-coated stem. Such a conservative approach is based on the trials with animals. Sheep were submitted to a hemiarthroplasty of the hip with a specially designed femoral component. The proximal two thirds of the stem had a circumferential, plasma-sprayed porous coating with hydroxyapatite. In the coated portion of the stem, apposition of woven immature bone was evident at 15–30 days. With time the gap between the endostem and the coated surface was filled by bridges of lamellar bone with a marked trabecular orientation. In the distal uncoated portion of the stem, the implant was initially surrounded by fibrous tissue that, with time, transformed into lamellar bone [62]. Though the rate of ingrowth of tissues into the implant and tested animals is dissimilar and runs by fibrous mechanism in men rather than osteous one, the restriction of bone motion relative to implant at the initial period of biological fixation of endorposthesis evidently makes sense. It is said in work [63] that bone remodelling in cementless hip stem replacement occurs as interface and periprosthetic bone remodelling. The conclusion is drawn on the basis of the results of radiological, histological, and bone mineral density findings. Atrophic, hypertrophic, and normotrophic bone reactions have been revealed. Osteodensitometry measurements have shown a bone density decrease of 20 % within the first 6 months in the region of the proximal femur. Measurements up to 4 years revealed no further loss of density.

When prescribing loads during the first month after operation it is advisable to have in mind the data from Table 4.6 [64].

Table 4.6.
Maximal permissible loads to artificial hip joint

Activity	*Max* force (% of body weight) several days after operation			
	3	6	16	31
Straight leg raising	–	1.0	1.5	1.8
Getting out of bed	0.8	1.0	1.2	1.4
Getting into to bed	0.8	1.0	1.5	1.5
Double-limb stance	0.5[1]	0.7[2]	0.9	1.0
Ipsilateral single-limb stance	1.2[1]	1.3[2]	1.4[2]	2.1[3]
Walking with aid	1.0[1]	1.5[4]	2.6[4]	2.4[5] –2.8[4]

Notes:
1 –using a walker,
2 –ipsilateral hand on crutch, contralateral hand in attendant's hand
3 – contralateral hand in attendant's hand,
4 – using crutches,
5 – between parallel bars.

Technically correct setting of endoprosthesis provides prophylaxis of postoperation luxations. However, the accuracy of free-hand cup and stem positioning is insufficient in most cases and can be considerably improved by navigation systems [21–23]. The same problem exists in total knee arthroplasty [24–26]. Stable fixation of components during first weeks is insufficient since a strong fibrous pseudocapsule is not formed yet. To reduce the risk of component displacement the patient should avoid some motions. One should know that the abduction of the hip aside is safe while its adduction is dangerous. The risk of luxation is the highest when excessive flexion and rotation of the hip inwards is added to bringing back.

Rehabilitation of patients on leaving the clinics should occur under orthopaedist's observation. The first clinical-radiological evaluation of the state of operated joint is done 6–12 weeks after operation. Further check visits commonly are scheduled for 6–12 months.

Modern views on patient's adaptation are based on pathogenetic dependence of early rehabilitation treatment after joint replacement. Patients with artificial hip joints can return to normal life except for the exercises giving intensive loads to the joint (running, jumps). Young men who ignored these recommendations had early fractures of prostheses. At the same time, implementation of the rational program of functional training allows reach 80 % of the initial strength of muscles providing the functioning of hip joint for 6...12 months [34]. Flexion and abduction of the leg is improved during a year after operation, adduction and rotation are improved during a longer period. In work [65] variations in functional behaviour of knee extensors and flexors were studied before and after implantation of uncemented knee prosthesis. A functional deficit was noted in the extensor apparatus, which increased in the first 6 moths following surgery. This strength was not improved 1 year following the arthroplasty, but it was more efficient as shown by the reduction of muscle fatigue. Flexor strength was conversed and a pathological muscle balance was maintained, reaching maximum efficiency over a small joint range. Commonly the bulk of motion in the joint is compared before and after operation using rating systems cited in 1.4. Each revision operation causes decreasing bulk of motions.

The patients with sedentary occupation may return to work within 6 weeks following total endoprosthetics. Those who must lift and carry objects in their jobs may have to delay returning to employment for 3–4 months and more. The patients operated at late stages of disease have fewer chances for early rehabilitation. This is an important reason for not delaying joint replacement in disabled persons.

As a rule, patients do not need supplementary walking aids when walking already 3 months following operation except for those suffering of arthritic pains in multiple joints, most commonly rheumatoid arthritis. Gentle swimming and water exercises are excellent. Patients can resume sexual activity already several weeks following operation. The position with abducted and flexed hip will be safe for women. The man should avoid hip adduction, at least during first weeks following operation

Special rehabilitation therapy is prescribed to the patients who had hip joint replacement accompanied with numerous lesions (polyarthritis).

Litter patients with different dysfunctions, in particular those who could not walk before operation need special rehabilitation measures both prior and following operation. Patients who need 3 or 4 joints to be replaced must have large vital forces, both physical and psychological. Total replacement of hip joint is performed before the knee joint endoprosthetics. If destruction and deformation are such that they do not permit performing operations during one hospitalization the second operation is planned in 2–3 months. Such tactics facilitates physical and psychological healing of the patient. Loss of motivation and deteriorated state of nervous system are the main causes of bad results of these operations.

The clinical and radiographic outcomes of total knee replacement in patients with body mass index (BMI – body weight divided by the square of the height) greater than 30 kg/m^2 were compared with the results of a matched group of patients with BMI < 30 kg/m^2. As shown in work [66] reporting the clinical and radiological outcome of obese osteoarthritic patients undergoing posterior cruciate retaining total hip replacement, revision and survivorship rates over the first decade are comparable to non-obese control.

The similar principles form the grounds for the rehabilitation programs of any joints to be replaced. It is attempted to optimize the procedure of therapeutic loading of operated joint using computer modelling by the criterion of recovering of the elasticity modulus of the bone after surgical intervention [67]. Rehabilitation of each joint certainly has its own features and numerous optimization criteria. This interesting topic is elaborated intensively in works on traumatology and endoprosthetics of joints [31, 34, 58, 61, 64, and 67].

Objective evaluation of the state of operated joints is a complex multi-factor problem.

In 1993 Sir Dennis Paterson wrote an editorial on the International Documentation and Evaluation System (IDES) [68]. He outlined the principles of IDES as consensus, hierarchical information, radiographic evaluation and acceptability. IDES was established by the International Society of orthopaedic Surgery and Traumatology (SICOT) Standing Committee on Documentation and Evaluation, which was founded in 1990 with Prof. M.E. Muller as chairman, and presented at the American Association of Orthopaedic Surgeons (AAOS) 61st annual meeting in 1994. The nomenclature used on the three IDES sheets for primary total hip arthroplasty (THA), revision THA and follow-up is based on the consensus paper by the Hip Society, the SICOT Commission on Documentation and Evaluation and the Task Force on Outcome Studies of the AAOS [69]. This consensus paper provided a terminology named CART (Clinical and Radiographic Terminology), in which each term, whether applying to a functional or radiographic parameter, was specifically defined to have a constant meaning. The initial impulse to create such a terminology was already given in 1985 when J. Galante [70] called for a uniform method of evaluating and reporting the results of hip-replacement surgery in order to compare the results on a common standardized basis.

Today, IDES represents one of the most valuable hip arthroplasty databases and documentation applications existing. During the past 4 decades, extensive information about 50,000 primary THA, 12, 000 revision THA and 77000 follow-ups was collected. By its means orthopaedic community is enabled to collect and compare more transparently and therefore finally improve the quality and efficiency of medical treatments.

4.4 COMPLICATIONS IN ENDOPRIOSTHETICS

With increasing number of joint replacement operations the number of errors and complications resulting sometimes in endoprosthesis failures and even fatal terminations has increased. Attention of orthopaedists to the issue encouraged them to develop methods for prophylaxis and treatment of possible negative outcomes of operations. That is why most works published recently on joint replacement are devoted to errors and complications whose frequency is from 7–9 to 16–30 % [69].

We will consider the complications occurring in joint replacement and methods for their treatment using hip joint replacement operations as an example. Complications following joint endoprosthetics are divided into somatic and specific.

Somatic complications are associated with the defects of somatic nervous system of patients, viz. the set of sensitive and motor nerve fibres innervating joints, muscles, and skin. Joint replacement is referred to higher-risk operations. The risk is due to not only traumatic surgical intervention (great duration of operation and blood loss, extended lesion of bony and soft tissues) but also to the fact that above 70 % of patients undergoing endoprosthetics are above 50 years old and have pathology of different systems and organs depending on their age. This is the main cause of general somatic complications occurring at intraoperation or at the closest postoperative period. The main condition for their avoidance is in scientifically justified selection of patients to be operated

Absolute contraindications for total hip replacement are [70]:

- heavy decompensated chronic diseases of cardiovascular system as well as respiratory, endocrine and excretory systems;
- psychic disorders (psychosis or pronounced dementia);
- inflammation in the region of hip joint, thrombophlebitis of lower limbs and untreated regions of chronic infection.

Relative contraindications to operation are artery hypertension, ischaemic disease of the heart and coronary insufficiency, cardiac rhythm breaks, dysfunctions of liver and kidneys, anaemia, paresis (partial paralysis) or pronounced hypotrophy of muscles near the hip joint, very young age, obesity, hormone osteopathy, drug intolerance, and previous hormone therapy. Revealing relative contraindications for endoprosthetics forms grounds for more thorough examination of the patients.

By statistical data of R.R. Vreden Research Institute of Traumatology and Orthopaedics (St.-Petersburg, Russia), general somatic complications amount to 60 % of all complications occurred after hip joint replacement.

Specific complications occur during operation, at early and distant postoperative periods.

Complication in intraoperation period are associated with technical aspects of the surgical procedure and depend on technical support of operations and surgeon experience. Below main complications are outlined.

Lesion of nerves is commonly due to non-optimal choice of access to the joint and its wrong implementation. To avoid intraoperation trauma of the sciatic nerve when performing operation it is exposed from backward position and treated thoroughly.

Lesion of vessels happens rarely at joint endoprosthetics. However, if it happens it leads to extremely bad results and even limb loss. Damages of vessels are divided into explicit that result in acute intraoperation blood loss, and latent that are revealed after operation. In work [71] it is reported on the causes of lesions of vessels, precautions, and vascular complications that can happen at hip joint replacement.

Bone fractures happen most often when preparing the bed for the femoral component and at its implantation. It can be fractures of the neck of the femur, which often occur at osteotomy with a chisel when the neck splits in the region of Adam's arc. When the diameter of the bony bed does not match the sizes and geometry of the stem its introduction into the medullary canal can cause spalling of greater or lesser trochanter and intertrochanteric or spiral fractures of the proximal part of the femur. In order to avoid these complications it is recommended to perform osteotomy of the neck with an oscillatory saw and the bony bed for the stem should be prepared using rasps of subsequently increasing size. In case of the fracture the fractured bone fragments are fixed with cerclage sutures or screws. Rare cases of impossibility to adapt bone fragments or their instable fixation cause abandoning immediate endoprosthetics and the operation is planned to be performed following fracture consolidation [72].

Longitudinal chopping of the upper third of diaphysis is caused by osteoporosis, enlarged anatomic curvature of the femoral bone, discrepancy in sizes of the medullary canal and endoprosthesis stem. If the stem length is sufficient to provide fixation of the fractured fragments it is used as an intramedullary fixture.

Complications in early postoperative period are hematoma and early (occurring during 1st–3rd week following operation) suppuration of soft tissues.

Early surface suppurations are the most probable with availability of hematoma in the region of endoprosthesis and large (above 2 hours) operation duration. Weakening of protective forces of the organism is aggravated by the effects of microflora which the implant contacts, namely conditionally pathogenic (available in the sites of latent infection) and pathogenic (coming from the environment). A fibrous capsule surrounds the implant being a foreign body in the human organism. If the implant is insufficiently biocompatible reactive aseptic inflammation occurs. With the infection penetrating the capsule an active inflammation process around the implant develops.

In the early period of endoprosthetics when there were no wide-spectrum antibiotics and efficient microbiological control of operating rooms the probability of infections was very high, viz. more than 11 %. Modern schemes of prophylaxis of suppurations by antibiotics permitted reducing their probability down to 1.3 %. Presently when performing endoprosthetics in "clean" operation sites it is 0.7 % and with using antibiotics – 0.6 %. Ultraviolet treatment of the operation operating room restricts the probability of infections at joint within 1% [71].

Thorough hemostasis (stoppage of bleeding) in surgical interventions and active draining of the operation wound with using wide-spectrum antibiotics are crucial prophylactic measures in prevention against early suppurations.

The risks associated with homologous blood transfusion, which include transmission of hepatitis viruses [73], have led to an increasing interest in autologous blood transfusion methods. Although transfusion reactions are rarely serious, and usually consist of febrile reactions, haemolytic reactions due to ABO (blood group) and Rh (Rhesus factor) incompatibility occur in 0.01 % of transfusions and are potentially fatal [74]. It has also been suggested that homologous blood transfusion has an immunosuppressive effect [75]. Recently, the theoretical risk of transmission of new-variant Creutzfeldt-Jacob disease through blood transmission has also forced health trusts to look into options that provide an alternative to homologous blood transfusion. The authors of work [29] have come to a conclusion that in operations like total joint replacement where a tourniquet is used intraoperatively and collection drains are used postoperatively, almost universally, reinfusion of drained blood would appear to be the most appropriate method of autologous transfusion. Wider use of this method should be considered, especially in operations like revision joint replacement where a substantial blood loss is anticipated.

Thromboembolism is an artery occlusion as a result of abruption of a part of thromb formed in veins, heart cavities, aorta, and its transfer with the blood stream. Symptomatic thromboembolic complications commonly take place up to 30 days after total hip replacement. In work [76] these data were analyzed to evaluate the incidence of symptomatic thromboembolic complications in patients undergoing total hip arthroplasty. It was concluded that the routine use of thromboembolic prophylaxis minimized the incidence of complications in orthopaedic patients receiving total joint arthroplasty under balance general anaesthesia.

Complications in the distant postoperative period are the most serious negative after-effects of joint endoprosthetics.

Deep suppurations in the region of endoprosthesis are due to: risk-factors availability (chronic infections, diabetes, diseases of urogenital organs, etc), insufficient quality of preoperative treatment, violation of conditions and technology of operation, insufficient biocompatibility of endoprosthesis, and its instability growth [69]. As has already been mentioned, postoperative monitoring of patients for hematoma and persistent drainage enables earlier intervention that may lower the risk of developing deep wound infection [37]. Preventive methods for avoiding infection after arthroplasty with methicillin-resistant Staphylococcus aureaus have been developed [38].

Methods of treatment of deep suppurations consist in the following: a purulent site is dissected, the cavity is washed with antiseptics and drained during 2–3 weeks. If the measures are not successful the endoprosthesis is removed. Contradictory opinions can be found about the terms of revision operations. Most often revision operation is performed 2–6 months following removal of primary implant while soft tissues are not reduced and marked osteoporosis is not developed because of the absence of loading [77]. Advocators of one-stage revision endoprosthetics believe that the infection previously brought into the organism gives rise to micro organisms that present danger even 30 years after primary flash and the absence of physiological load in the waiting period impedes reendoprosthetics [34]. In work [78] a three-stage operation is outlined: 1) resection arthropasty and 4-week antibiotic therapy, 2) osseoplasty in the proximal part of the femur and acetabulum 3–6 months following operation, and 3) implantation of revision porous-coated endoprosthesis 9–12 months following the second stage.

Bone fractures after total replacement of hip joint happen with a probability of 0.1 % within a period from several days to many years after operation [71]. Early fractures are most often initiated intraoperationally or result from stress concentration in the bone-implant contact. The fractures can be caused by tensile stresses that occur at tight setting of the stem in the medullary canal of the femoral bone. The change in the proximal loading pattern stimulated the formation of new trabeculae streaming up to the level of the lateral flare [79]. The new bone apposition is typical of every case examinated and may represent the reaction of the physiologically loaded bone as opposed to the unloaded situation (proximal stress shielding). The change on the geometry of the stem and the mechanical stimulus imposed on the loaded femur accounts for the presentation of the bone mass and the increased density in the periprosthetic areas [80].

The second type of fractures is generated by trauma and unhardening of bones. Septic failure of total hip arthroplasty is often associated with extensive osteolysis on both the acetabular and the femoral side. The essential treatment for this devastating complication begins with debris elimination and removal of all foreign material [81].

The complications are treated by revision endoprosthetics.

Different length of legs can occur if prior to operation the patient had congenital incomplete dislocation or low dislocation of femur and as the young patient grew up the soft tissues surrounding hip joint get shortened. The attempts to elongate the leg intraoperationally can cause nerve-vascular lesions, flexion and adduction contractures in hip joint. The problems can be avoided if planning operation thoroughly.

Heterotopic ossification is formation of the bone tissue in the surrounding muscles and in the capsule of the operated joint. This is a significant complication of total hip arthroplasty. Etiology of heterotopic ossifications is not clearly understood, though it is believed that they have traumatic nature. Classification of these formations suggested by Brooker is given in Table 4.7 [82].

Table 4.7
Grading system for postoperative heterotopic bone formation

Grade	Formation
I	Islands of bone within the soft tissues about the hip
II	Bone spurs from the pelvis or proximal end of the femur leaving at least 1 cm between opposing bone surfaces
III	Bone spurs from the pelvis or proximal end of the femur reducing the space between opposing bone surfaces to less than 1 cm
IV	Apparent bone ankylosis of the hip

Some patients feel pain or heat in the region of the joint when heterotopic formations are generated. In the patients having these symptoms 2–3 weeks after operation early thin formations are revealed radiographically. As the signs appear the prophylaxis measures are ineffective. An incidence, without prophylaxis, has been reported of between 8 % and 90 % [83]. To stop the growth of formations moderate activity of the patient is recommended. At degrees II and III the bulk of motions in the joint should be decreased.

Prophylactic treatment (non-steroid anti-inflammatory medicines and deep radiographic therapy) is prescribed by the at risk patients having had trauma of the hip joint, long complicated endoprosthetic operations, suffering hypertrophic osteoarthritis and Bekhterev's disease. The incidence of ossification was lower in the patients of O blood group [84].

Aseptic instability of endoprostheses arises when inflammation is absent in the artificial joint and is fatigue by nature. Main causes of instability are the following [85]:

A. Considerable difference in deformation-strength characteristics of cartilage-cartilage and metal-UHMWPE contacts. Elasticity moduli of hyaline cartilage (1–3 MPa) and UHMWPE (1400 MPa) differ by about three orders of magnitude. Because of these kinetic dependences of stiffness and lubricity of friction units of endoprosthesisis and natural joint are absolutely different. New loads to the bones forming the reconstructed joint cause increased a probability of osteolytic lesions. Recently [86] the phenomenon of bone remodelling of the acetabulum after the insertion of prosthesis has been revealed, and this can be considered as retroacetabular stress shielding. However, this does not eliminate the problem of higher loosening rate of the acetabular component, particularly in the patients with rheumatoid arthritis [87].

B. Artificial joint has an endoprostheisis-bone interface, which is not characteristic of natural joint. The latter is characterized by gradual transition from cancellous to cortical bone. The similar transition is absent both in cement and cementless endoprostheses. In the latter the load is transferred from the metallic stem with a 100–200 thousand MPa to the cortical (15–20 thousand MPa) or cancellous (100–1000 MPa) bone. This produces significant shear stresses and stress concentration spots at the interface.

Atrophy of bone related to stress shielding has been observed around uncemented femoral components in total hip replacement [88]. The extent of the atrophy depends on the elasticity: the stiffer and thicker the stem, the greater the atrophy [89]. A coating of high modulus increases the atrophy [90]. Bone density seems to be presented better around isoelastic stems than around stiff stems [91]. New designs of stem have different patterns of strain distribution within the proximal femur. Whilst improvements in cementing technique have led to improvement in the survival in cemented femoral components [92], similar improvements in the survival of cemented acetabular implants have not been achieved [93].

C. The interfaces in endoprostheses are broken under the effect of loads that are lower than the yield point of the cement conjunction of endoprosthesis and bone. Such a type of failure is named fatigue one. Fatigue failures occur because all materials have defects, which under repeating loading are developed into cracks. Fatigue strength is characterized by the breaking stress as a function of the number of loading cycles, Fig. 4.22. The lower limit of the stress under which the material does not break at $N \to \infty$ is named fatigue life σ_f. As shown in Chapter 5 the friction surface of polymer liners of endoprostheses is developed by the fatigue mechanism.

Loosening of endoprostheses is a fatigue process whose kinetics is affected by fatigue strength of materials, number of loading cycles, patient's weight and his/her activity, implantation quality, and engineering efficiency of endoprosthesis. Loosening is accelerated at:

- wrong setting of the components of endoprosthesis,
- early (earlier than following 3 months) full load to the operated limb and patient's obesity,
- progressing dystrophic changes in bony tissue,
- hormono-, radio-, and chemotherapy as accompanying indication [72].

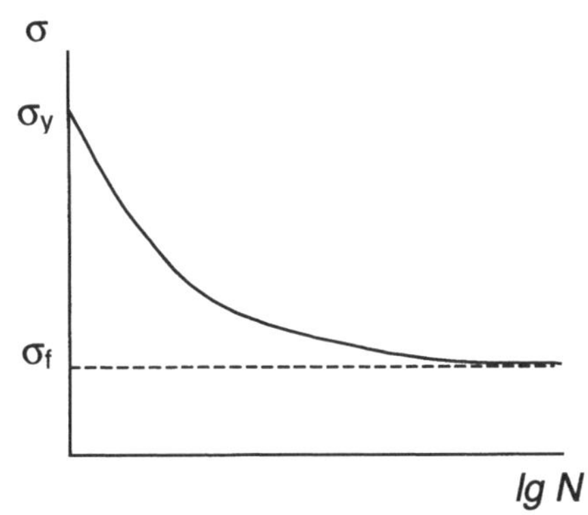

Fig. 4.22. Breaking stress (σ) vs. logarithm of number (*N*) of loading cycles: σ_y – yield point, σ_f – fatigue life

Radiologically initial instability is characterized by the availability of insignificant (up to 1.5 mm) light zones around the stem and cup. With pronounced and deep instability the width of the zones reaches 3–5 mm, also typical displacements can be seen, viz. of the cup to the small pelvis or its proximal-lateral displacement, that of the stem–medial deviation of the upper part and lateral deviation of the lower one.

In work [94] a surprisingly accurate method by using radiographs for custom hip stem design is presented. To optimize the prediction of canal geometry, they successfully used a model of average sectional shape of femur. However, simple radiological methods continue to be the basic tool in clinical series to evaluate the canal fill of the femoral stems. At present, computer tomography is carried out by means of an extended scale that allows a satisfactory differentiation between bone and implant. Further improvements in artefact reduction by special image reconstruction methods using raw data will give additional information, particularly at the bone-implant interface [95].

In work [96] the following procedure for avoiding and treatment of aseptic instability of endoprostheses is proposed:

- at the initial stage with moderate pains and good general patient's state – restricted walk, usage of cane, and refusal of carrying weights are recommended; with severe pains – joint replacement;
- at pronounced instability if pain is tolerable and patient's general state is good, endoprosthesis replacement and reduced activity is recommended to young patients; restricted loads, usage of cane and crutches – to elderly people; if the measures are not effective – revision endoprosthetics should be done;
- at deep instability (great cavity in the acetabulum, sharp thinning of the walls, resorption and break of diaphysis of the femoral bone) – revision endoprosthetics.

The goal of work [96] was to investigate whether the clinical variables of non-steroid anti-inflammatory drugs usage and cigarette smoking are possibly linked to aseptic loosening around total hip arthroplasty. The following information was recorded: age, gender, primary and revision surgery details, radiographic parameters, smoking history and drugs usage history. Logistic regression analysis was used to determine if there is any statistically significant association between anti-inflammatory drugs usage or smoking habit and aseptic loosening. No such association was found.

Study [97] was based on data from the Finnish Arthroplasty Register. The incidence of periprosthetic fractures was calculated separately for the years 1990-1994 and 1995-1999. The incidence of periprosthetic fractures in the first period was greater than in the latter. It was shown that gender, type and age were of no significance as risk factors for periprosthetic fractures.

Luxation of endoprosthesis head results from errors in planning and performing operations; wrong space setting of endoprostheses; discrepancy in the lengths of the neck of endoprosthesis and resection segment of the femoral bone; excessive dissection of the bony tissue when forming the bed of the femoral component resulting in stem collapse, approach of the points of muscle fixing, and weakness of gluteus muscles.

Violation of optimal parameters of resection of the femoral bone and implant setting requires revision operation. At isolated weakness of gluteus muscles "closed" setting of the head can be attempted under narcosis and radiological control with further fixation of the joint with plaster bandage.

Stem fractures are caused by the following factors [72]:

1) absence of close contact of stem and bone in the proximal part of the femur, which results to fracture at the boundary of dense and loose setting of the stem;

2) nonconformity of the endorposthesis design to biomechanical conditions of its loading that arise at degenerative-dystrophic lesions of other joints and backbone and cannot be predicted;
3) electrochemical corrosion aggravated by cyclic loading of the stem, engineering defects of its manufacture.

Probability of fractures is less at optimal selection of type and size of endoprosthesis and standard preparation of the bony bed. Combination of the components of different endoprostheses or different series of industrial production increases the probability of their electrochemical heterogeneity.

Complication due to endoprosthesis wear will be discussed in the next chapter.

Thus, considerable part of complication following joint arthroplasty is due to not only objective but also subjective causes (errors when choosing endoprosthesis design, violation of the operation procedure, treatment methods at postoperative period). The number of complications reduces when operations are performed in endoprosthetics centres equipped by high-expertise orthopaedists, modern implants of various design and tools for their implantation, as well as all drugs required.

4.5. BONE TRANSPLANTS

Transplantation of organs and tissues as a surgical method has been known since antiquity. Modern ideas of joint endoprosthetics assume a wide application of osseoplasty with using the bone tissue belonging to the patient (autotransplant), tissue belonging to other people (allotransplant) or tissues of some other origin (xenotransplants).

The history of bone transplantation as treatment method in orthopaedics has started in the middle of 1800-ies. In 1878 the first successful allopasty of the bone was reported and in 1907 Axhausen presented the study results on the biology of bone transplants [98]. Later Albee F. H. reported the clinical results of bone trasplantation to more than 3,000 patients [99]. Lexer E. reported 50 % of successful operations on joints with using allotransplants [100]. High probability of unfavourable results of alloplasty was associated with resection, infection, fixation and fracture problems of transplants

In 1950-ies weakening of immune reactions to bone allotransplants prepared under deep cooling was revealed [101, 102]. Then it was shown that despite some unsolved problems allotransplants could be successfully used in reconstruction of extensive bone defects in oncological patients. Advances in endoprosthetics made osseoplasty an indispensable procedure in joint replacement. With increasing number of revision operations the need in additional source of bony tissue has become evident. Banks of tissues, development of the network of transplant delivery to clinics and better understanding of immune aspects of bone transplantation specify its present state as one of the main surgical techniques at joint endoprosthetics [105].

Grafting of bone transplants occurs in the following way.

Autotransplants of cancellous and cortical bones behave similarly at first two weeks. Inflammatory site of the organism corresponds to ingrowth of vessels and cell diffusion from the autotransplant surface. Then rates, mechanisms and degree of fusion of cancellous and cortical bones become different.

In transplant – host bone contact new formation of vessels begins, after which osteogenic cells form a new bone on necrotic trabeculae of the transplant. The process is accompanied by growing density and strength of the transplant and registered radiologically [106]. Reconstruction of cancellous autotransplant in the course of which new cells replace it, goes on rapidly enough. Axhausen introduced a new term «creeping substitution» to denote the process.

New formation of vessels in the cortical autotransplant occurs more slowly than in the cancellous bone and remains uncompleted. Therefore the transplant is a mixture of reconstructed viable and necrotic bony tissues. Such factors as patient's age, diseases, distribution of vessels in transplant, infections affect the reconstruction of cells [107].

Allotransplants cause inflammatory response of the organism during which lymphocytes penetrate the "foreign" bone. New formation of vessels is restricted by immune reaction of the organism and therefore it only takes place in the bone layer of several millimetres thick. Grafting of allotransplants goes on more slowly and not so completely as that of autotransplants. Small fragments of cancellous bone are reconstructed more fully in transplantation than large ones. Cortical transplants are intergrown with vessels 2–3 mm in depth. The inner part of cortical fragments is deprived of vessels and is not reconstructed even in response to dosed mechanical loading.

Immune response of the organism to inserted transplant is governed by their genetic difference and causes the intensity of transplant rejection reaction. Antigens are organic substances that coming from the transplant into the organism cause response immune reaction, viz. antibodies formation. The difference by the antigens between the individuals of one kind causes the rejection of allotransplants. Therefore, impaction bone grafting causes a significant number of complications. In work [108] the results evidence that if adequate measures are taken to avoid complications in the perioperative period, the long-term results will be excellent.

Antigens of bone transplants are localized mainly on the surface of bone cells (osteocytes) and marrow [107]. Compact bony substance – matrix is less antigenic since the chemical structure of collagens, hydroproteins and its other components is the same in all persons. Xenotransplants differ significantly from the host bone by the composition of the components of bone matrix, which causes strong antigen-antibody reactions.

Rejection of new allotransplant is caused by the fact that each of its components (plasma, viable bone, marrow, blood cells) has its own antigens. Blocking or reducing immune response of the organism treating the bone tissue by boiling, decalcination, protein extraction, or local agents failed. All these attempts affect contradictorily immunology suppressing reconstruction and deteriorating mechanical properties of bone allotransplants [98].

Nowadays radioactive irradiation, freezing, lyophilization (freezing out aqueous components) and irradiation combined with lyophilization.

Gamma irradiation dosed (2–3) $\cdot 10^{-2}$ Gy provides sterilization and somewhat decreasing antigen ability of bone transplants. Freezing and lyophilization noticeably decrease their rejection reaction compared to untreated bony tissue. Freezing enables preserving and keeping transplants. The strength of bone transplants depends on the treatment method (Table 4.8) deep freezing affects insignificantly its strength. Lyophilization decreases flexion and torsion strength but insignificantly affects the tensile strength. Irradiation results in some decrease in strength under all types of testing and irradiation combined with lyophilization makes the effect more pronounced.

Table 4.8

Effect of treatment methods on relative strength λ of bone transplants [107] ($\lambda=\sigma/\sigma_0$, where σ and σ_0 are ultimate strength prior to and after treatment)

Treatment	λ, %, under testing at:		
	tension	torsion	flexion
Lyophilization	120 or unvaried	39	55–90
Gamma-irradiation dose (Gy):			
0.01	unvaried	90	unvaried
0.03–0.04	unvaried	90	60–90
0.06	80	65	70
Lyophilization + gamma-irradiation	decreased or unvaried	70	20–80

Other ways for decreasing fails of bone transplants need working out immune-biological methods for choosing donors and prescribing immunodepressants (substances suppressing immunity) to patients. The best up-to-date method for choosing histocompatible pairs by human leukocyte antigens (HLA) is named serological or leukocytic typing. Its efficiency in endoprosthetics is ambiguous though it is noticed that the more dissimilar the tissues of host and donor the more frequently rejections occur [109]. Temporary prescription of immune depressants is aimed at weakening immune response of the organism in the period of allotransplant fusion. It is shown experimentally [110] that in this case the results of application of allotransplants can approach those when using autotransplants. The main disadvantage of the method is in toxicity of immune depressants and long-term side effects.

Tissue banks appeared in 1970-ies when application of bone allotransplants increased drastically. The banks provide clinics with the transplants that were given as gifts and removed surgically under aseptic conditions or obtained in some other juridically irreproachable circumstances however being unsterile and sterilized later. When in 1975 American Association of Tissue Banks was established and in 1980 the criteria of donor selection were published the activity and structure of the tissue banks has become professional [111]. A law regulating the procedure of making inquires about donors was also helpful. Despite the fact that the number of donors owing to banking establishment has increased significantly the need for bone-joint transplants exceeds their availability. The issue is aggravated by high requirements to dimensions, quality, and safety of bone allotransplants [105].

Banks differ by infrastructure and efficacy of technological facilities. They include institutions with small freezing chambers to keep hip heads as well as branched structures which are able to give any bone transplants with cryopreserved cartilage, tendons, and joint capsules (Fig. 4.23). During last decade a new reconstructive method has being developed in acetabular roof reconstruction, which uses a pedicled iliac graft. This technique provides good primary stability and allows local bony remodelling and incorporation under load [112].

Fig. 4.23. Typical workpieces of bone allotransplants

Successful activity of banks is based on thorough quantitative control, application standardization techniques and high-efficient methods for allotransplant preparation. The latter if taken under unsterile conditions undergo secondary sterilization by γ-irradiation or treated by ethylene oxide in autoclave.

Hepatitis, AIDS, cortical-striospinal degeneration (Kreitzweld–Yakob's disease), syphilis, and some other lesions can be transferred through allotransplants of blood and tissues. Though the risk of infection with bone allotransplants is very low the recipients should be subjected to total examination. Enhanced hazard of AIDS and hepatitis causes the necessity of revealing of infected donors and their exclusion of the reserve.

Crushed bone transplants are obtained from the tissue banks or a surgeon prepares them by crushing heads of femoral bone and other cancellous fragments. It is desirable to place transplants on the vessel bed of the host bone of a sufficiently large area to provide a new formation of vessels and transplant ingrowth. Formation of trabecular arcs between transplant and host bone and increased density of the cancellous bone of graft is a radiographic sign of fusion. On the computer tomography scan, the fusion is marked by increased intensification of the contact zone and periosteum formation on the transplant.

Engineering of bone transplants is a reasonable alternative to the practice existing in orthopaedics when bone autotransplants are taken from one part of patient's body (e.g. upper part of pelvis being not the main one) and transplanted to the bone being reconstructed.

Presently a strategy of cultivating the bony tissue by multiplying of alive cells is developed (Fig.4.24). Its essence is in the following [113]. The cells of marrow taken from the recipient by respiration method or dissected by biopsy cells of connective tissue are sifted out in the cultivated medium. Then they can be used to cultivate specific tissues, be secondarily sifted out in the pores of load-carrying matrix, and affected by bioactive factors encouraging cells multiplication.

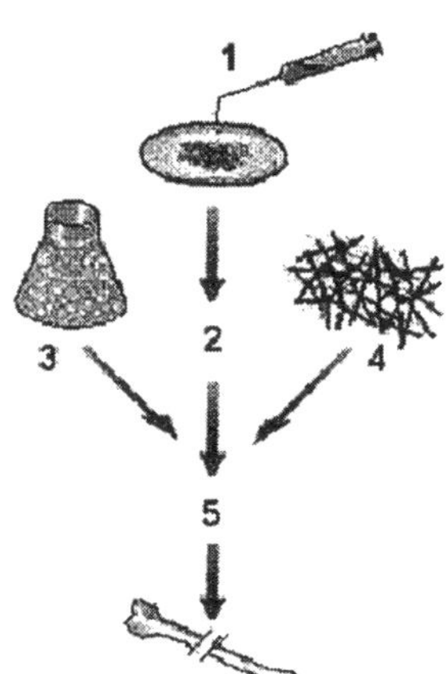

Fig. 4.24. Diagram of cultivating *in vitro* of bony tissue, which can be used in orthopaedics for reconstruction of skeletal structures: *1* – cells of connective tissue and/or embryonic bone cells, *2* – bioactive transforming factors, *3* – special tissue, *4* – carrying matrix, *5* – bioprosthesis

In California University (San-Diego, US) a cartilage tissue was cultivated *in vitro* having structure and properties of natural cartilage (Fig. 4.25). Properly made mixture of cartilage cells (chondrocytes) of different types was suspended in a gel where the cartilage matrix was formed layer-by-layer. The idea of experts is that the cartilage embryo would grow up adapting to the specific geometry of individual joint [114].

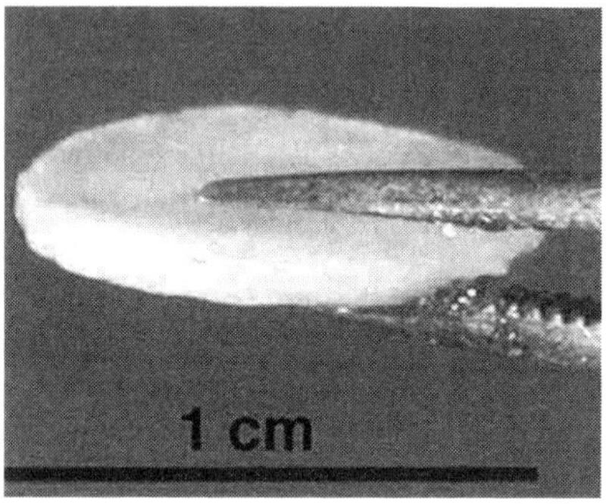

Fig. 4.25. Plate of the cartilage tissue cultivated in the lab for 14 days

In work [115] the technique is reported of cultivating *in vitro* of bony tissue by affecting osteogenic cells with transforming growth factor (TGF) - β1 for which steroid preparation Dexamethasonum was used. As a result, in 24–28 hours spherical cells arise that after 3–7 days are transformed into specific bony proteins and the latter are transformed into crystalline microspicules (bony needles) available in the alive bony tissue.

The third trend in artificial osteogenesis consists in filling porous (hundreds of micrometers diameter of pores) the matrix of sea corals [116] with mesenchymal cells (cells of embryonic tissue) taken from the bone trunk. Sea corals consist of calcium carbonate and have a structure simulating that of the cancellous bone. The composite transplant cultivated in such a way was tested in sheep. Following 4 months after its implantation into the metatarsal bone the coral matrix was resolved and the transplant was transformed in radiographically registered bone that turned out to be totally identical to the bony tissue in the contact zone.

Principally different way to replenish the tissue banks is proposed by genic engineering. It is a division of molecular genetics dealing with purposeful obtaining *in vitro* of new combinations of genetic material, which is capable of multiplying in the host cell and synthesizing final products of exchange. Obtaining implants by cloning (viz. vegetative, cell multiplication) consists in using so-called "trunk" cells, which are found in human ovocytes at early stages following fertilization. DNA spiral is extracted from the ovocyte and replaced by the donor cell DNA, after which the ovocyte is developed into an embryo. Thus various human tissues genetically compatible to the tissues of a person who "gave" his DNA can be grown.

The important step is to deliver the target gene into the patient's cells in a way that is safe, efficient, and specific. The vehicles that encapsulate therapeutic genes for delivery are called vectors (Fig. 4.26). Many vectors currently in use are modified or attenuated viruses. These modified viruses cannot replicate but can efficiently deliver genetic materials to the cells. Genes can be delivered in two ways: ex vivo or in vivo. With the former, the gene is transferred to a cell culture, which is later returned to the host, while with the latter, the gene is delivered directly into the host. In spinal research the viral vectors most commonly used are retroviruses, adenoassociated virus, and adenadenovirus. Other viral vectors being developed include herpes simplex, lentivirus, and papillomavirus [117].

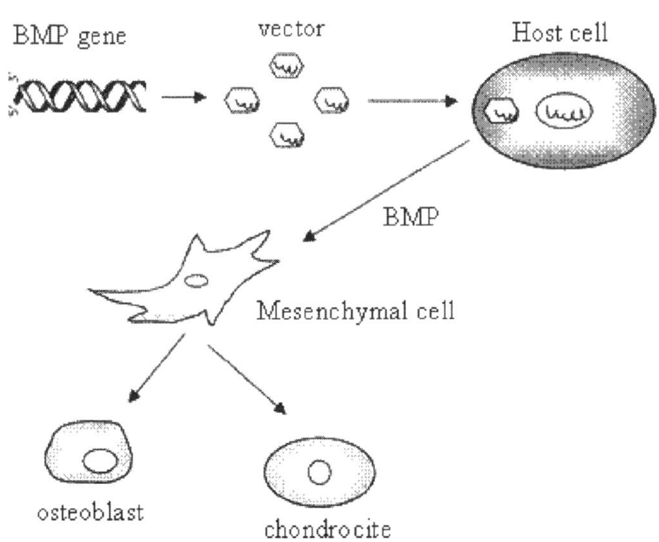

Fig. 4.26. Diagram showing how the genes are introduced to the host cell and stimulate the mesenchymal stem cells differentiation by expressing specific proteins.

Genic engineering has revealed the issue of responsibility of researchers on the consequencies of their work. Its methods differ from the manipulations with microbes by the fact that the spread of "designed" organisms can lead to serious problems. With this in mind numerous experts as far as in 1974 summoned to self-restriction in genic engineering trials. However, the moratorium was soon broken. Therefore, in 1982 in Strasburg at the meeting of European Parliament the recommendations were approved on the legality of genic engineering methods in reference to people. It was the first attempt to seek legal protection against sequences of genic manipulation at the international level. Nowadays high-quality popularization of scientific achievements using mass media makes possible working out responsive attitudes and objective opinions on advantages and risks of genic engineering. Legislation mechanisms are produced to check on the consequences of cloning of human organs and realzation of the products. In the beginning of 2001 Britain Parliament approved a law draft allowing cloning human cells and regulating donors' rights.

The data given in the Chapter evidence that joint endoprosthetics is a multi-factor process that despite the availability of medicinal traditions and standard surgical solutions requires creative approach essentially to all details of operation procedure. It supports the idea that endoprosthetics is one of the most science-consuming methods in surgery. The authors comprehend that it is impossible to outline medical – engineering aspects of performing operations and postoperative monitoring of patients in one chapter. Nevertheless, though aiming at presenting medical-biological features of joint replacements to engineers we assumed the data to be helpful for practicing surgical orthopaedists.

The review of endoprosthetic methods and technical facilities to implement them has led us to the conclusion that traditional methodology has reserves both in terms of improvement of engineering provision of operations and in terms biological reserves of human organism. The following chapters are devoted to discussing these potentialities.

References:

1. Singh M. Changes in trabecular pattern of the upper end of the femur as an index of osteoporosis. *J. Bone Joint Surgery*, 1970, V. 52 A, p. 456–467.
2. De Lee J. and Charnley J. Radiological demarcation of cement socket in total hip replacement. *Clin. Orthop.*, 1976, V. 121, p. 20–32.
3. Spotorno L., Grappiolo G., and Mumenthaler A. Das CLS-system. In: Endoprothetik, Ed. by E. W. Morscher. Berlin: Springer, 1995, S. 291–318.

4. Konstantoulakis Ch., Anastopoulos G, Papaeliou A., et al. Uncemented total hip arthroplasty in the elderly. *Int. Orthop.,* 1999, V. 23, p.334-336.
5. Voronovich A.I. Endoprosthetics of hip joint with acetabular defects, Ph.D. Thesis, Minsk, 1999.
6. Pitto R.P., and Schikora N. Acetabular reconstruction in developmental hip dysplasia using reinforcement ring with a hook. *Int. Orthop.,* 2004, V. 28, p.202-205.
7. Carlos E.R. Displaced intracapsular hip fractures: hemiarthroplasty or total arthroplasty? *Clin. Orthop.,* 2002, V. 399, p.72-77.
8. Hasan M.Y., Shinomiya F., Osada M., et al. Intracapsular hip fracture in patients with rheumatoid arthritis. *Int.orthop.,* 2003, V. 27, pp.294-297.
9. Stulberg B., and Hupfer T. Indications and preoperative planning for total hip arthroplasty. In: Total Joint Replacement, Ed. by W. Petty. Philadelphia, Saunders Co., 1991, p. 229–238.
10. Muller M.E., and Jaberg H. Total hip reconstruction. In: Surgery of the Musculoskeletal System / 102. Ed. by E. McCollister, 2nd ed. New York, Churchill Livingstone, 1990, p. 2979–3017.
11. Spotorno L., and Romagno S. Preoperative planning. In: The CLS Uncemented Total Hip Replacement System. Berne, SULZER medica Co., 1991, p. 8–11.
12. Widman J., and Isacson J. Lateral position reduces blood less in hip replacment surgery: a prospective randomized study of 74 patients. *Int. Orhtop.,* 2001, V. 25, p. 226-227.
13. Berry D.J., Berger R.A., Callaghan J.J., et al. Minimally invasive total hip arthroplasty. Development, early results, and a critical analysis. *J. Bone Joint Surg. (Am),* 2003, V. 85, p. 2235-2246.
14. Nakamura Sh., Matsuda K., Arai N., et al. Mini-incision posterior approach for total hip arthroplasty. *Int. Orthop.*, 2004, V. 28, p. 214-217.
15. Higuchi F., Gotoh M., Yamaguchi N., et al. Minimally invasive uncemented total hip arthroplasty through an anterolateral approach with a shorter skin incision. *J. Orthop. Sci.,* 2003, V. 8, p. 812-817
16. Lester D.K., and Helm M. Mini-incision posterior approach for hip arthroplasty. *Orthop. Traumatol.*, 2001, V. 4, p. 245-253.
17. Berger R.A. Total hip arthroplasty using the minimally invasive two-incision approach. *Clin. Orthop.*, 2003, V. 417, p. 232-241.
18. Surgical technique. Lubinus classic plus®. Modern Alternative. Hamburg, WALDEMAR LINK GmbH, 1999, 6 p.
19. Lewinnek G.E., Lewis J.L., Tarr R., et al. Dislocation after total hip-replacement arthroplasties. *J. Bone Joint Surg. (Am)*, 1978, V. 60, p. 217-220.

20. Tannast M. The measurement of anterversion and inclination with respect to the pelvic frontal plane. M.D. Thesis, University of Bern, 2000.
21. Saxler G., Marx A., Vandervelde D., et al. The accuracy of free-hand cup positioning – a CT based measurement of cup placement in 105 total hip arthropasties. *Int. Orhtop.*, 2004, V. 28, p.198-201.
22. Borner M., Bauer A., and Lahmer A. Rechnerunterstutzter robotereinsatz in der huftendoprothetik. *Orthopade*, 1997, V. 26, S. 251-257.
23. Jerosch J., Peuker E., von Hasselbach C., et al. Computer assisted implantation of the femoral stem in THA – an experimental study. *Int. Orthop.*, 1999, V. 23, p. 224-226.
24. Jenny J.Y., and Boeri C. Computer–assisted implantation of total knee prostheses: a case-control comparative study with classical instrumentation. *Comput. Aided Surg.*, 2001, V. 6, p. 217-220.
25. Hart R., Janecek M., Chaker A., and Bucek P. Total knee arthroplasty implanted with and without kinematic navigation. *Int. Orhtop.*, 2003, V. 27, p. 366-269.
26. Perlick L., Bathis H., Tingart M., et al. Minimally invasive unicompartmental knee replacement with a nonimage-based navigation system. *Int. Orthop.*, 2004, V. 28, p.193-197.
27. Keating E.M. Current options and approaches for blood management in orthopaedic surgery. *Inst. Course Lect.*, 1999, V. 48, p. 750-762.
28. Earnshaw P. Blood conservation in orthopaedic surgery: the role of epoetin alfa. *Int. Orthop.*, 2001, V. 25, p. 273-278.
29. Sinha A., Sinha M., and Burgert S. Reinfusion of drained blood as an alternative to homologous blood transfusion after total knee replacement. *Int. Orthop.*, 2001, V. 25, p. 257-259.
30. Petrenko A.M. Clinic, diagnosis and treatment of disorders of locomotor system in combination with cranial-brain trauma, Ph.D. Thesis, Minsk, 1985
31. Stillwell W.Th. The art of total hip arthroplasty. New York, Grune and Statton, Inc., 1987, 507 pp.
32. Herberts P., Ahnfelt L., and Malchau H. Multicenter clinical trials and their value in assessing total joint arthroplasty. *Clin. Orthop.*, 1989, V. 249, p. 48–55.
33. Kuzmenko V.V., and Fokin V.A. Endoprosthetics of hip joint: state-of-the-art and promises of development of the method. *Orhtopaedics and Traumatology (Moscow),* 1991, No.10, p.74-78.
34. Neverov V.A., and Zakari S.M. Revision endoprosthetics of hip joint. St.-Petersburg, Education Publ., 1997, 112 pp.

35. Trotsenko V.V. Dynamics of protective-adaptive reactions in endoprosthetics of hip joint. In: Endoprosthetics in traumatology and orthopedics. Moscow, Medicine, 1993, p. 24-31.

36. Malchau H., Herberts P., Ahnfelt L., and Johnell O. Prognosis of total hip replacement. Goteborg, University, 1993, 10 pp.

37. Resig S., Saleh Kh.J., and Bershadsky B. The outcome of perioperative wound infection after total hip and knee arthroplasty. *Int. Orthop.,* 2002, V. 26, p. 257.

38. De Lucas-Villarubia J.C., Lopez-Franco M., Granizo J.J., et al. Strategy to control methicillin-resistant *Staphylococcus aureus* postoperative infection in orthopedic surgery. *Int. Orthop.*, 2004, V. 28, p. 16-20.

39. Cailouette J.T., Gorab R.S., Klapper R.C., and Anrel S.H. Revision arthroplasty facilitated by ultrasonic tool cement removal. Part 1. *In vitro* evaluation. *Orthop. Rev.,* 1991, V. 20, No. 4, p. 353–357.

40. Petty W. Revision total hip arthroplasty. In: Total Joint Replacement, Ed. by W. Petty. Philadelphia, Saunders Co., 1991, p. 371–430.

41. Perner K. Revision operations with using endoprostheses of Zweimuller system. *Advances in Traumatology and Orthopedics by name of N.N. Priorov (Moscow),* 1998, No. 3, p. 33-38.

42. Hartwig C.-H., Bohm P., Czech U., et al. The Wagner revision stem in alloarthroplasty of the hip. *Arch. Orthop. Trauma Surg.*, 1996, V. 115, p. 5–9.

43. Katthagen B. Fermurdefectklassification. Arbeitskreis knochenersatz der DGOT. *Orthop. Mitteilungen DGOT*, 1993, No. 3, S. 236.

44. Pak J., and Paprosky W. Femoral strut allografts in cementless revision total hip arthroplasty. *Clin. Orthop.*, 1993, V. 295, p. 172–178.

45. Grady-Benson J. Revision hip replacement surgery. *Current Orthopedics*, 1995, No. 9, p. 9–20.

46. Voronovich A.I., and Nikolaev V.I. Revisions of hip joint in patients with extended destruction of the acetabulum resulting from instability of previously inserted endoprostheses. Materials of Scientific practical Conf. of traumatologists-orthopedists of Belarus, Minsk, 1998, p.47-50.

47. Van Koeveringe A.J., and Ochsner P.E. Revision cup arthroplasty using Burch-Schneider anti-protrusio cage. *Int. Orthop.,* 2002, V. 26, p. 291-295.

48. Nikolaev V.I. Aseptic instability of acetabular component of endoprostheses: biophysical aspects, diagnosis, treatment, and prophylaxis (clinical and experimental study). Ph.D. Thesis, Minsk, 2000.

49. Effenberger H., Ramsauer T., Dorn V., and Imhof M. Factors influencing the revision rate of Zweimuller acetabular cup. *Int. Orthop.,* 2004, V. 28, p.155-158.

50. Gore D.R. Impaction bone grafting for total hip revision. *Int. Orhtop.,* 2002, V. 26, p.162-165.

51. Gross A.E., Allan D.G., Catre M., et al. Bone grafts in hip replacement surgery: The pelvic side. *Orthop. Clin. N. Am.*, 1993, V. 24, p. 679–695.

52. Delimar D., Cicak N., Klobucar H. et al. Acetabular roof reconstruction with pedicled iliac graft. *Int. Orhtop.*, 2002, V. 26, p. 344-348.

53. Volkman R., Bretschneider C., Eingartner Ch., and Weller S. Revision arthroplasty – femoral aspect: the concept to solve high grade defects. *Int. Orthop.* , 2003, V. 27 (Suppl.1), S24–S28.

54. Pulliam I.T., and Trousdale R.T. Fracture of ceramic femoral head after a revision operation. A case report. *J. Bone Joint Surg. (Am),* 1997, V. 79, p.118-121.

55. Arenas A., Tejero A., Garbayo A., et al. Ceramic femoral head fractures in total hip replacement. *Int. Orthop.,* 1999, V. 23, p. 351-352.

56. Abadu A., Sivardeen K.A.Z., Grimer R.J., et al. The overcome of perioperative wound infection after total hip and knee arthroplasty. *Int. Orthop.*, 2002, V.26, p. 40-43.

57. Pitto R.P., and Spika I.A. Antibiotic-loaded bone cement spacers in two-stage management of infected total knee arthroplasty. *Int. Orthop.,* 2004, V. 28, p.129-133.

58. Insall J., and Scott N. Surgery of the knee. Philadelphia, Churchill Livingstone, 2001, 500 pp.

59. Domingo L.J., Calallero M.J., Cuena J., et al. Knee arthrodesis with the Wichita fusion nail. *Int. Orthop.*, 2004, V. 28, p. 25-27.

60. Spotorno L. Postoperative treatment. In: The CLS uncemented total hip replacement system. Berne, ZULZER *medica* Co., 1991, No. 1, p. 17–19.

61. Chan Y.K., Chin K.Y., Yip D.K.H., et al. Full weight bearing after non-cemented total hip replacement is compatible with satisfactory results. *Int. Orthop.,* 2003, V. 27, p. 94-97.

62. Doria C., De Santis V., Falcone G., et al. Osseointegration in hip prostheses: experimental study in sheep. *Int. Orthop.,* 2003, V. 27, p. 272-277.

63. Braun A., and Reiter A. The periprosthetic bone remodeling process – signs of vital bone reaction. *Int. Orthop.,* 2003, V. 27 (Suppl. 1), S7 – S10.

64. Davy D.T., Kotzar G.M., Brown R.H., at al. Telemetric force measurements across the hip after total arthroplasty. *J. Bone Joint Surg.,* 1988, V. 70, No. 1, p. 45–50.

65. Anchuela J., Gomes-Pellico L., Ferrer-Blanco M., et al. Muscular function and bone mass after knee arthroplasty. *Int. Orthop.*, 2001, V. 25, p. 253-256.

66. Spicer D.D.M., Pomeroy D.L., Badenhausen W.E., et al. Body mass index as a predictor of outcome in total knee replacement. *Int. Orthop.,* 2001, V. 25, p.2 46-249.

67. Akulich Yu.V., Denisov A.S., Nyashin Yu.I., et al. The influence of the scheme of loading variations on the recovering of the bone tissue elastic modulus. *Russian Journal of Biomechanics*, 1999, V. 3, No. 3, p. 63–72.

68. Paterson D. The International Documentation and Evaluation System (IDES). *Int. Orthop*., 1993, V. 11, p. 303-307.

69. Jonston R.C., Fitzgerald R.H., Harris W.H. et al. Clinical and radiographic evaluation of total hip replacement. A standard system of terminology for reporting results. *J. Bone Joint Surg. (Am),* 1990, V. 72, p. 161-168.

70. Galante J. The need for a standardized system for evaluating results of total hip surgery. *J. Bone Joint. Surg*., 1985, V. 67, p. 511-512.

71. Petty W. Total hip arthroplasty: complications. In: Total Joint Replacement, Ed. By W. Petty. Philadelphia, Saunders Co., 1991, p. 287-314.

72. Karptsov V.I., Vorontsov S.A., Epstein G.G., et al. Specific complications in operation of hip joint replacement. *Tramatology and Orthopedics in Russia,* 1994, No.5, p. 91-98.

73. Barbara J.A.J., and Contreras M. Infectious complications of blood transfusion: viruses. *Br. Med. J.*, 1990, V. 299, p. 450-453.

74. Wallace J. Blood transfusion for clinicians. Edinburgh, Churchill Livingstone, 1977, 380 pp.

75. Murphy P., Heal J.M., and Blumberg N. Infection or suspected infection after hip replacement surgery with autologous or homologous blood transfusions. *Transfusion,* 1991, V. 31, p. 212-217.

76. Borghi B., and Casati A. Thromboembolic complications after total hip replacement. *Int. Orthop*., 2002, V. 26, p. 44-47.

77. Salvati E.A., Cherofsky K.M., Brause B.D., and Wilson Ph.D. Reimplantation in infection. *Clin. Orthop.,* 1982, V. 170, p. 62–75.

78. Joint Replacement Arthroplasty, Ed. by B.F. Morray. New York, Churchill Livingstone, 1991, 1205 pp.

79. Walker P.S., Culligan S.G., et al. The effect of a lateral feature on uncemented hip stem. *Hip Int.,* 1999, V. 9, p. 71-80.

80. Leali A., Fetto J.F. Preservation of femoral bone mass after total hip replacement with a lateral flare stem. *Int. Orthop.,* 2004, V. 28, p.151-154.

81. Colyer R.A., and Capello W.N. Surgical treatment of the infected hip implant: two stage Reimplantation with a one month interval // *Clin. Orthop.,* 1994, V. 298, p.75-79.

82. Brooker A.F., Bowerman J.W., Robinson R.A., and Riley L.N. Ectopic ossification following total hip arthroplasty. *J. Bone Joint Surg.*, 1973, V. 55 A, p. 1629–1632.

83. Ahrengart L. Periarticular heterotopic ossification after total hip arthroplasty. Risk factors and consequence. *Clin. Orthop.,* 1991, V. 263, p. 49-58.

84. Toom A., Haviko T., Rips L. Heterotopic ossification after total hip arthroplasty. *Int. Orthop.,* 2001, V. 24, p. 323-326.

85. Hollister S.J. Mechanical factors influencing the outcome of total joint replacement. *Current Orthopaedics*, 1995, V. 9, No. 1, p. 2–8.

86. Wright J.M., Pellicci P.M., Salvati E.A., et al. Bone density adjacent to press-fit acetabular components. A prospective analysis with quantitative computer tomography. *J. Bone Surg. (Am),* 2001, V. 83, p. 529-536.

87. Van der Lugt J.C.T., Onstenk R., and Nelissen R.G.H. Primary Stanmore total hip arthroplasty with increased cup loosening in rheumatoid patients. *Int. Orthop.*, 2003, V. 27, p. 291-293.

88. Oh I., and Harris W.H. Proximal strain distribution in total loaded femur // *J. Bone Joint Surg. (Am),* 1978, V. 60, p. 75-83.

89. Engs C.A., McGovern T.F., Bobyn J.D., and Harris W.H. A quantitative evaluation of periprosthesis remodeling after cementless total hip arthroplasty. *J. Bone Joint Surg. (Br),* 1992, V. 74, p. 1009-1020.

90. Kilgus D.J., Shimaoka E.E., Tipton J.S., and Eberle R.W. Dual-energy X-ray absorptiometry measurement of bone mineral density around porous-coated cementless femoral implants. *J. Bone Joint Surg. (Br),* 1993, V. 75, p. 279-287.

91. Niimimaki T.J., and Jalovaara P.A.K. Bone loss from the proximal femur after arthroplasty with an isoelastic femoral stem. *Acta Orthop. Scand.,* 1995, V. 66, p. 347-351.

92. Herberts P., and Malchau H. How outcome studies have changed total hip arthroplasty practices in Sweden. *Clin. Orthop.*, 1997, V. 344, p. 44-60.

93. Mulroy W.F., Estok D.M., and Harris W.H. Total hip arthroplasty with use of so-called second-generation cementing techniques. *J. Bone Joint Surg. (Am),* 1995, V. 77, p.1845-1852.

94. Iguchi H., Hua J., and Walker P.S. Accuracy of using radiographs for custom hip stem design. *J. Arthroplasty,* 1996, V. 11, p. 312-321.

95. Schmidt R., Muller R., Kress L., et al. A computed tomography assessment of femoral and acetabular bone changes after total hip arthroplasty. *Int. Orthop.,* 2002, V. 26, p. 299-302.

96. Malik M.H.A., Gray J., and Kay P.R. Early aseptic loosening of cemented total hip arthroplasty: the influence of non-steroidal anti-inflammatory drugs and smoking. *Int. Orthop.,* 2004, V. 28, p. 211-213.

97. Sarvilinna R., Huhtala H.S.A., Puolakka T.J.S., et al. Periprosthetic fractures in total hip arthroplasty: an epidemiologic study. *Int. Orthop.,* 2003, V. 27, p. 359-361.

98. Burchardt H. The biology of bone graft repair. *Clin. Orthoped.,* 1983, V. 174, p. 28–42.

99. Albee F.N. Fundamentals in bone transplantation. Experiences in three thousand bone graft operation. *J.A.M.A.*, 1923, V. 81, p. 1429–1432.

100. Lexer E. Joint transplantation and arthroplasty. *Surg. Gynecol. Obstet.,* 1925, V. 40, p. 782–809.

101. Herndon C.H., and Chase S.W. The fate of massive autogenous bone grafts including articular surfaces. *Surg. Ginecol. Obstet.*, 1954, V. 98, p. 273–290.

102. Burwell R.G., Gowland G., and Dexter F. Studies in the transplantation of bone. VI. Further observations concerning the antigenicity of homologous cortical and cancellous bone. *J. Bone Joint Surg.*, 1963, V. 45 B, p. 597–608.

103. Ottolenghi C.E. Massive osteoarticular bone grafts. Technic and results of 62 cases. *Clin. Orthop.,* 1972, V. 87, p.156–164.

104. Volkov M. Allotransplantation of joints. *J. Bone Joint Surg.*, 1970, V. 52 B, p. 49–53.

105. Tissue Enegineered Medical Products (TEMPs). Ed. by E. Schutte, G.L. Picciolo, and D. Kaplan. Philadelphia, ASTM Int., 2004, 280 pp.

106. Nikolsky M.A. Outcomes of osseoplasty on the bodies of lumbar vertebrae. Ph.D. Thesis, Novosibirsk, 1971.

107. Scarborough N., and Van der Griend R.A. Allografts. In: Total Joint Replacement, Ed. by W. Petty. Philadelphia, Saunders Co., 1991, p. 43–50.

108. Gore D.R. Impaction bone grafting for total hip revision. *Int. Orthop.,* 2002, V. 26, p. 162-165.

109. Muscolo D.L., Caletti E., and Schajowicz F. Tissue-typing in human massive allografts of frozen bone. *J. Bone Joint Surg.,* 1987, V. 69 A, p. 583–595.

110. Bone graft substitutes. Ed. by C.T. Laurencin. Philadelphia, ASTM Int., 2003, 250 pp.

111. Tomford W.W., Doppelt S.H., Mankin H.J., and Friedlaender G.E. 1983 Bone banking procedure. *Clin. Orthoped.*, 1983, V. 174, p. 15–21.

112. Delimar D., Cicak N., Klobucar H., et al. Acetabular roof reconstruction with pedicled iliac graft. *Int. Orthop.*, 2002, V. 26, p. 244-348.

113. Niclason L.E. Engineering of bone grafts. *Nature Biotechnology*, 2000, V. 18, No. 9, p. 929–930.

114. Artificial cartilage. *Materials today*, 2001, V. 4, No. 6, p. 14.

115. Kale S., Biermann S., Edwards C., et al. Three-dimensional cellular development is essential for *ex vivo* formation of human bone. *Nature Biotechnology*, 2000, V. 18, No. 9, p. 954–958.

116. Petite H., Viateau V., Bensaid W., et al. Tissue engineered bone regeneration. *Nature Biotechnology*, 2000, V. 18, No. 9, p. 959–963.

117. Li H., Zou X., and Bunger C. Gene therapy and spinal disorders. *Int. Orthop.*, 2001, V. 25, p. 1-4.

Chapter 5. TRIBOLOGICAL ASPECTS OF ENDOPROSTHETICS

The goal of joint endoprosthetics is considered achieved if artificial joints function in the human body without failure (breaking of normal operation) for a long time. Main reasons for failures of implanted endoprostheses are related to tribology, a science about friction. The most often aseptic complications of endoprostheses result from the wear of endoprostheses and accumulation of wear debris in the surrounding tissues that are not removed through periarticular (joint-surrounding) lymphatic system. This results in osteolysis and loss of bearing capacity and, consequently, in artificial joint failure. With due post-operation observation of the patients the ultimate wear is a less probable reason for joint failure. Therefore, the wear resistance of implanted endoprostheses governs essentially the terms of revision operations and becomes presently one of the most urgent challenges in joint endoprosthetics. Natural joints differ essentially from artificial ones by the availability of microporous cartilage functioning as an antifrictional material and a reservoir for the lubricating synovia. In diseases disturbing natural mechanisms of joint lubrication unfavourable tribological situations arise that need endoprosthetics.

The methodology for studying friction and wear of endoprostheses *in vivo* has certain limitations. The most complete and valid information is obtained by analyzing endoprostheses removed in revision operations and histological examination of joint capsules, and surrounding tissues.

New-type endoprostheses are subjected to tribological testing prior to clinical approbation. They consist of laboratory trials, tests in animals, joint simulators, etc. Therefore, it will be possible to assess new endoprosthesis efficiency 7–10 years after completing their elaboration.

The tribojoints of modern endoprostheses possess a satisfactory tribological resource. Polymeric friction parts of endoprostheses made by J. Charnley have the lowest wear resistance and provide 20- years-and-more service life of artificial joints. Under long-term *in vivo* operation some problems arise due to metallic, polymeric, and other wear debris accumulated in the human body.

The present chapter deals with this and some other challenges related to the tribology of artificial joints.

5.1 FRICTION IN SYNOVIAL JOINTS

Friction in a natural joint occurs between the layers of hyaline cartilage fixed at the conjugated ends of the bones and lubricated by the synovial fluid. The elastic layers of the cartilage at the joint ends of the bones have different thickness providing a uniform pressure distribution on the contact surface under joint loading. The cartilage damps dynamic loads and regulates pressure on the friction surface of the joint.

The load-carrying cartilage matrix is formed by huge aggregates of proteoglycane molecules comprising hyalurone acid, glycoproteins, and core proteins with numerous side chains of polysaccharides [1, 2]. The chains comprise many electrostatically interacting anion groups and form “brushes” on the macromolecules of the core protein. In the vacancies between the chains the water and the ingredients of protein-polysaccharide complex are accumulated selectively. Therefore, the joint cartilage is a highly hydrated tissue and water is one of its main components [3]. Under loading the water is squeezed out from the cartilage tissue and comes back under load relieving, which makes cartilage deformation reversible. A non-linear dependence exists between cartilage tissue deformation and its hydrodynamic resistance [4].

Synovia is a nutrient medium for the cartilage matrix cells (chondrocytes) producing protein components necessary for cartilage regeneration [5]. It is an aqueous solution of proteins and biopolymers including hyalurone acid. The latter produces in the synovia polymolecular space structures capable of binding significant water amounts. Because of this the synovia is a thixotropic medium whose viscosity depends on the shear rate. Macromolecules of the complex compounds containing hyalurone acid molecules are adsorbed on the cartilage surface producing the protective layer having gel constituency and some elasticity [6].

Work [3] gives a retrospective review of lubrication models of natural joints. The views have been developed from the hydrodynamic model according to which the geometry of the cartilage surface and synovia viscosity cause the hydrodynamic flow of the lubricating synovia in joint operation to the weeping lubrication model when the liquid squeezed out from the cartilage in compression wets directly the friction surfaces [7]. A more complex lubrication model, i.e. “squeezed film lubrication” provides the cartilage ends of the bones to rest on the liquid interlayer whose carrying capacity prevents them from contact [8]. The “boosted lubrication” model is realized by filtration of low-molecular fraction of the lubricating synovia interlayer being under total squeezing into the pores of the cartilage matrix. As a result, the concentration of high-molecular components increases in the interlayer. This improves its carrying capacity and provides better mechanical load bearing by the joint [9]. The boundary lubrication friction model [10] taking into account polarization-electrical interactions of the tribosystem components is shown in Fig.5.1. On the negatively charged cartilage surface an adsorption of electrically non-equilibrium molecules of glycoproteins – complex proteins of synovia – occurs. Following the mechanism hydrophobic-hydrophobic interaction they attract other glycoprotein molecules thus forming strong bonds. The third chain is bound to these chains by weak hydrophilic-hydrophilic bonds. The friction in the system is governed by the breaking energy of weak bonds responsible for low shear resistance in the boundary layer consisting of zones 4-7.

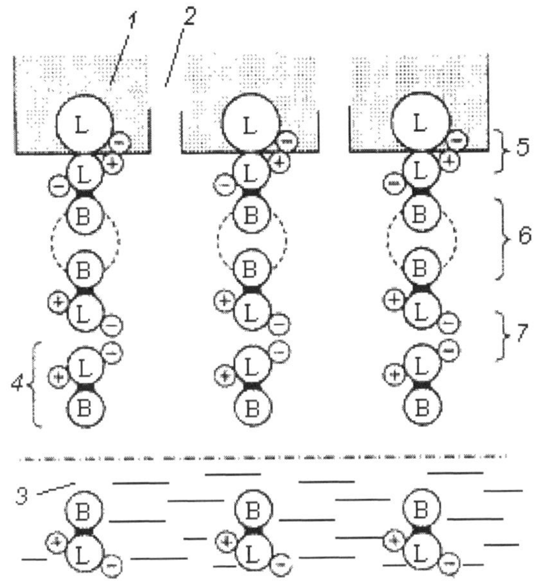

Fig. 5.1. Diagram of boundary lubrication of natural joints: *1* – cartilage, *2* – pore, *3* – synovia, *4* – glycoprotein molecule, *5* – zone of electrostatic attraction of glycoproteins to the cartilage surface, *6* and *7* – zones of formation of strong hydrophobic-hydrophobic and weak hydrophilic-hydrophilic bonds between glycoprotein molecules, L and B – hydrophilic and hydrophobic zones

In 1980ies liquid-crystalline components were discovered experimentally in the synovia [11] (liquid crystals having anisotropy of properties which is due to the ordered orientation of molecules). It has been supposed that etherified cholesterol derivatives whose content of synovia is 0.16–0.20 wt % are responsible for the liquid-crystalline state [12]. They are characterized by phase transitions both in the range of physiological temperatures (31–43 ^{0}C) and at higher temperatures arising in friction. A hypothesis has been proposed that low friction in healthy joints (0.005–0.025 friction coefficient) is due to liquid-crystalline state of the synovia [13]. Then it was found experimentally that biophysical friction mechanisms in joints were mainly governed by the specific structure of liquid-crystalline ethers of cholesterol. They are responsible for the antifriction behaviour of the joints through realization in the friction zone of the mesomorphic (intermediate between liquid and solid) nematic state of the lubricating film (molecules of nematic liquid crystals are located in parallel to the friction surface) with the layer-by-layer oriented structure [14].

The data obtained give grounds to compare the tribological conditions for the functioning of natural joints and endoprostheses and are helpful in finding ways to improve the efficiency of artificial joints.

5.2 FRICTION AND WEAR OF ENDOPROSTHESES

Wear of endoprostheses and tissue inflammation initiated by wear debris are nowadays the main challenges in joint endoprosthetics. They are the wear debris, their migration on the bone-implant interface, and cell reaction to them that cause aseptic instability of endoprostheses. Under the effect of the increasing pressure of juridical aspects of endoprosthetics on medical practice the requirements to the implant quality get more severe. The number of young patients leading active lifestyle rises, which makes necessary prolonging tribological resource of endoprostheses. Below we will consider factors affecting the wear of implanted joint endoprostheses.

Lubrication of artificial joints occurs by principally different mechanisms and is considerably less efficient than that of synovial joints.

Friction pairs of endoprostheses are traditionally made of structural rigid and nonporous materials that do not allow for realizing biophysical mechanisms of lubrication characteristic of the cartilage. In endoprostheses used nowadays it is impossible to regulate electric potential of friction surfaces. That is why unique tribological properties of the synovia remain unclaimed and the lubrication mechanism whose diagram is shown in Fig.5.1 does not work. The unavailability in the traditional endoprostheses of the sources of electrical and magnetic fields that simulate the natural biophysical field of the joint does not make possible realizing low friction mechanisms related to the optimization of the structure and tribological characteristics of the liquid-crystalline component of the synovia. Loads and sliding velocities characteristic of human joints cause insufficient synovial lubrication of endoprostheses. In the best case the implanted endoprostheses work under the conditions corresponding to the region of mixed lubrication on the Stribeck curve showing friction coefficient as a function of the dimensionless load parameter. It means that the hydrodynamic effect of lubrication occurs only on small spots of the friction surface. On the larger rest area the contact of microasperities is not provided by the lubricating interlayer or they are divided by such a fine (shares of micrometers) synovia film that its viscosity does not affect essentially the friction.

In work [15] the thicknesses of the synovia film in natural and artificial joints are calculated by the method of finite elements (Fig.5.2). It is seen that in a healthy hip joint the film thickness is 2 μm on average while in an arthrosis-damaged one it diminishes down to 0.5–1.0 μm, and in hip joint endoprosthesis with the UHMWPE friction pair the lubricating film is even finer, i.e. 0.2–0.5 μm. In metal-metal and ceramics-ceramics friction pair involving endoprostheses the rated thickness of the synovial film does not exceed 0.1 μm. Therefore, the tribological state of implanted endoprostheses is in the best case similar to the arthrosis-damaged joint.

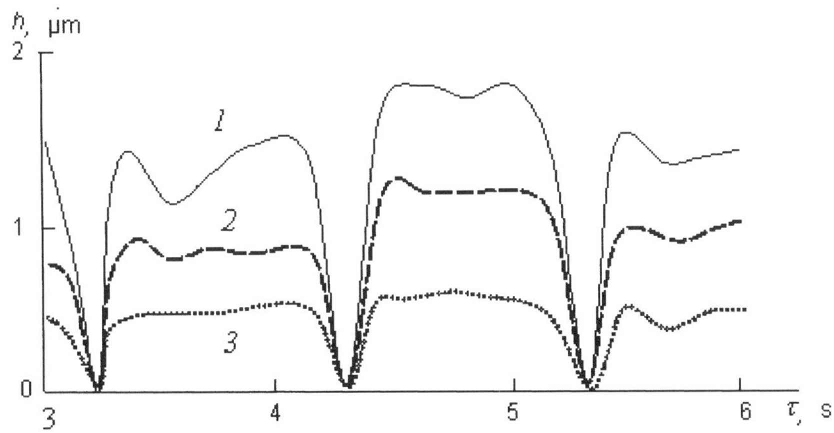

Fig. 5.2. Kinetic dependence of lubricating synovial film when walking: ***1*** **– healthy joint,** ***2*** **– arthrosis-affected joint,** ***3*** **– UHMWPE-metal endoprosthesis**

Table 5.1
Application of friction pairs in hip joint endoprostheses [16]

Head material	Cup material						
	Polymers					Metal	Ceramics
	PTFE	UHMWPE	PTFCE	POM	PETF	CoCrMo	Al_2O_3
FeCrNiMo	×	++	×	•	•	–	–
FeCrNiMoNbN	•	++	–	•	•	–	–
CoCrMo	•	++	×	×	×	++	–
TiAlV	•	×	×	–	–	–	–
Al_2O_3	•	++	•	•	•	–	++
ZrO_2	•	+	•	•	•	–	–

++ long-term clinical application; + clinical approbation; – unsuitable by technical reasons;

× clinically unsuitable; • not studied.

Combining of friction materials in joint endoprostheses is based on the long-standing world experience in endoprosthetics. Table 5.1 presents the data on the tribological effectiveness of engineering materials under usage in orthopaedics [16].

Polytetrafluoroethylene (PTFE) is one of the best antifriction materials. However, it produces wear products that cause inflammation of soft tissues and complications long after endoprosthetics [17]. After it had been found out in friction pairs with stainless steel further clinical experiments with PTFE stopped. Up to now the traditional UHMWPE–CoCrMo friction pair for endoprostheses proposed by J. Charnley has remained actual. UHMWPE works excellently in pair with ceramics (wear rate *in vivo* 0.05–0.13 mm/year [18]) and all metallic alloys except titanium ones. The latter are characterized by good biological compatibility but are unsuitable for working in moving joints of endoprostheses. Friction damage of the passivating oxide titanium film occurring *in vivo* in the biological fluid initiates corrosive-mechanical wear of the titanium heads [19]. Reduction of UHMWPE wear and elimination of wear-induced osteolysis are essential to improve the longevity of endoprostheses. Cross-linked UHMWPE and ceramic heads are currently used for wear reduction [20]. These materials have been applied to total hip arthroplasty based on the encouraging results of a number of wear simulator tests [21]. However, the latter do not always accurately reflect *in vivo* wear. In work [22] three types of femoral heads working in pair with UHMWPE sockets more than 9 years after total hip arthroplasty were evaluated to determine the influence of socket and head tribosurface quality on wear. The retrieved alumina heads showed significantly better surface roughness and roundness than heads of CoCr and stainless steel. Nevertheless, the volumetric wear rates of the retrieved UHMWPE sockets were significantly greater in those coupled with an alumina head. However, no significant difference was found in socket quality demonstrated as fusion defects among three different groups. Wear products of polytrifluorchlorethylene (PTFCE) have the same disadvantages as PTFE. Biologically compatible polyoxymethylene (POM) and polyethyleneterephthalate (PETF, Dacron) used as artificial tendons have a low wear resistance.

Many orthopaedists hope that metal-metal friction pairs can improve significantly endoprosthetic results. Currently the endoprostheses involving CoCrMo–CoCrMo friction pairs hold promise since they have shown wear rate less than tens micrometers per year [23].

It is advisable to use a ceramic counterbody in pair with Al_2O_3–ceramics for tribological criteria. Endoprostheses with these friction pairs have the lowest wear rate (less than 5 μm/year) among all tribojoints shown in Table 5.1. However, when ceramic cups were set in a wrong way the avalanche-like wear resulting in instant failure of endoprostheses was reported [24].

Summing up the numerous data [16, 18–25] on distant results of hip joint endoprosthetics (a large scatter in the data is due to different endoprosthesis designs) one can suggest that *in vivo* the wear rates shown in Fig.5.3 characterize the friction pairs within a series based on the wear resistance criterion.

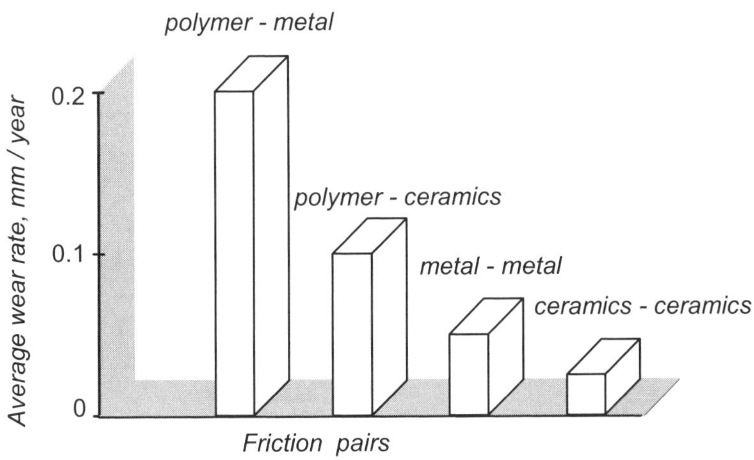

Fig. 5.3. Average wear rate *in vivo* of hip endoprosthesis with different friction pairs

Nowadays UHMWPE (ISO 5834/1, ASTM F 603) is the main material for manufacturing pelvic components of hip joint endoprostheses and wrought stainless steel FeCrNiMoMn (ISO 5832/1, ASTM F 648), alloy CoCrMo (ISO 5832/4, ASTM F 75), and Al_2O_3–ceramics (ISO 6474, ASTM F 603) are used for heads of the femoral component [25].

Hydroxyapatite (HA) coatings are commonly used for fixation of uncemented femoral components, and several studies have shown satisfactory results [26]. However, it has been suggested that HA particles detached from the stem surface may enhance osteolysis either by stimulation bone less or by migration to the joint space producing third-body wear [27]. A retrospective matched-pair study has been carried out of the patients who had a primary hip replacement with insertion of either porous-coated (PC) or a HA-coated Omni flex femoral component [28]. The prevalence of radiographic osteolysis was 16 % in hips with HA-coated stems and 43 % in hips with PC femoral stems. In hips with HA-coated stems, osteolysis was limited to Gruen zones 1 and 7 (Fig.5.4). In contrast, distal osteolysis was present around 26% of the PC stems. At 7 years, the survival-free rate of distal osteolysis was 100 % in hips with HA-coated stems but 90% in hips with PC stems. Circumferential HA coating of the femoral component reduced the occurrence of osteolysis and eliminated distal osteolysis at 5–10 years of follow-up. In addition, HA coating did not alter the wear rate.

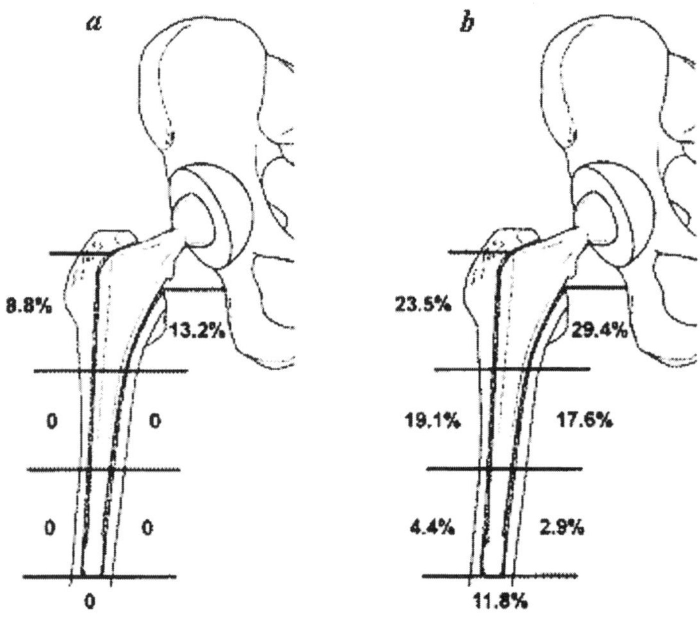

Fig. 5.4. Distribution of the osteolytic lesions according to the frequency of involvement of each femoral zone: *a* – HA-coated, *b* – PC femoral stems

Wear of metallic components of endoprostheses developed in 1990ies is characterized by the following data.

Wear of the new-generation heads made of stainless steel FeCrNiMnMoNbN (ISO 5832/9) is comparable to that of the heads made of alloy CoCrMo. The steel is polished well yet is scratched easily. With displacement of endoprosthesis during operation and occurrence of cement particles in the friction unit as well as "close" setting of luxations the head can be damaged irretrievably, which increases its wear. Alloy CoCrMo is significantly more resistant to mechanical damages and abrasive wear. European manufacturers of endoprostheses while improving the quality of alloy CoCrMo heads have gained the wear rate of matched UHMWPE inserts commensurable to that of UHMWPE-Al_2O_3 ceramics friction couple [25].

Tribological characteristics of titanium heads are unsatisfactory and worse than those of common stainless steel (ISO 5832/1, ASTM F 648). However, the efforts are made to improve the wear resistance of the heads made of titanium alloys by strengthening oxide layer by diffusive saturation with oxygen or applying diamond-like titanium nitride coatings. Low wear resistance of titanium and its alloys is the cause of their limited usage in producing stems and cups of cement endoprostheses. As limited micromobility arises between the stem and the bed the cement particles scratch the thin oxide film imparting chemical inertness to the titanium implant [29]. As a result, already after 3-4 years a so-called "titanium granuloma" is developed.

Studies are carried out on applying metal zirconium as a head material. Its disadvantages, among which treatment difficulties, high raw material cost, lack of data on biological compatibility, as well as residual radiation can be mentioned, are compensated by its high wear resistance.

To improve lubrication by the synovia and reduce endoprosthesis wear, grooves are made on the surface of metallic heads with a diamond needle to provide a regular microrelief (Fig.5.5). It improves head wetting with the synovia and decreases friction torque of the metallic head matched with UHMWPE cup by 25–35 % in simulator testing [30].

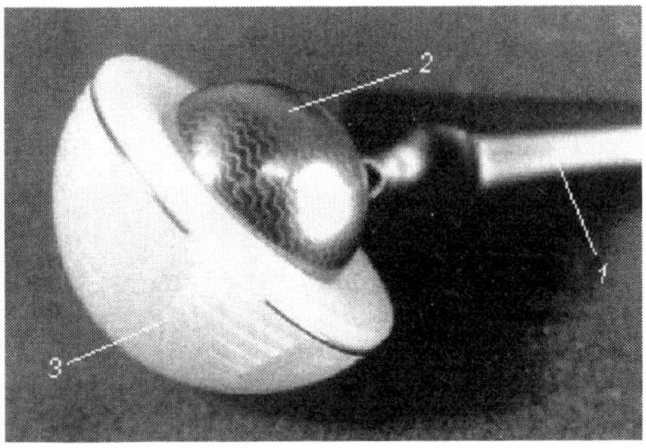

Fig. 5.5. Friction unit of total hip endoprosthesis with a micro relief on the metallic head [30]: *1* – stem, *2* – head, *3* –UHMWPE cup

Several leading European orthopaedists including B.G. Weber, H. Wagner, R. Kotz, and K. Zweymuller pin their hopes of improving distant results of joint endoprosthetics on metal-metal friction couples considering them as most suitable for young and active patients. In 1999 Zweymuller and his collaborates determined chrome and cobalt concentration in patient's blood serum after implanting endoprosthesis with the CoCrMo–CoCrMo friction pair by plasma ion spectrometry. Exponential reduction of the elements content of serum depends on the post-operation time and endoprosthesis quality. It can be said that accumulation of chrome and cobalt ions in the organism corresponds to the run-in period when the equilibrium roughness of the friction surfaces is established and intensive wearing conditions are changed by the stationary ones. In the run-in period the compensatory capabilities of the human organism and capacity of periarticular lymphatic system are significantly less than the efficiency of the endoprosthesis friction unit as wear product generator. As the friction unit begins working under stationary conditions at a constant and low wear rate, the capability of macrophages of utilizing wear products and capabilities of lymphatic system of removing them from the organism become prevailing. Element content of serum is stabilized or decreases exponentially. The level of decrease depends on the quality of endoprostheses whose production in different companies differs by precision and technological effectiveness.

There are only a few publications evaluating bone-resorbing cytokines in patients with aseptic loosening. These reports have suggested that some cellular products – interleukin 6 (IL-6), granulocyte-macrophage colony-stimulating factor (GM–CSF) – may be useful as early markers of implant failure [31, 32]. In synovial-like membranes from failed total hip prostheses, an increased level of cytokines and cellular mediators has been identified. In work [33] two matched groups of patients with total hip arthroplasty have been compared. One group was with a surgically proven component loosening and one without. Serum levels of IL-6, GM-CSF, and elastase were measured. Soluble interleukin 2 receptor was also measured to exclude any hypersensitivity reaction. No significant difference was found in serum values between groups. Neither were there any differences with respect to implant material, mode of fixation, and periprosthetic osteolysis. In contrast to previous reports these results suggest that serum levels of cytokines and cellular mediators may not be affected in aseptic loosening.

The ultimate wear resistance of endoprostheses with metal-metal friction pairs is reported in work [34]. The cup and the head of hip joint endoprostheses produced by Mathys Medical Ltd are made of forged alloy CoCrMo with high carbon content (more than 0,2 %) and have good friction surface quality. The endoprostheses were tested in collaboration with the Laboratory of independent orthopaedic studies in Los Angeles (USA) and Orthopaedic Fund of professor R.Mathys in Betlakh (Switzerland). Proceeding from the fact that the hip joint experiences *in vivo* 1 mln loading cycles per year it was found that the wear rate of Mathys endoprostheses was only 2 μm/year.

Wear of UHMWPE is an urgent issue in the tribology of artificial joints since the majority of implanted endoprostheses comprises friction parts made of UHMWPE (Fig.5.6) [35]. It can be attributed to such advantages of metal-polymer endoprostheses as their comparatively low cost, low friction coefficient, technological effectiveness, good impact load damping and running-in abilities.

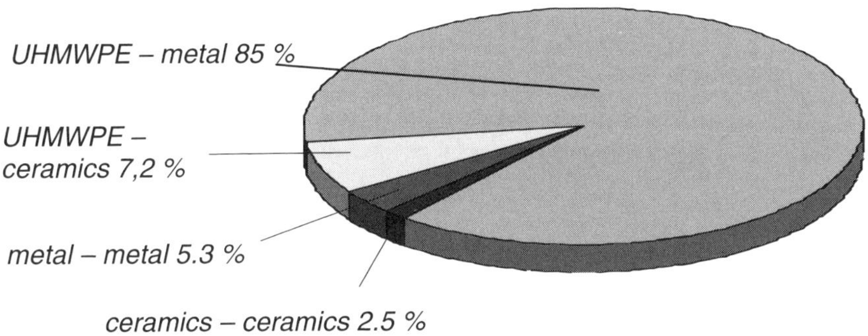

Fig. 5.6. Volumes of usage of hip endoprostheses vs. material combination in the friction pair of endoprostheses

Recently, wear debris of the UHMWPE cup as a component of the tribosurface of joint prostheses has been identified as a serious problem and one of the causes of osteolysis around, and loosening of, the prosthesis [36]. Several authors reported on "adhesive", "abrasive", and "fatigue" wear, but various terminologies have been used to describe the micromode changes on the tribosurface with the progress of UHMWPE wear, and many factors of the mechanisms are unknown [37, 38].

An English tribologist D. Dowson with collaborates have found out [39] that the wear of UHMWPE in pair with a smooth metallic counterbody (R_a<0,02 μm) occurs *in vivo* by fatigue mechanism (Fig.5.7). At the run-in period the microasperities on the friction surface of the polymer part contact the counterbody. Beneath the contact microareas of the polymer part the stresses are concentrated. As some time passes microcracks originate in the polymer part in the zones of stress concentration at a depth of 10–40 μm from the friction surface. Under dynamic loading of the artificial joint the subsurface cracks grow leading to the spalling of the microasperities and considerable damage of the friction surface. Therefore, the wear rate of endoprostheses *in vivo* is higher by two and more orders than that recorded in laboratory tests of UHMWPE-metal pairs by the rod-disc scheme under similar constant load.

The authors of [40] presume the UHMWPE wear process to take three steps: (1) foldings are generated first, (2) ripples are formed, and (3) when the UHMWPE cup is exposed to more severe wear condition, ripples decrease and fibrils increase. The amount of wear on cross-linked UHMWPE cup decreased as the radiation dose increased.

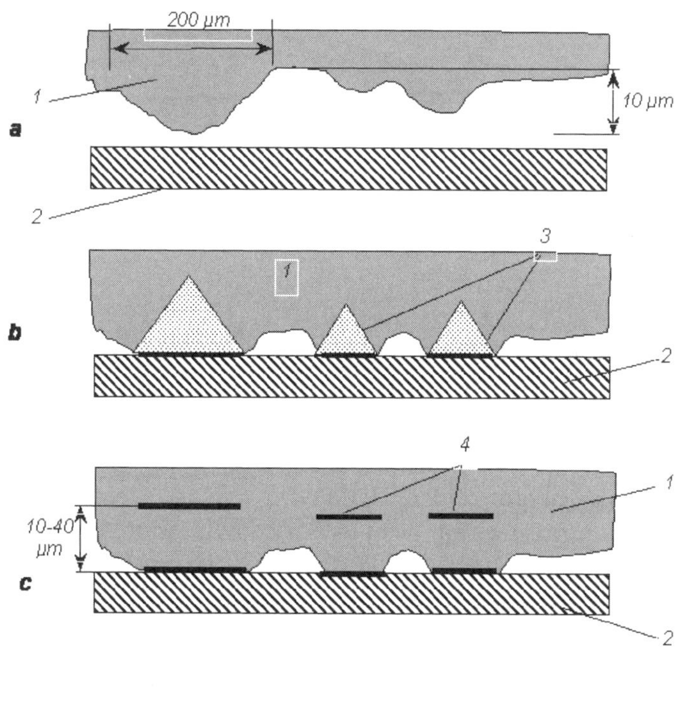

Fig. 5.7. Mechanism of UHMWPE wear in implanted joint endoprostheses: *a* –contact onset; *b* – loading of the friction pair and wear of microasperities; *c* – initial phase of UHMWPE failure. *1* – polymer, *2* – metal, *3* – zone of stress concentration, *4* – crack

It is suggested [41] that conventional UHMWPE gamma sterilized in air is liable to progressive oxidative degradation and prolonged shelf-life time before implantation. Oxidative aging of UHMWPE makes the liner susceptible to severe wear. The reason for the catastrophic wear of the modular Hexloc liner was thus, according to [41], an unfortunate synergism of many inferior material and design factors. The authors of work [42] question this finding assuming that the statistical data do not allow for those conclusions. Other recently published studies [43, 44] present contradictory results and further investigations are necessary to evaluate the reported assumption.

In work [45] tribological characteristics of Charnley endoprostheses are presented, currently the best hip joint endoprostheses with a UHMWPE-metal friction pair. The simulator tests have shown the friction coefficient of new endoprostheses to be $\mu = 0.11 \pm 0.025$, and under liquid lubrication at viscosity $\eta = 0.01$ Pa·s $\mu = 0.04 \pm 0.001$. For the set of the tested new endoprostheses we have Gaussian distribution of μ while for the endoprostheses working *in vivo* and removed in revision operations the Gaussian curve is distorted because of increasing number of endoprostheses exposed to increased wear. Under unlubricated friction 30% of removed endoprostheses have shown $\mu \geq 0{,}16$ and under lubricated friction 39% of endoprostheses have had $\mu \leq 0.07$. *In vivo* the linear wear rate was 0.21 mm/year over the first 2 years, followed by 0.14 mm/year thereafter [46].

In view of the risk of loosening and of a dramatic increase in the wear rate when the femoral head articulates with a metal backing, it is important to plan revision of this cup in good time and to ensure regular radiographic surveillance of these patients [47].

The design of polymer cups affects greatly the osteolysis progress. The cups without holes for screws, which can allow migration of UHMWPE wear debris, which might induce osteolysis of the adjacent ilium, are preferable [48].

The technology of UHMWPE modification by electron flow radiation has been developed [49]. The treatment causes the formation of cross-links between macromolecules and increase in wear resistance of the new material called WIAM. It was exposed to friction testing on the hip joint simulator under 20 mln cycles of alternating stresses. In testing, wear debris were not revealed though J. Charnley follower B.M. Wroblewski recorded only somewhat less wear rate of the new material *in vivo* (10-12 year observation period) than that of conventional UHMWPE in pair with the ceramic head [25].

The latest advance in development of alternative polymeric friction materials for endoprostheses is polyether-etherketone (PEEK) reinforced by carbon fibre. In simulator testing of the cups made of the material (in pair with ceramic head) the wear is 30 times less than that of the UHMWPE-ceramics pair [25]. Its advantage is high biological compatibility and unique in polymers chemical stability owing to which it does not essentially age in biological media. However, high rigidity of PEEK reduces endoprosthesis damping of impact loads, which accelerates mechanical loosening of endoprosthesis components.

Wear of ceramic components of endoprostheses occurs at a rather low rate (1–5 μm/year). Wear debris of Al_2O_3–ceramics are biologically inert and cellular reaction does not develop on them.

Figure 5.8 shows the plots demonstrating tribological characteristics of the friction units of endoprostheses comprising ceramic parts [50]. It is seen that a high friction torque under unlubricated conditions (~ 5 Nm) is reduced essentially (0.5–0.8 Nm) when rubbing in Ringer's solution simulating physiological medium of the organism. It is characteristic that in rubbing without lubrication the pair temperature rises up to 40°C because of friction heat release. In Ringer's solution, viz. glucose-salt aqueous solution heated up to the human body temperature, when friction occurs under mixed lubrication regime, the ceramic-ceramic friction pair exceeds the ceramic-UHMWPE one by low friction criterion.

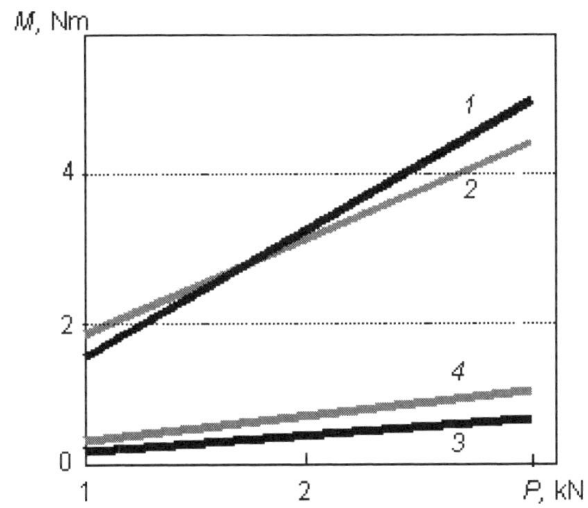

Fig. 5.8. Friction torque (*M*) in cup-head friction pairs vs. load (*P*): *1, 3* – BIOLOX®/ BIOLOX® friction pair; *2, 4* – BIOLOX®/UHMWPE; *1, 2* – unlubricated friction, *3, 4* – friction in Ringer's solution at 37°C

Owing to advances in technology the leaders in production of Al_2O_3–ceramics could decrease the wear rate of endoprostheses with ceramic-ceramic pairs down to 10–20 μm/year [25]. This is similar to the wear of zirconium ceramics heads whose commercial production is a complex and expensive process. By breaking stress ZrO_2–ceramics exceeds Al_2O_3–ceramics 2 and more times, which ensures reliable operation of small-diameter heads (22 mm). However, the ZrO_2–ZrO_2 pair has poor tribological characteristics and yields a great amount of wear debris. In the beginning of XX century best manufacturers of heads having the high-quality surfaces CeramTek (Germany), Saphirwerk Industrieprodukte (Switzerland), Ceraver Osteal (France), Matroc (England) reduced the prices of the Al_2O_3–ceramics heads down to those of cobalt-chrome ones.

Unfortunately, negative experience of application of ceramic heads, particularly some cases of their failure in implantation and operation of endoprostheses has repelled numerous orthopaedists from using ceramics on the whole.

The data cited above indicate that operation of implanted endoprostheses in biological medium differs essentially from that of natural joints under more severe friction conditions than on the rigs simulating *in vivo* conditions. This, being the matter of concern, encourages orthopaedists to improve tribological parameters of joint endoprostheses.

5.3 WEAR DEBRIS OF ENDOPROSTHESES

Analysis of wear debris occurring in operation of the tribosystem is a popular investigation method in tribology. It allows for establishing wear mechanisms and predicting engineering resource of friction units. Its significance in studying efficiency of endoprostheses increases since wear debris of endoprostheses initiate the resorption of the bone bed of endoprosthesis that results in its loosening.

In machine construction the methodology of studying wear debris has been developing beginning from 1970ies from ferrographic analysis (regulated deposition of steel wear debris suspended in the lubricating oil on the substrate by the magnetic field and optical control of the deposited debris) to more refined ones making possible distinguishing wear mechanisms by the wear products and simulating debris formation process [51]. Methods of on-line control of wear in machines and devices have been developed, among them the optical-magnetic detector that allows determining tribosystem wear index by express analysis of the worked-out oil [52].

In orthopaedics of late of 1970ies – early of 1980ies it was found out that pathological joints were worn abnormally. Severe wear of the joint cartilage is a symptom of arthritis and abnormal wear debris initiate joint inflammation [53]. Size, shape, and composition of wear debris of human joints characterize the wear pathology in which the debris are generated [54]. The efforts were undertaken to relate the visual evaluation of wear debris to the mechanism of their formation [55]. Phenomenological approach to wear debris analysis and classification of wear modes on the basis of wear debris formation were proposed [56]. A PC-aided analysis of synovial joint wear debris was worked out using fractal parameters of debris morphology [57]. It allows for overcoming disadvantages of visual observations yielding wrong findings about the spatial outline of wear debris. The authors believe that the method can be helpful in diagnosing and predicting arthritic diseases as more sensitive and less aggressive than arthroscopy.

Wear debris formation in endoprostheses in their long-term operation (about 10 years) was always the matter of concern of orthopaedists. It is particularly true about polymeric components of endoprostheses whose fatigue wear mechanism (Fig.5.7) gives rise to a great amount of UHMWPE wear debris (the wear rate reaches 0,29 mm/year [58]).

Efficiency of endoprostheses as wear debris generators depends essentially on the design parameters of the friction pair. Thus, in work [59] statistically justified data are reported that the number of wear debris recorded in the tissues surrounding the head of hip joint endoprosthesis turns out to be twice as large as near the cups fixed immovably in the acetabulum. In 1999, Japanese orthopaedists H. Oonishi and N. Amino found a correlation between thickness h of the UHMWPE liner and its wear rate v by an aluminium-ceramic head [25]:

h, mm	7	8	9	10
V, mm/year	0,15	0,12	0,08	0,06

However, other researchers have not revealed any similar correlations [58].

The correlation is found between the wear rate of the polymeric liner and the diameter of the endoprostheses metallic head. The studies have been performed for 9.5 years on 385 implanted hip joint endoprostheses [60]. The least wear of polymeric liners was observed in pair with heads 28 mm in diameter while endoprostheses with heads 22 mm in diameter demonstrated some higher wear. However, the difference was within an error. The highest wear was revealed in liners coupled with heads 32 mm in diameter (Fig.5.9).

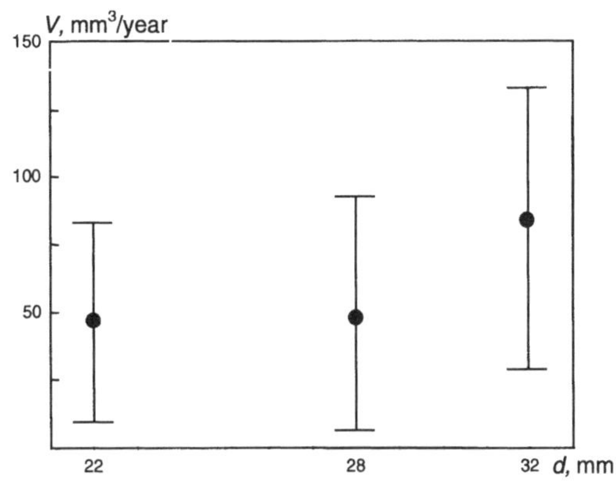

Fig. 5.9. Volumetric wear rate of UHMWPE liners vs. diameter of the metallic head of hip endoprostheses

The largest amount of UHMWPE wear debris is formed during the run-in of endoprostheses, which, as the authors of work [58] believe, goes on *in vivo* within 5 years. Then the wear rate of the liners decreases and the contamination rate of the joint capsule with polymeric debris are stabilized.

Deposition of wear debris in the tissues surrounding the endoproshtesis occurs since the periarticular lymphatic system does not manage to remove them.

Studies performed by K.M. Sherepo in 1990ies in CITO (Moscow) in revision operations on replacement of total metallic Sivash's endoprosthesis have brought him to the following conclusions. Comparison of wear results of endoprostheses *in vivo* and on simulator shows the run in of metal–metal tribojoints of implanted endoprostheses to continue not less than 3 years. The capsules around the necks of titanium-chrome endoprostheses have a pronounced brown colouring during 1–4 year period and then (5-15 years) they have deep black colouring sometimes yellowish. If the tribojoints is made of stainless steel the tissues are grey-blue or blue-brown (9–17 years). The metal content of capsules (found by emission spectral analysis) decreases with time insignificantly. This differs from the cited in 5.2 conclusion drawn by K. Zweimuller on exponentially decreasing metal concentration in blood serum of patients after completing the run-in of endoprostheses. The capabilities of the human body of neutralizing metallic wear products might be different in tissues and physiological fluids. Black friable particles imbibing (absorbing) in the surrounding tissue as well as "fields" of decomposed macrophages with fine black crumbs have been revealed histologically inside the cells and outside them.

An American orthopaedist W. J. Maloney while doing revision operations on 35 hip joints examined histologically the joint capsule tissues [59]. Wear debris were washed out from tissues sections by papain solution and then examined by optical and electron microscopy as well as X-ray microanalysis using an automatic particle analyzer. UHMWPE wear particles size was measured 0.5 μm and metallic – 0.7 μm on the average. More 90% of particles were less than 0,95 μm by size. Particle concentration was within 1.5–2.0 mlrd particles in 1 g tissue. Metallic particles were revealed in 46 % of studied tissue samples mainly taken in the vicinity of titanium alloy stems and less often in the vicinity of Co–Cr alloy stems. The titanium alloy wear debris was more often found in the vicinity of the porous coated stems than without. Fine (< 1 μm) particles were registered by X-ray microanalysis since conventional histological examination could not determine such a fraction of particles. It explains a widely spread in clinical practice under-estimation of endoprosthesis wear debris concentration in surrounding tissues.

The reaction of the organism to endoprosthesis wear debris was noticed by J. Charnley when he tested Teflon cups of endoprostheses and revealed their accelerated *in vivo* wear. He related the destruction of bone tissues surrounding the cup to infections resulted from Teflon wear debris [61]. This indicates that the clinical osteolysis problem (bone tissue destruction) is not new but was one of the first issues in joint endoprosthetics.

Pathomorphological studies provided a unique opportunity to investigate the structure of tissues contacted *in vivo* with the implant [62]. Wear debris proved to cause the organism reaction to a foreign body thus giving rise to characteristic soft-tissue capsule between the implant and the bone. Immune histochemical and biochemical studies have shown the capsule to consist mainly of macrophages and fibroblasts capable of secreting a variety of factors including cytokines and other metabolism products destructing the bone tissue. Therefore, the quantitative parameters of lysis of the proximal part of femoral bone correlate with the wear rate of the polymeric liner of the hip joint endoprosthesis [62]. Presently the relation between biological reaction of the organism to the endoprosthesis wear debris and osteolysis responsible for the instability of the latter is beyond question. Analysis of the tissues being in contact with endoprosthesis, study of cellular and tissue samples, and experiments with animals have given grounds for such relation to exist.

Many details of the process need further understanding. The capability of wear debris of initiating granuloma formation and encouraging cell metabolism is not differentiated by the nature of the worn materials. The mechanisms responsible for the kind and spread of osteolysis due to individual features of the organism are not found. It is still not understood how granuloma formation depends on the weight, composition, and size of wear debris as well as mechanisms of their delivery to the bone-implant interface.

A mode of endoprosthesis fixation influences essentially the resorption of the bone tissue. W.H. Harris found the correlation between the localized osteolysis of the femoral bone and instability of cemented endoprosthesis [63]. Cement debris arise from breakage of the cement mantle under alternating endoprosthesis loading. As they appeared on the friction surface and the bone-implant interface they initiated intensive wear of endoprostheses and lysis of the bone tissue. The phenomenon of bone damage received the name "cement disease" [64].

Correlation between the cement mantle damage and bone resorption at the bone-cement interface was studied in work [65]. When studying sections of the bone tissue being in contact with cement it was found that fine (1–12 μm) cement particles were the crucial factor for bone resorption. They are subjected by phagocytosis (active seizure and absorption) with macrophages. The *in vivo* experiments carried out on three kinds of PMMA-cement couples have shown the macrophages to release newly generated necrosis factors which "start" the bone damage mechanism. The result can give scientific grounds for developing pharmacological agents suppressing necrosis factors and seeking bone cement materials being an alternative to PMMA.

"Cement damage" of bones has become one of the reasons for developing cementless fixated endoprostheses. However, it soon became clear that the term was incorrect since cement fixation evidently enhanced the bone resorption but was not the main cause for osteolysis. The latter accompanies all modern cementless endoprostheses, which is found out within first two years after replacement. With time X-ray signs of osteolysis around endoprostheses are becoming ever more evident.

Meloney studied pelvic bone osteolysis after cementless implantation of cups with UHMWPE liners [66]. All used standard endoprostheses had holes or gaps over which the wear debris could migrate to the cup-bone interface. The correlation between the availability of the holes in the UHMWPE cup and migration of its wear debris is reported in work [48]. The time between the operation and pelvis osteolysis identification was 53–84 months (65 months on average). Many patients felt good and the estimation of their artificial joint functioning by Harris scale (Table 1.2) exceeded 90 points. Nevertheless, the liner wear and osteolysis were revealed by X-ray examination and in some cases wide destruction of the pelvic bone was reported. Histological analysis made possible revealing numerous macrophages in joint capsules subjected to revision operation. UHMWPE wear debris was present in all joints investigated. Intraoperation examination of joints indicated underestimation of bone losses determined by X-ray analysis. This supports the necessity for regular checking of the patients after endoprosthetic operation by using good-quality X-ray apparatuses.

Comparative evaluation of osteolysis degree accompanying implantation of cement and cement-less hip joint endoprostheses was undertaken in work [67]. At cement fixation of the cups the authors attribute linear propagation of osteolysis pelvis damages to the loosening of implants. Pelvis damages at cementless fixation of the cups are characterized as localized and extensive. They occur more rarely than at cement fixation and are not associated with the loosening of the endoprosthesis components. However, cementless fixation of cups causes more extensive development of osteolysis in the femoral bone than cement fixation. At any fixation of endoprstheses, osteolysis is localized in the zones of wear debris accumulation (Fig.5.10). As was noted earlier [28] hydroxyapatite coating of the femoral component reduced the occurrence of osteolysis and eliminated distal osteolysis at 5–10 years of follow-up.

The following types of bone tissue reaction to cementless fixation of endoprostheses (2–4 year post-operation period) are distinguished [68]. In the metaphysis part of the femur the bone is commonly integrated with alloy Ti-6Al-4V stems. This is due to bone fusion caused by primary stable press-fit of the stems and biocompatibility of titanium alloy. In the proximal part, two types of reactions were revealed by histological examination of the tissues being in contact with stems: a connecting capsule appeared from one side of the stems, and, from the other – biological reaction ran directly in the bone-metal contact.

In this connection there are two questions to be answered: 1) which of two reactions is most favourable for providing long-life stability of endoprosthesis, 2) whether the type of reaction is related to migration of the wear debris to the bone-implant interface, and 3) whether the modification of friction unit design cause the prevalence of this or that type of reaction?

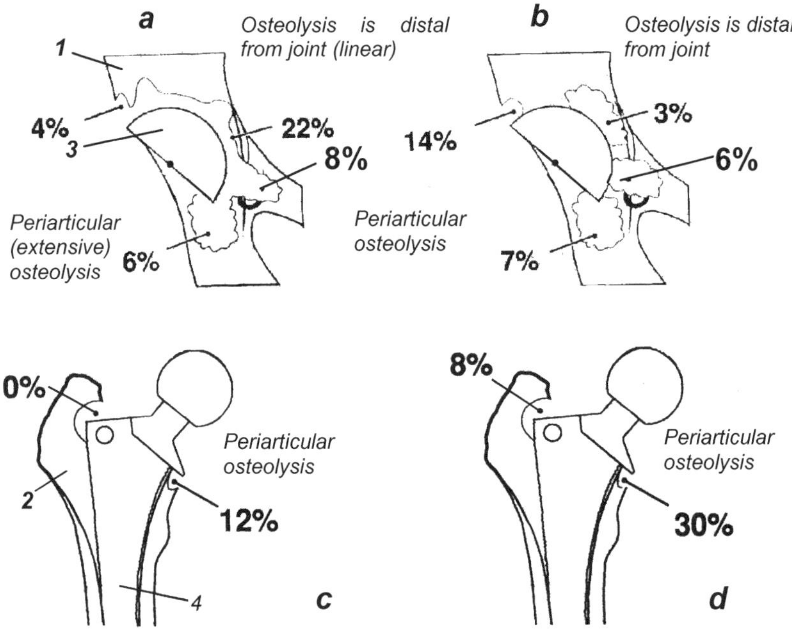

Fig. 5.10. Probability of osteolysis (%) around cups (*a*, *b*) and stems (*c*, *d*) of endoprostheses with cement (*a*, *c*) and cementless (*b*, *d*) fixation: *1* and *2* – pelvic and femoral bones, *3* – cup, *4* – stem

Thus, an up-to-date idea of "starting" biological mechanisms of aseptic instability of joint endoprostheses is based on the fact that inflammation in the tissues surrounding the endoprosthesis is the response to "the contamination" of tissues with endoprosthesis wear debris that integrates the cell reactions resulting in osteolysis. Osteolysis initiates instability of fixation of endoprosthesis components, which is intensified by the development of fatigue processes in the zones of fixation of the components in the bones. One can assume that an advanced level of endoprosthetics would be based on new principles of friction unit designing that make possible utilization of wear products and application of antifriction materials whose dispersion does not cause tissue inflammation.

5.4 ANALYSIS OF REMOVED ENDOPROSTHESES

Revision operations give us a unique opportunity to correlate tribological damages of joint endoprostheses *in vivo* with osteolyisis propagation and endoprosthesis loosening. In 1990, the orthopaedists of CIS-countries were forced to use low quality home-made endoprostheses. They were not unified and did not match international standards. In clinical practice, most of them proved to be unreliable and were removed in revision operations. The analysis of errors made in their development and implantation gives us an opportunity to follow up post-operation complications due to tribological conditions for endoprosthesis functioning.

The total endoprosthesis of the hip joint manufactured in Ivano-Frankovsk (Ukraine) is made of stainless steel of austenite class 12X18H10T (according to the results of metallographic analysis and X-ray spectrometry). A conical neck and a support ring are welded to the hardened stem at angle 135^0. A ball head is inserted on the neck taper. The pelvic component of the endoprosthesis shaped as a polymer cup was fixed in the acetabulum with the bone cement (Fig. 5.11). Aseptic inflammatory process and instability of the pelvic component were indications for the revision operation.

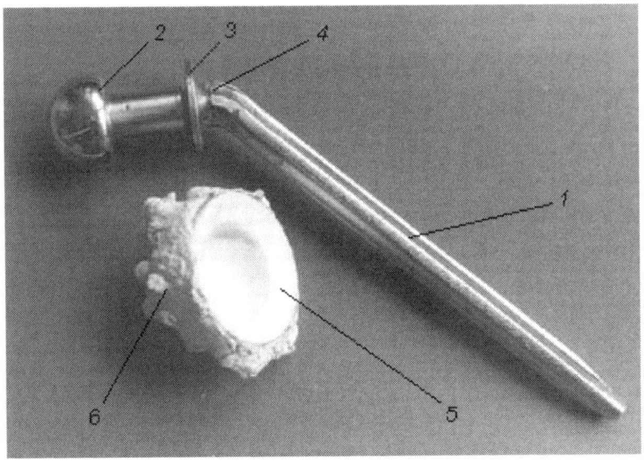

Fig. 5.11. External view of endoprosthesis manufactured in Ivano-Frankovsk: *1* – stem, *2* –head, *3* – ring, *4* – welding joint, *5* – polymer cup, *6* – cement mantle

The inflammatory reaction of soft tissues to the foreign body and extensive osteolysis of the hip bone around the cup that results in its loosening were noticed when removing endoprosthesis. An insignificant wear of the cup (Fig.5.12) made us to conclude that the run-in of the endoprosthesis tribosystem had been completed within a year after primary endoprosthetics.

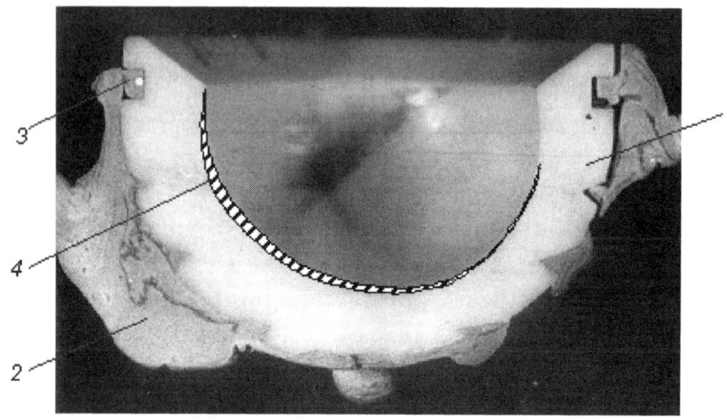

Fig. 5.12. Cross-section of the pelvic component: *1* – polymer cup, *2* – cement mantle, 3 – metallic ring for X-ray evaluation of the cup state, *4* – worn part of the cup

The polymer cup was studied by IR-spectroscopy and DTA. Clear characteristic peaks of amide groups were revealed on the IR-spectra. Fitting spectra of the cup material and the reference one corresponding to caprolone-V has given us grounds to identify the cup material to this proper polyamide. This was confirmed by the melting peak at 220–225 °C available on the DTA curve for the cup material, which coincided with the melting temperature of caprolone-V. Visual examination proved the cup to be produced by turning from a block workpiece and have characteristic signs of a cutter. The manufacturing technique is typical of caprolon-V.

World-wide medical practice refused from using polyamides in tribojoints of endoprostheses more than 40 years ago. As is noted in 2.3 and 5.2, the polyamide wear debris had lead to chronic aseptic inflammation of soft tissues and lysis of the bone tissue. These damages were revealed in removed Ivano-Frankovsk endoprosthesis. Periarticular osteolysis was found in the femoral bone around the welding on the stem. Ca salt deposit was revealed in the welding zone. This indicates the difference in the electrode potentials of the stem and the welding joint as well as electrochemical reactions running on the implant surface with biological fluids participation.

Endoprosthesis "Foenix" (St.-Petersburg) consists of four parts (Fig.5.13)

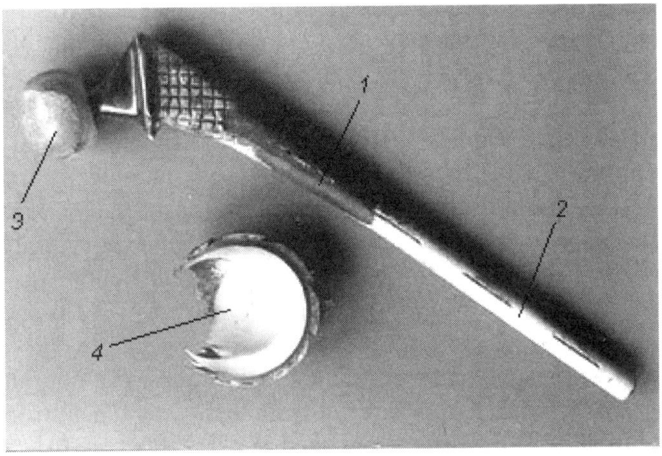

Fig. 5.13. Endoprosthesis "Foenix" removed in revision operation: *1* – stem, *2* – bone tissue layer, *3* – ceramic head, *4* – polymeric liner, *5* – metallic cup

One end of the metallic stem (titanium alloy VT-5 by the data of X-ray spectroscopy) is a cylindrical rod with the grooves for the bone to grow through. The other end is provided with an oval collar transforming into the conical neck with a ceramic cap on it. It matches the polymeric liner mechanically fixed in the metallic screwed cup. It is seen that the liner has experienced catastrophic wearing. Even the metallic cup is worn with the distortion of its initial geometrical shape. These points to the fact that the patient was not subjected to X-ray or clinical examination after arthroplasty.

The analysis of the endoprosthesis materials makes grounds for the following conclusions.

After removing the cementless stem from the medullar canal of the femoral bone, a layer of the sponge bone tissue remains on its distal part (is seen in Fig.5.13). Clear peaks are seen on X-ray spectra taken from the proximal stem having the metallic lustre that correspond to Ca, K, and Cl. This indicates the growth of the bone tissue into the titanium oxide layer on the stem surface [69].

The oscillation bands of CH_2-groups typical of high density polyethylene (HDPE) are identified on IR-spectra of the polymeric liner. They are relevant to the crystalline and amorphous HDPE phases. The melting temperature found by endothermic peak maximum on the DTA curve corresponds to 128–129° C. It is a strong evidence of initial suggestion that the liner was made of some other material rather than UHMWPE. As was mentioned above, already in the beginning of 1960es, J. Charnley abandoned HDPE in favour of more wear-resistant UHMWPE. A low wear resistance of the liner and no control over endoprosthesis condition resulted in participation of the metallic cup in friction after the total wear of the liner. This caused catastrophic wear of the ceramic head and cup. In line with up-to-date views on biological compatibility of friction units [16, 29, 50], the ceramic-metal pair, especially with the titanium counterbody, is not recommended for usage in endoprostheses. During catastrophic wear of the pair a great amount of wear debris was formed that were withdrawn from the friction zone and deposited in the tissues forming a "titanium granuloma".

Figure 5.14 presents the microphotograph of the worn-out liner edge. One can suggest that the wear particles result from their detachment from the fibrous structures (called in [40] fibrils) originated under friction and shear stresses in HDPE. The surface of the worn part of the cup (alloy VK-1) was work hardened as a result of its rubbing against the ceramic head and impregnated by abrasive wear particles.

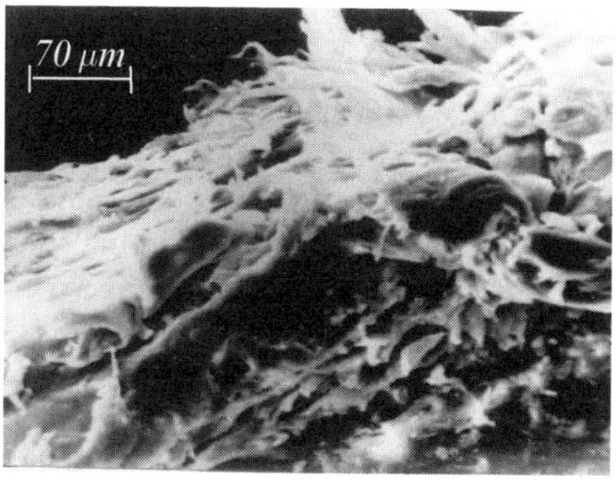

Fig. 5.14. Structure of the worn part of insert *4* (See Fig 5.13) prior to formation of HDPE wears debris

The head structure was studied by the elemental X-ray spectral analysis. Its main component is Al_2O_3, however, there are lines corresponding to silicates, alkali oxide, and calcareous inclusions on the spectra. These admixtures form the hyaloid phase with a low chemical stability, which decreases the mechanical strength and durability of ceramics. The head material did not meet stringent requirements of ISO 6474 concerning the purity of alumina ceramics for medical implants. There are lines on the spectra taken from the head friction surface corresponding to titanium compounds. This indicates the friction transfer of titanium which gives rise to "titanium granuloma".

Accumulation of a great amount of polymeric, metallic, and ceramic wear debris in tissues caused acute inflammations leading to lysis of acetabulum and proximal part of the femur. This resulted in subluxation and then to luxation of the cup followed by revision operation.

Endoprosthesis "Arete" (St.-Petersburg) had a design similar to endoprosthesis "Foenix" (Fig.5.15). When implanting pelvic component the metallic cup was screwed into the acetabulum and the polymeric insert was set in it. The signs of bone integration in the form of the white layer of the spongy bone on the cup surface facing the pelvis testify a reliable fixation of the pelvic component. The stem was fixed in the bone marrow canal of the femoral bone with the cement compensating the lack of the bone tissue in the proximal canal. The cement mantle was formed only in the upper part of the stem rather than over the whole length. Aseptic instability of the endoprosthesis stem was an indication for the revision operation.

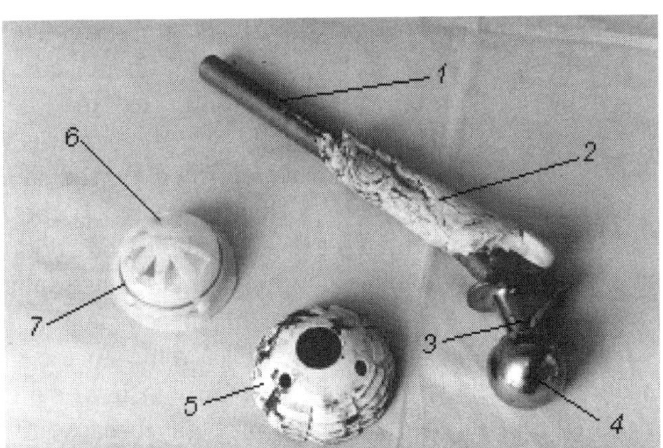

Fig. 5.15. Endoprosthesis "Arete" explanted in revision operation: *1* – stem, *2* – cement mantle, *3* – neck, *4* – head, *5* – cup, *6* – insert, *7* – metallic ring

By the data of IR-spectroscopy and DTA the endoprosthesis liner was made of HDPE. It is shown in Fig. 5.16 that the liner wear was close to ultimate after a 3-year operation. One can suggest that polyethylene wear debris appearing on the cement-bone interface caused lysis of the bone tissue of the femoral bone. The process was aggravated by the effect of "cement damage" of the femoral bone. Rapidly progressing instability of the stem stipulated the necessity of the revision operation.

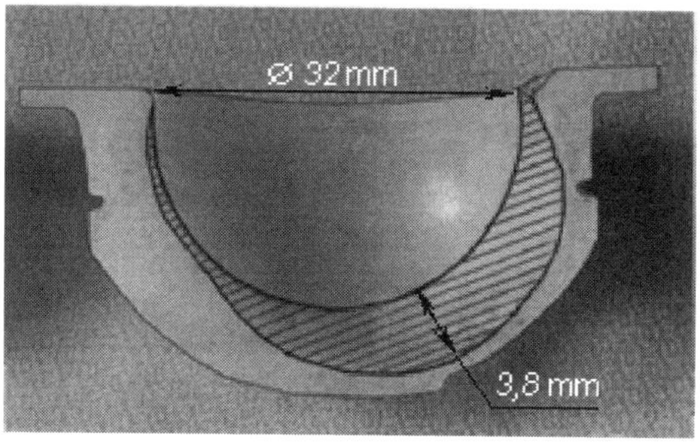

Fig. 5.16. Cross-section of the worn polymeric liner. The cross-section of the worn part is dashed

Mechano-chemical (corrosion) wear of joint endoprostheses occurs when electrochemical similarity of their parts is not provided. Figure 5.17 shows the endoprosthesis removed in revision operation caused by mechanical failure of the femoral component. According to the certificate the endoprosthesis was made of certified materials, viz. the cup was made of pure titanium, the liner – UHMWPE (trade mark Chirulen P), the stem – alloy Ti-6Al-4V, the porous insert on the stem – titanium powder, the head – alloy CoCrMo (Endocast SL). Despite the fact that the implanted endoprosthesis has served less than a year the visual examination testified catastrophic wear of the metallic head whose initial spherical friction surface acquired a shape of polyhedron. The wear of some spots exceeded 1.5 mm. At the same time, the polymeric liner did not have any signs of wear. Only insignificant squeezing of the polyethylene by the head was registered probably because of the errors in endoprosthesis assembling. This gives us grounds to assume that the head was subjected to mechano-chemical wearing.

Presently, a simplified understanding of mechano-chemical wear as a two-stage process of corrosion products formation on the friction surface of metallic parts and their detachment under mechanical effect of the counterbody has been developed. A friction pair is a multi-electrode system because of different energetic states of newly formed and film-coated areas of the friction surface, contact discreteness, micro-heterogeneity of the structure of the friction metallic material. Electrode potential φ of the part under attrition is the basic characteristic of the friction pair [70].

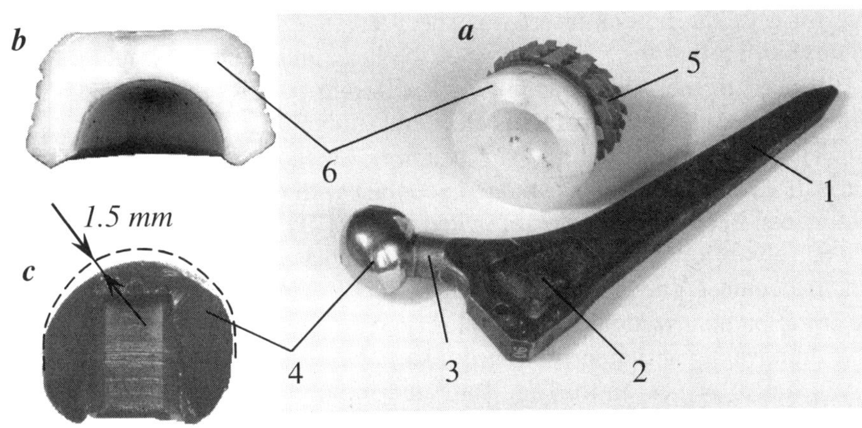

Fig. 5.17. External view (*a*) and cross-sections of liner (*b*) and head (*c*) of electrochemically heterogeneous endoprosthesis: *1* – stem, *2* – porous insert for bone integration, *3* – neck, *4* – head, *5* – cup, *6* – liner

The results of searching for electrodes in the implanted endoprosthesis that form an electrochemical cell whose electromotive force causes the mechano-chemical wear of the head are given in Fig.5.18.

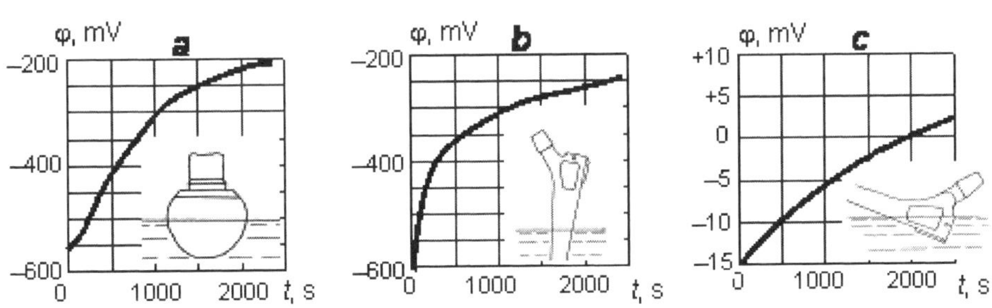

Fig. 5.18. Kinetic dependencies of electrode potentials and diagrams of immersion of endoprosthesis parts into electrolyte (0,9 %- aqueous solution of NaCl): *a* – head, *b* – distal stem, *c*– porous insert

It is seen that the stationary electrode potentials of the head and the wedge-shaped part of the stem are essentially equal (–200 and –230 mV), whereas φ of the proximal porous-coated stem (*c*) corresponds to +5 mV. Since kinetic curve $\varphi(t)$ in Fig. 5.18, *c* did not show the saturation, the steady-state value for φ of the porous insert was specified by extrapolation, viz. +10 mV. Even expecting that the values for φ of the head and the stem would be close, we still need further understanding of a different polarity of the head and the porous insert made of the materials close by nature.

Titanium is close by φ to stainless steel. It is capable of strong polarization, and serves as a cathode in contact with most metals. Nevertheless, titanium can intensify corrosion of mated metals [71]. Thus, at a low velocity of the electrolyte flow ($v < 0.15$ m/s), titanium increases the corrosion rate of the steel in contact (negative protective effect). This can be caused by a concentration polarization of the titanium parts of endoprosthesis.

Immovable and poorly mixed electrolytes that surround the implant contain much oxygen so they are saturated rapidly by the products of anodic reaction. The removal of the latter and delivery of fresh electrolyte to the implant are impeded. The electrode potential of the titanium insert depends on more numerous factors. Electrochemical heterogeneity of the insert is due to macroelements of the "slit – open surface" type in which the edges of the slits work as cathodes and their middle part – as an anode [71]. Apparently the porous insert located in the essentially immovable electrolyte (the implant-surrounding liquid) has a pronounced electrochemical heterogeneity. It exerts a negative protective effect thus considerably increasing the rate of contact corrosion of the endoprosthesis head.

Hence, a closed electrochemical circuit operated in the implanted endoprosthesis (Fig.5.19) in which the head (φ_1 = –200 mV) and the porous insert (φ_2 = +10 mV) were electrodes and the liquid contacting the implant functioned as an electrolyte.

Only a part of the spherical head contacting the insert is involved in friction at implanted endoprosthesis operation. The nominal contact area at which mechano-chemical wear is localized is less than a half of the sphere area. Formation of the wear spots on the head that make it polygon-shaped can be attributed to the fact that the head-insert contact spots were located differently under different physiological loads (when sitting, standing, walking, etc). A paradox situation when the polymeric liner of the endoprosthesis was worn more slowly than the ball head made of a noble metallic alloy resulted from an unfavourable combination of metallic materials in the endoprosthesis.

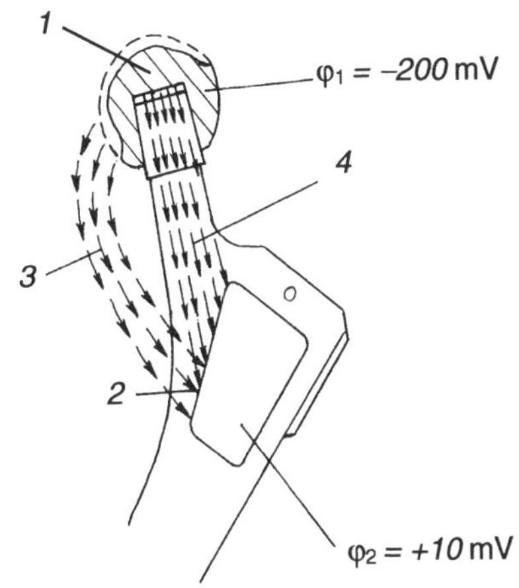

Fig. 5.19. Diagram of electrochemical dissolution of the head of implanted endoprosthesis: *1* – head surface on which metal oxidation occurs; *2* – porous insert; *3* – flow of metal ions; *4* – flow of electrons; φ_1 and φ_2 – electrode potentials of the head and the insert, respectively

It follows that electrochemical homogeneity and wear resistance of joint endoprostheses as well as biological inertness of their wear debris are the main factors governing long-life efficiency of artificial joints. Formation of wear products "unfriendly" in respect to the organism causes chronic inflammations of soft tissues and lysis of bone tissues. It gives rise to progressing aseptic instability of the modern joint endoprostheses and is the main cause of their failure.

5.5 TRIBOLOGICAL TESTING OF ENDOPROSTHESES

The data given in this chapter indicate that tribological state of joint endoprostheses essentially determines their serviceability *in vivo*. Therefore producers of endoprostheses since J. Charnley's times have paid a great attention to the investigation into friction and wear processes in endoprostheses. In the modern medical engineering a 4-stage system of tribological testing of joint endoprostheses is accepted:

- tribological tests of materials under usage in endoprosthetic friction units by the schemes and standards accepted in machine construction;
- laboratory tests of the same materials in unconventional friction units simulating frictional loading of natural joints;

- simulator tests of joint endoprostheses using test machines simulating the motion and loading that implanted joint endoprostheses experience;
- clinical tests of implanted enodprostheses.

The experience gained in tribotesting of joint endoprostheses [74] has shown the advisability of their performance in "clean" rooms equipped with sluices and boxes that create a laminar air flow above the test friction unit. It is aimed at decreasing the effect of impurities contained in the room air on the test results. To avoid artefacts (processes that are commonly uncharacteristic of the object under investigation and arising in its study) the lubricant is sterile-filtered by membrane filters with 0.2 μm – fineness and the test friction unit is sealed by a polyurethane membrane. A buffer stabilized mixture of Ringer's solution (pH = 7.2) with calf serum (30 %) is recommended for usage as an endprosthesis lubricant in the experiments.

Tribological tests of materials are conducted in laboratories under accelerated conditions, i.e. under more severe conditions for contact pressure, velocity, or temperature, to reduce the initial stage of materials selection. The main goal of the tests is to determine principle suitability of the materials for usage in the endoprosthesis friction unit.

Laboratory tests are carried out on commercial friction machines. A disc-rod friction pair with the discs made of harder materials are used most often in the initial assessment of friction materials for endoprostheses. The test conditions are identified to have an opportunity to compare the results obtained in different laboratories. For this reason, the criteria and requirements of ISO TR 9325 and 9326, ASTM F 732, Russia State Standard 23.204-78, 23.215-84 and 11629-75 have been determined.

Usage of multi-positioned friction machines allows for augmenting the database that provides the statistical validity of the results. Simplification of the friction conditions compared to those available *in vivo* certainly makes the conclusions based on measurement results less valuable. Nevertheless, laboratory tests are simulating tribological behaviour of materials under conditions rarely met *in vivo* but useful to predict failures of artificial joints. We will show a number of the results being of evident interest for orthopaedists.

Figure 5.20 presents the parameters of the CoCrMo–UHMWPE pair versus lubrication conditions and sliding velocity [16]. Increasing sliding velocity is seen to result in exponentially decreasing wear rate. Under unlubricated friction the friction coefficient passes through a minimum. Under lubricated friction with increasing velocity hydrodynamic effects may occur that decrease the UHMWPE wear rate by an order and the friction coefficient – by half. In the experiments a standard deviation of some results from their average value is within 10–50 %.

A decrease in UHMWPE wear by 30–75 % is registered in the Al_2O_3 ceramics–UHMWPE couple in comparison to metal-UHMWPE couples, all other conditions being the same. It is interesting that according to the results of laboratory tests with irregular friction pairs the above value is only about 20 %.

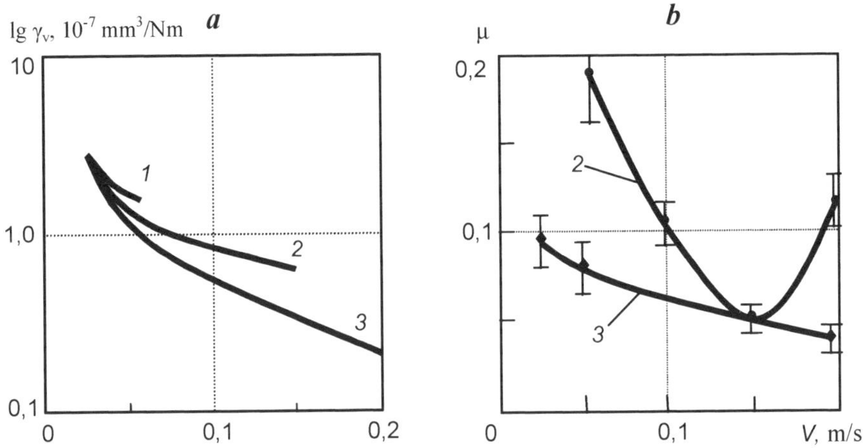

Fig. 5.20. Wear rate of UHMWPE (*a*) and friction coefficient (*b*) in the metal-UHMWPE couple vs. sliding velocity: *1* – unlubricated steel, *2* and *3* – CoCrMo without lubrication and under serum lubrication

Irregular friction pairs are tested to simulate the scheme of frictional loading available in the artificial joint. A very popular in orthopaedics pendulum tribometer used even by J.Charnley is one of the examples. The pendulum represents motions in the hip, knee, and some other joints in a simplified form. It provides high sensitivity and accuracy of tribological measurements since it comprises only one friction pair, viz. that under testing (Fig.5.21). The specimen is secured to the tribometer base. The pendulum indenter resting on the specimen and shaped as cylinder, sphere, or bar with a curved support surface serves as a counterbody. A kinetic dependence of the pendulum oscillation amplitude is recorded by which the friction coefficient in its support is found. The linear dependence is characteristic of unlubricated friction or boundary lubrication friction and the exponential dependence happens when friction parts are deformed plastically or the lubricant has high viscosity.

The experiments involving a friction pair in the form of "ball-shaped pendulum indenter – support with a spherical hollow" were carried out for CoCrMo–CoCrMo, CoCrMo–UHMWPE, and Al_2O_3 ceramic–UHMWPE combinations. All of them are characterized by a linear mode of pendulum damping when lubricated by blood serum. It is seen in Fig. 5.22 that the minimal friction occurs in ceramic–UHMWPE friction pair and maximal friction is characteristic of the metal–metal pair. Reduced head diameter causes decreasing friction.

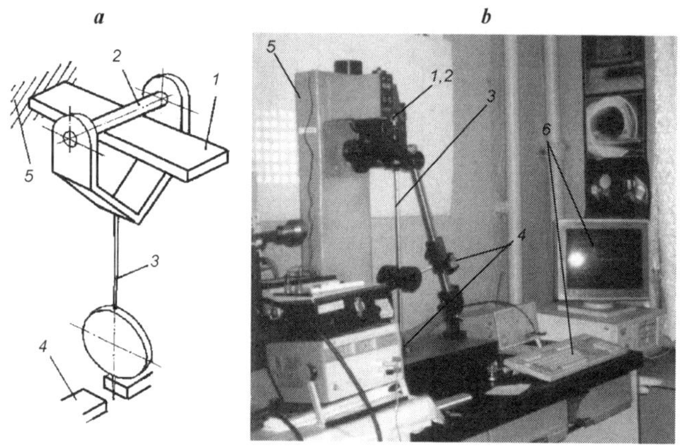

Fig. 5.21. Diagram (*a*) and external view (*b*) of pendulum tribometer: *1* – specimen, *2*– counterbody, *3* – pendulum with weight, *4* – oscillation gauge, *5* – frame, *6* –recorder

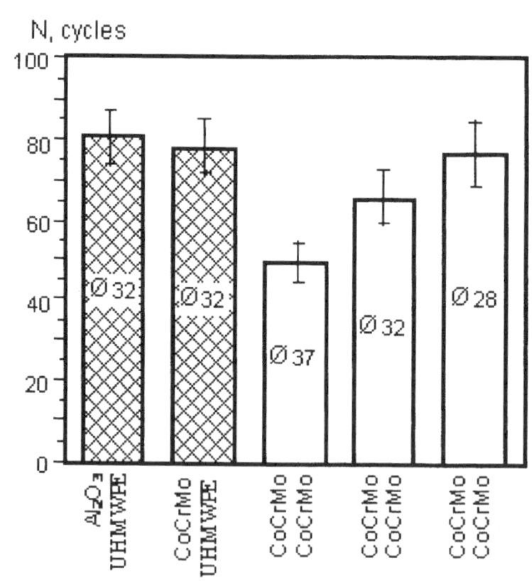

Fig. 5.22. Maximal number of cycles of pendulum oscillations up to its stop vs. combination of indenter and support materials and the diameter of the ball indenter of pendulum tribometers. Lubrication by Ringer mixture solution with calf serum

The friction in hip joint endoprosthesis is simulated in Bialystok Technological University (Poland) by a tribometer whose diagram is shown in Fig. 5.23.

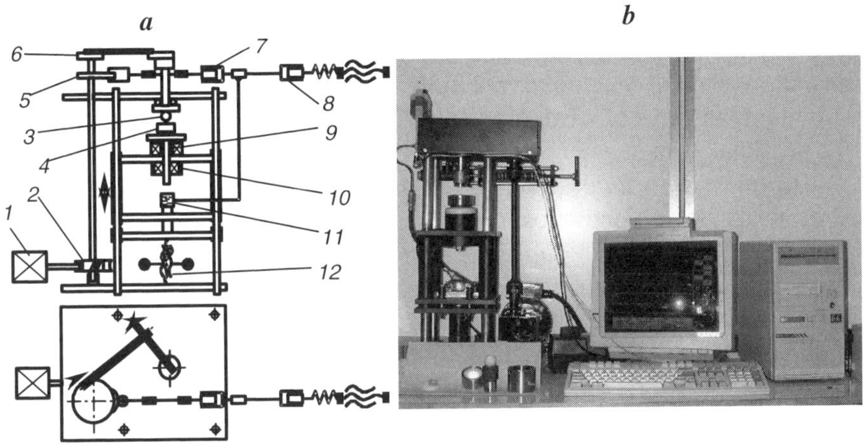

Fig. 5.23. Diagram (*a*) and external view of the tribometer (*b*) operating under impact cyclic loading: *1* –engine, *2* –transmission, *3*and *4* – specimens simulating the endoprosthesis head and cup, *5* – cam, *6* – crank gear, *7* – hydraulic pump, *8* –load regulator, *9* and *10* – strain gauges, *11* – hydraulic plunger, *12* – lifting device

Specimens simulating the head and the cup of endoprosthesis are secured to a table and the tribometer spindle. The loading cycle consists of application of a normal specific load N = 1.2 kN, head turning by 50°, its returning to the initial position under the same load, and removal of the normal load. The cycle continues 1 s. The tests can be carried out without or with the lubricant. The computer unit serves to record kinetic dependencies of N and friction torque during head turning.

When choosing the irregular procedure for the laboratory tests the following rule is observed [16]: reduced number of the parameters to be studied and simplification of the test procedure lower the significance of the indices specific of orthopaedics (volume of wear debris, ways of their migration, etc.) but increase the reliability of general ones (wear velocity and intensity of endoprosthesis part, etc.). To obtain reliable information on tribological parameters of endoprosthesis statistical processing of the results and performance of the complete test program including simulator tests are needed.

Simulator tests are carried out using joint endoprosthesis designed for implantation as specimens. The simulator is a test rig equipped with mechanical, pneumatic, and hydraulic drives by which motion and loading range is specified for the endoprostheses similar to those they experience *in vivo*. It is important that the endoprosthesis position in the simulator's clamps corresponds to the anatomic position of the implant. It makes possible determining the ways of wear debris migration and their effect on the efficiency of the artificial joint.

Hip joint simulators Stanmore Mk III specify the cyclic load with two loading peaks to the endoprosthesis. It corresponds to a natural cycle of leg motion when walking [73]. This is achieved by respective kinematics of driving cams. The test duration on the basis of several million loading cycles was 2–6 months. It is believed that the wear of the hip joint endoprosthesis after 1 million loading cycles is identical to the wear of the endoprosthesis in its operation *in vivo* for a year. When testing on Stanmore simulator the endoprostheses comprising CoCrMo–CoCrMo friction pairs, the wear of each component during the run-in period (0.5–1.0 mln cycles) achieved 20–30 μm and then decreased down to 2–4 μm / mln cycles. The wear rate set at this stage corresponded to a maximal wear rate of the endoprosthesis functioning *in vivo* for several years [16]. After testing on the simulator the wear groves having a certain direction appeared on the endoprosthesis friction surface. It is not characteristic of the endoprostheses removed in revision operations.

Professor D. Dowson, Leeds University (England), used the hip joint simulators equipped with an electrohydraulic loading system [74]. The loading cycle was produced by applying to the endoprosthesis the forces directed normally to each other, viz. vertically (3 kN), reciprocating (0.525 kN), and to centre – outside (0.525 kN). The figures in the brackets show the peak loads in a cycle. Under the effect of the loads the endoprosthesis was subjected to flexion-extension (from +35° to –20°), abduction – adduction (±7°), and rotation (±7°). The frequency of loading cycle repetition was 1 Hz. The simulator was equipped with a coordinate measuring device to record the cup wear.

In work [75] the test results are reported of hip joint endoprosthesis with the ceramic-ceramic (Al_2O_3) friction pair. The simulator [76] simulated loads and motions of the endoprosthesis components common for walking. Five endoprostheses were tested simultaneously on the simulator. The loading cycle consisted of extension-flexion by 60° and loading with a vertical force 3.5 kN. "The support phase" during which the loading took place was 2/3 of the cycle duration, "the oscillating phase" simulating the support-free position of the leg (the load was absent in this position) was 1/3.The loading frequency was 65 cycles per minute. The endorpostheses were tested under water lubrication. To simplify the water supply the endoprostheses were fixed in a cup-down position (Fig.5.24). The test basis was 5 mln cycles, the temperature was 37± 1 °C.

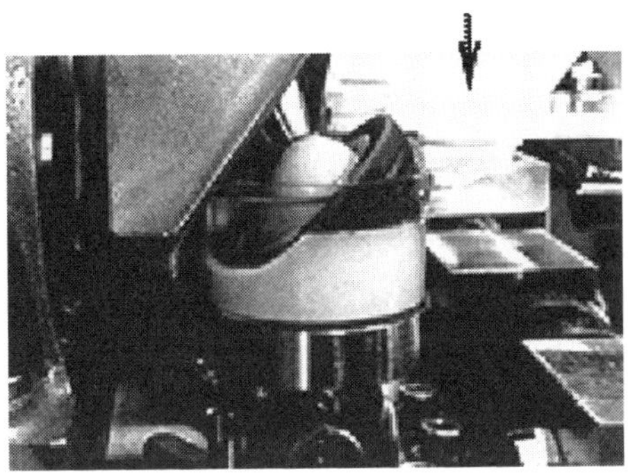

Fig. 5.24. Endoprosthesis installed on a simulator in the cup-down and ultimate "flexion" position. The arrow shows the reservoir with water to automatically fill the cup

Visual examination of the friction surfaces and weighing of the endoprostheses did not reveal any signs of wear. Rare microcraters (~1 μm by size) were registered by the scanning electron microscopy that resulted from wear particle pitting and unlikely directed scratches formed by the particles moved in the clearance of the friction pair. Since the endoprosthesis was fixed in the simulator in the upturned position the wear debris were accumulated in the cup's bottom affecting it abrasively.

Knee joint simulator [77] provides flexion-extension of the endoprosthesis (60° maximal flexion) under specified loads and sliding velocities

German sledge knee joint endoprosthesis «Schublade» consists of the UHMWPE part fixed in the tibia and the CoCrMo part fixed in the femoral bone. Its friction in blood serum (37 °C) has shown [16] the results presented in Fig. 5.25. It is seen that after run-in the steady-state wear rate of UHMWPE is 60 μm / mln cycles. During the run-in, the metallic counterbody with the ceramic nitride coating produces less wear debris than the uncoated one.

A simulator of model 1115 Instron Testing Machine [78] allows for evaluating the stability of unconstrained endoprostheses of the knee joint in any directions within the loads corresponding to normal walking. The tool allows simulating subluxation of the artificial joint at beyond-ultimate displacement in all directions of the components of the tibia and femur as well as recording the friction torque in rotation of the endoprosthesis components about the vertical axis.

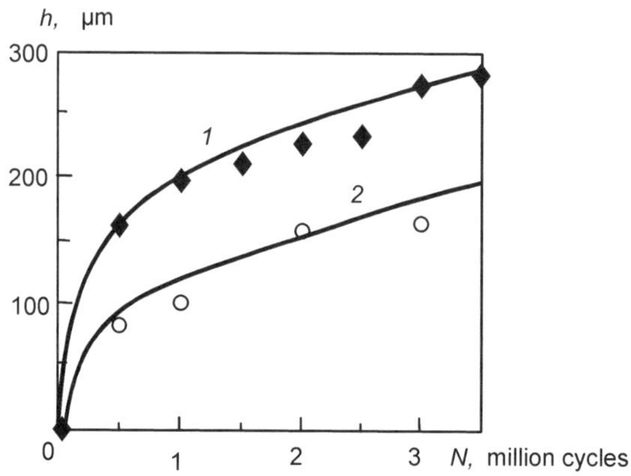

Fig. 5.25. Linear wear of the buttress UHMWPE plate of sledge endoprosthesis of the knee joint vs. number of testing cycles and the conjugate part material: *1* – CoCr, *2* – the same coated with titanium nitride

A «TE 89 Hip and Knee Friction Simulator» has been developed in England for friction and wear testing of endoprostheses of hip and knee joints [79]. The tool comprises two fixing blocks (for each kind of endoprostheses) which are connected to the electromagnetic drive. The load is recorded by piezoelectric gauges. The experiments can be carried out without and with lubrication keeping the endoprosthesis in a reservoir filled with water, synthetic or synovial fluid.

Clinical testing of implanted endoprostheses is a unique exploration whose results are extremely valid. Nowadays the only method that allows for actual measuring the *in vivo* wear of endoprostheses is the X-ray analysis. The results of X-ray analysis have a large scatter since the resolving capacity of clinical X-ray apparatuses is about 0.2 mm and positioning of the patient in front of the screen as well as the X-ray image analysis affect greatly the results. As a rule, the result of direct measurement of the wear after endoprosthesis removal in revision operation turns to be higher than the result of the X-ray analysis made before the operation. Thus, a 10-year monitoring of *in vivo* operation of 134 UHMWPE cups of hip endoprostheses conjugate with the CoCrMo heads (32 mm- diameter), has shown the wear rate 0.23 mm/year and total volume of the worn UHMWPE 180 mm^3. Direct measurement of the removed cups shows that really the wear is by 25 % greater [16]. Three zones can be distinguished on the surface of the spherical cavity of the cups, viz. the zone of load application with the polished surface; the unloaded, and the transition one.

Examination of endoprostheses removed in revision operations with the metal-metal friction pairs has shown the following. During operation for 25 years, the average wear rate of components (head and cup) was 2–8 μm/year. With increasing service life the roughness of endoprosthesis friction surface decreases since the scratches arising in the run in period and from abrasion by osseous-cement particles are polished. The wear is accelerated significantly if the spherical surfaces of the endoprosthesis friction surfaces are produced inaccurately [16].

Thus, the main goal of the tribotesting of endoprostheses is forecasting their wear rate *in vivo*. The tests allow for determining when the run-in finishes and operation of endoprostheses under steady-state wear conditions begin. It is needed to quantitatively assess wear debris arising in these periods and find out the regularities for their migration in artificial joints. Still there are issues to be solved, i.e. the absence of convenient and valid methods for registration of *in vivo* wear of endoprostheses and impossibility to control the migration of wear debris so as to avoid their appearance at the bone-implant interface.

◈ ◈ ◈

Summing up the results of achievements made in the tribology of artificial joints we should admit that just a few methods have been involved for solving a global tribological issue whose name is instability of joint endoprostheses. The hardest and strongest engineering materials as well as reinforcing coatings are now used in orthopaedics. Biocompatible and wear resistant polymers alternative to UHMWPE are being searched for. The tribological parameters of UHMWPE are improved by cross-linking and filling with wear-resistant components and with recently-avaiable nanoparticles [80]. Lubrication is improved by the methods adopted in machines [30]. Artificial synovia fluids containing hyalurone acid and liquid-crystalline components such as “Diasinol” [13, 14], “Hyalgan” (Fidia Pharmaceutical, Inc., Italy) [81], «Synvisc» (Biomatrix, Inc., England) [82], «Orthovisc» (Anika Therapeutic, Inc., USA) [83], which are called the synovia prostheses. However, their application is not referred to joint endoporsthetics. The results gained in long-term joint endoprosthetics are realized in recommendations on choosing traditional endoprosthesis designs with account for the age and individual features of a patient shown in Fig.5.26 [84]. It is seen that the most wear-resistant endoprostheses with ceramic-ceramic and metal-metal friction pairs are chosen mainly by young people having raised individual demands.

Nevertheless, the tribological issue of joint endoprosthetics is far from being solved optimally. When comparing the structure, lubrication and functioning of natural joints and those of endoprostheses one can say that the latter resemble friction units of machines that suit poorly for operating in a human body.

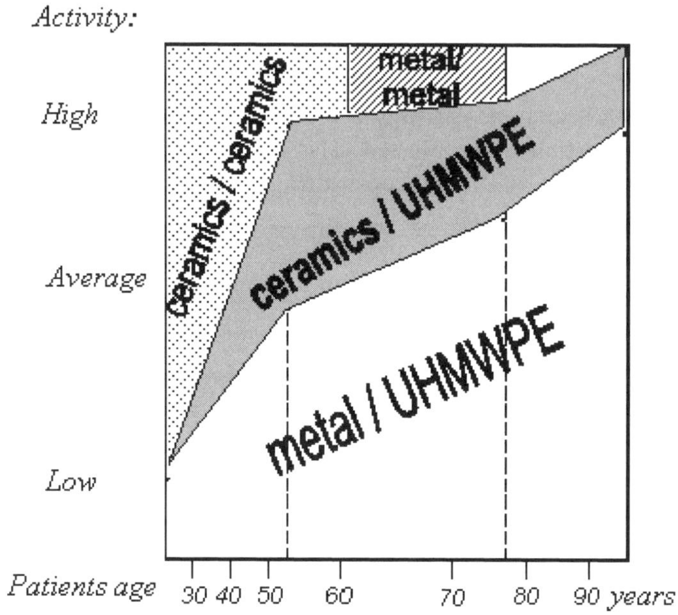

Fig. 5.26. Recommended allocation of friction pairs of hip endoprostheses depending on age and physical activity of patients

Study of implanted joints has shown "contamination" of the organism with wear debris of endoprosthesis friction pairs to be the main issue impeding their reliable and long-life operation. The problem lies at the interface of medicine, biophysics, tribology, and material science and creative collaboration of orthopaedists, immunologists, and experts in engineering is needed for the issue to be resolved. In work [84] it is assumed that the age of talented individuals who create endoprostheses has gone. Only by uniting the efforts of the experts of different fields of knowledge one can produce novel joint endoprostheses with enhanced tribological resource. Further we will report the data on the development of endoprostheses simulating various physiological functions of natural joints.

References:

1. Pavlova V.N.,.Kopieva T.N, Slutski L.I., and Pavlov G.G. Cartilage. Moscow, Medicine, 1988, 320 pp.
2. Fulkerson J.P., Edvards C.C., and Chrisman O.D. Articular cartilage. In: The Scientific Basis of Orthopaedics. 2-nd ed. Los Altos, 1987, p. 347–371.
3. Ermakov S.F., Rodnenkov V.G., Beloenko E.D., and Kupchinov B.I., Liquid crystals in engineering and medicine. Minsk, "Asar Ltd."; Moscow, "ChePo", 2002, 412 pp.
4. Mansour J.M., and Mow V.C. The permeability of articular cartilage under compressive strain and at the high pressures. *J. Bone Joint Surg.*, 1976, V. 58 A, p. 509–516.
5. Pavlova V.N. Synovia medium of joints. Moscow, Medicine, 1980, 296 pp.
6. Walker P.S., Sikorski J., Dowson D., et al. Features of the synovial fluid film in human joint lubrication. *Nature*, 1970, V. 225, No. 5236, p. 956–957.
7. Mc Cutchen C.W. The frictional properties of animal joints. *Wear*, 1962, V. 5, p. 1–17.
8. Fein R.S. Are synovial joints squeezed-film lubricated? *Proc. Inst. Mech. Eng.*, 1966–67, V. 181, p. 125.
9. Walker P.S., Dowson D., and Longfield M.D. Boosted lubrication in synovial joints by fluid entrapment and enrichment. *Ann. Rheum. Dis.*, 1968, V. 27, p. 512–520.
10. Davis W.H.J., Lee S.L., and Sokoloff L. A proposed model of boundary lubrication by synovial liquid: Structuring of boundary water. *Trans. ASME J. Biomech. Eng.*, 1979, V. 101, No. 3, p. 185–192.
11. Kupchinov B.I, Rodnenkov V.G., Bobrysheva S.N, et al. On the mechanism of operation of joint as a rubbing body, Belarus AS Proceedings, 1985, V. 29, No. 5, p. 463–465.
12. .Kupchinov B.I, Ermakov S.F, and Beloenko E.D. Biotribology of synovial joints. Minsk, Vedy, 1997, 272 pp.
13. Beloenko E.D. Optimization of orthopedic-surgical treatment of rheumatoid arthritis based on the methods of biophysical control of antifriction properties of the joint synovia (clinical-experimental research), D. Sc. Med. Thesis, Kiev, 1992.
14. Ermakov S.F. Tribophysics of liquid-crystalline materials in metallic and biopolymeric joints. Dr. Sc. Eng. Thesis, Gomel, 2001.

15. Knoll J. Analyse von druckverteilung und schmierfilmbildung im kunstlichen huftgelenk. *Schmiertechnik und Tribologie*, 1978, V. 25, No. 2, S. 43–46.
16. Streicher R.M. Tribologie künstlicher gelenke. In: Endoprothetik. Ed. by E.W. Morscher, Berlin, Springer, 1995, S. 38–53.
17. Movshovich I.A., and Vilenski V.Ya. Polymers in traumatology and orthopedics. Moscow, Medicine, 1978, 320 pp.
18. Zichner L.P., and Willert H.G. Comparison of alumina-polyethylene and metal-polyethylene in clinical trials. *Clin. Orthop.*, 1992, V. 282, p. 86–94.
19. Zagorodnyi N.V., Ilyin A.A., Karpov V.N., et al. Titanium alloys in hip joint endoprosthetics. *N.N.Priorov Traumatology and Orthopedics Proceedings,* 2000, No. 2, p.73-75
20. Wroblewski B.M., Siney P.D., and Fleming P.A. Low-friction arthroplasty of the hip using alumina ceramic and cross-linked polyethylene. A ten-year follow-up report. *J. Bone Joint Surg. (Br)*, 1999, V. 81, p. 54-55
21. Chiesa R., Tanzi M.C., Alfonsi S., et al. Enhanced wear performance of highly cross-linked UHMWPE for artificial joints. *J.Biomed. Res.,* 2000, V. 50, p. 381-387
22. Haraguchi K., Sugano N., Nishii T. et al. Influence of polyethylene and femoral head surface quality on wear: A retrieval study // *Int. Orthop.*, 2001, v.25, p.29-34
23. Semlitsch M.F., Streicher R.M., and Weber H. Verchleibverhalten von pfannen und kugeln aus CoCrMo – gublegierung bei langzeitig implantierten ganzmetall – huftprothesen. *Orthopädie*, 1989, V. 18, S. 36–41.
24. Plitz W., and Hoss H.U. Untersuchungen zum verschleibmechanismus bei revidierten huftendoprothesen mit gleitflachen aus Al_2O_3–ceramic. *Biomed. Tech.*, 1980, V. 25, S. 165–168.
25. Fokin V.A. Friction pairs for total hip joint endoprostheses and wear issues. *Margo Anterior*, 2000, No. 4, p. 1–4.
26. McNally S.A., Shepperd J.A., Mann C.V., Walczak J.P. The results of nine to twelve years of the use of a hydroxyapatite-coated femoral stem. *J. Boone Joint Surg. (Br),* 2000, V. 82, p. 378-382
27. Bloebaum R.D., Beek D., Dorr L.D. et al. Complications with hydroxyapatite particulate separation in total hip arthroplasty. *Clin.Orthop.,* 1994, V. 298, p.19-26
28. Sanchez-Sotelo J., Lewallen D.G., Harmsen W.S., et al. Comparison of wear and osteolysis in hip joint replacement using two different coatings of the femoral stem. *Int. Orthop.,* 2004, V. 28, p. 206-210

29. Mc Kellor H.A., Sarmiento A., Schwinn C.P., and Ebramzaden E. *In vivo* wear of titanium - alloy hip prostheses. *J. of Bone and Joint Surgery*, 1990, V. 72A, No. 4, p. 512–517.
30. Levitin M. Friction reduction in hip implants treated with regular microrelief. Haifa, Regmi-Tech, 1997, 12 pp.
31. Heringou P., Intrator L., Bahrami T., et al. Interlinkin-6 in the blood of patients with total hip arthroplasty without sossening. *Clin. Orhtop.*, 1999, V. 366, p. 147-154
32. Granchi D., Verri E., Ciapetti G., et al. Bone-resorbing cytokines in serum of patients with aseptic loosening of hip prostheses. *J. of Bone Joint Surg. (Br),* 1998, V. 80, p. 912-917
33. Streich N.A., Breusch S.J., and Schneider V. Serum levels of inetrleukin-6 (IL-6), granulocyte–macrophage colony-simulating factor (GM-CSF) and elastase in aseptic prosthetic loosening. *Int. Orhtop.*, 2003, V. 27, p. 267-271
34. RM – cup metal/metal. *Margo Anterior*, 2000, No. 4, p. 5.
35. Kadolich Zh.V. Physical modification of polymer –metal joint to increase wear resistance based on simulation of biophysical properties of natural joints, Ph.D. Thesis, Gomel, 2002.
36. Harris W.H. The problem is osteolysis. *Clin.Orthop.*, 1995, V. 311, p. 46-53.
37. Dowling J.M., Atkinson J.R., Dowson D., and Charnley J. The characteristics of acetabular cups worn in the human body. *J. of Bone Joint Surg. (Br)*, 1978, V. 60, p. 375-382.
38. Walker P.S., and Bullough P.G. The effects of friction and wear in artificial joints. *Orthop. Clin. North Am.*, 1973, V. 4, p. 275-293.
39. Cooper J.R., Dowson D., and Fisher J. Macroscopic and microscopic wear mechanisms in ultra-high molecular weight polyethylene. *Wear*, 1993, V. 162–164, p. 378–384.
40. Yamamoto K., Imakiire A., Masaoka T., et al. Wear mode and wear mechanism of retrieved acetabular cup. *Int. Orthop.*, 2003, V. 27, p. 286-290.
41. Puolakka T.J.S., Keranen J.T., Juhola K.A., et al. Increased volumetric wear of polyethylene liners with more than 3 years of shelf-life time. *Int. Orthop.*, 2003, V. 27, p.153-159.
42. Heisel C., and Schmalzried T.P. Letter to the Editor. *Int. Orthop.*, 2003, V. 27, p. 318.
43. Hopper R.H.J., Young A.M., Orishimo K.F., and Engh C.A.J. Effect of terminal sterilization with gas plasma or gamma radiation on war of polyethylene liners. *J. of Bone Joint Surg. (Am)*, 2003, V. 85, p. 464-468.

44. Oonishi H., Kadoya Y., and Masuda S. Gamma-irradiated cross-linked polyethylene in total hip replacement – analysis of retrieved sockets after long-term implantation. *J. of Biomed. Mater.Res.*, 2001, V. 58, p. 167-171.
45. Hall R.M., Unsworth A., Wroblewski B.M., and Burgess I.C. Frictional characterization of explanted Charnley hip prostheses. *Wear*, 1994, V. 175, p. 159–166.
46. Wilkinson J.M., Gordon A., and Stockley I. Experiences with the Plasmacup – early stability, wear, remodeling, and outcome. *Int. Orthop.*, 2003, V. 27, (Suppl. 1), S16-S19.
47. Van Haaren E.H., and Heyligers I.C. Implant wear and osteolysis with a hydroxyapatite-coated screw cup. *Int. Orthop.*, 2003, V. 27, p. 282-285.
48. Farizon F., de Lavison R., Azoulai J.J., and Bousquet G. Results with a cementless alumina-coated cup with dual mobility. *Int. Orthop.*, 1998, V. 22, p. 219-224.
49. US Patent 6242507, C 08 J 3/28. Process for medical implant cross – linked ultrahigh molecular weight polyethylene having improved balance of wear properties and oxidation resistance. K.A. Saum, W.M. Sanford, W.G. Di Mao, and E.G. Howard, 2001.
50. Blomer W. Design aspects of modular inlay fixation. In: Performance of the wear couple BIOLOX forte in hip arthroplasty. *Proc. 2-nd Symp. on Ceramic Wear Couple*, Stuttgart, 1997, p. 95–104.
51. Hunt T.M. Handbook of wear debris analysis and particle detection in liquids. London, Elsevier, 1993, 370 pp.
52. Myshkin N.K., and Petrokovets M.I. Tribology. Principles and Applications. Gomel, MPRI NAS of Belarus, 2002, 310 pp.
53. Evans C.H., Mazzochi R.A., Nelson D.D., and Rubash H.A. Experimental arthritis induced by ultra-articular injection of allogenic cartilaginous particles into rabbit knees. *Arth. Rheum.*, 1984, V. 27, p. 200–208.
54. Swanson S.A.V. Friction, wear and lubrication. In: Adult Articular Cartilage, London, Pitman Publ., 1979, p. 401–414.
55. Implant Wear: The future of total joint replacement. New York, American Academy of Orthopaedic Surgeons, 1996, 230 pp.
56. Cho U. Phenomenological approach to wear debris analysis. *J. of Friciton and Wear*, 2000, V. 21, No. 3, p. 23-31.
57. Panzera D., Kirk T.B., and Anamalay R.V. Computer image analysis of debris from normal and arthritic synovial joints. *Proc. of Int. Tribology Conf. "AUSTRIB'94"*, Perth, 1994, p. 407–414.
58. Woolson S.T., and Murphy M.G. Wear of the polyethylene of Harris-Galante acetabular components inserted without cement. *J. of Bone and Joint Surg.*, 1995, V. 72 A, No. 9, p. 1311–1314.

59. Maloney W.J., Smith R.L., Schmalzried T.P., et al. Isolation and characterization of wear particles generated in patiens who have had failure of a hip arthroplasty without cement. *J. of Bone and Joint Surg.*, 1995, V. 77 A, No. 9, p. 1301–1310.
60. Livermore J., Ilstrup D., and Morrey B. Effect of femoral head size on wear of the polyethylene acetabular component. *J. of Bone and Joint Surg.*, 1990, V. 72 A, No. 4, p. 518–528.
61. Charnley J. Low friction arthroplasty of the hip. Theory and practice. New York, Springer, 1979, 320 p.
62. Maloney W.J., and Smith R.L. Periprosthetic osteolysis in total hip arthroplasty: the role of particulate wear debris. *J. of Bone and Joint Surg.*, 1995, V. 77 A, No. 9, p. 1448–1461.
63. Harris W.H., Schiller A.L., Scholler J.M., et al. Extensive localized bone resorption in the femur following total hip replacement. *J. of Bone and Joint Surg.*, 1976, V. 58 A, p. 612–618.
64. Jones L.S., and Hungerford D.S. Cement disease. *Clin. Orthop.*, 1987, V. 225, p. 192–206.
65. Horowitz S.M., Doty S.V., Lane J.M., and Burstein A.N. Studies of the mechanism by which the mechanical failure of polymethylmethacrylate leads to bone resorption. *J. of Bone and Joint Surg.*, 1993, V. 75 A, No. 6, p. 802–811.
66. Maloney W.J., Peters P., Engh C.A., and Chandler M. Severe osteolysis of the pervis in association with acetabular replacement without cement. *J. of Bone and Joint Surg.*, 1993, V. 75 A, No. 11, p. 1627–1635.
67. Zicat B., Engh C.A., and Gokkcen E. Pattern of osteolysis around total hip components inserted with and without cement. *J. of Bone and Joint Surg.*, 1995, V. 77 A, No. 3, p. 432–439.
68. Lintner F., Zweymuller K., Bohm G., and Brand G. Reaction of surrounding tissue to the cementless hip implant Ti-6Al-4V after an implantation period of several years. *Archives of Orthopaedic Traumatic Surgery*, 1998, V. 107, p. 357–363.
69. Nikolaev V.I., Tsvetkova E.A., Pinchuk L.S., and Beloenko E.D. Biocompatibility of hip joint endoprostheses removed in revision operation. *Public Health*, 1999, No. 7, p. 42–44
70. Preis G.A., and Dzyub A.G. Electrochemical phenomena in metal friction. *Soviet J. Friction and Wear*, 1980, V. 1, No. 2, p.18-31.
71. Rozenfeld I.L. Corrosion and protection of metals. Moscow, Metallurgia, 1970, 448 pp.

72. Streicher R.M., Schon R., and Semlitsch M. Untersuchung des tribologischen verhaltens von metall/metall – kombinateonen fur kunstliche huftgelenke. *Biomed. Tech.*, 1990, V. 35, No. 5, S. 107–111.
73. Wright K.W.J. Friction and wear of materials and joint replacement prostheses. In: Biocompatibility of Orthopaedic Implants. Ed. by D.F. Williams, V. 1, Boca Rato, CRC, 1993, p. 141–195.
74. Dowson D., Jobbins B., and Seyed-Harraf A. An evaluation of the penetration of ceramic femoral heads into polyethylene acetabular cups. *Wear*, 1993, V. 162–164, p. 880–889.
75. Saikko V., and Pfaff H.-G. Wear of alumina-on-alumina total replacement hip joints studied with a hip joint simulator. *Proc. of 2-nd Symp. on Ceramic Wear Couple*, Stuttgart, 1997, p. 117–122.
76. Saikko V., Paavolainen P., Kleimola M., and Slatis P. A five-station hip joint simulator for wear rate studies. *J. Eng. Med.*, 1992, V. 206, p. 195–200.
77. Stallforth H., and Ungethum M. Die tribologische testung von knieendoprothesen. *Biomed. Tech.*, 1978, V. 23, No. 12, S. 295–304.
78. Postak P.D. Stability characteristics of the natural congruent and ultra-congruent total knee systems. Bern, SULZER*medica*, Lit. No. 1934e, Ed. 08/93.
79. Hall R.M., Unsworth A., Wroblewski B.M., and Burgess I.C. Frictional characterization of explanted Charley hip prostheses. *Wear*, 1994, V. 175, p. 159–166.
80. Okhlopkova A.A., Vinogradov A.V., and Pinchuk L.S. Plastics filled with ultradisprsed inorganic compounds. Gomel, MPRI NAS of Belarus, 1999, 164 pp.
81. Richter W., Ryde E.M., and Zatterstrom E.O. Non-immunogenicity of a purified sodium hyaluronate preparation in man. *Int. Arch. Appl. Immunol.*, 1979, V. 59, p. 45–55.
82. Scale D., Wobig M., and Wolpert W. Viscosupplementation of osteoarthritic knees with hylan: a treatment schedule study. *Curr. Ther. Res.*, 1994, V. 55, p. 220–232.
83. Peyron J. Intra-articular hyaluronan injections in the treatment of osteoarthritis: state-of-the-art review. *J. of Rheum.*, 1993, V. 20, Sup. 39, p. 10–15.
84. Black J. Prospects for alternate bearing surfaces in total replacement arthroplasty of the hip. *Proc. of 2-nd Symp. on Ceramic Wear Couple*, Stuttgart, 1997, p. 2–10.

Chapter 6. SIMULATION OF THE CARTILAGE TISSUE

Modern endoprosthetics of joints attaches much importance to elaboration of the cartilage-simulating material. Great efforts of professionals in medicine, biomechanics and materials science imparting non-traditional ideology to creation of a new generation of endoprostheses are directed to solution of this crucial problem. Along with the cartilage-simulating material endoprosthetics of today decides a series of other major problems touched upon in the previous chapters, namely: i) starvation lubrication of friction joints *in vivo*, ii) prevailing fatigue wear of friction members resembling machine joints, iii) generation of large amount of debris that can not be removed fully by the periarticular lymphatic system and deposit on the bone-implant interface thus adding aseptic instability to endoprostheses. Artificial cartilage is able to solve in part named problems, deliver drugs to the operation wound, regulate bio potentials of the post-operative joint and realize some other functions, e.g. simulate biological mechanisms of a natural cartilage. This step will be of a revolutionary importance for the methodology of the novel generation of endoprostheses, showing the possibility of modelling biological functions of the artificial joint.

The present chapter presents a survey of numerous endeavours in the development of artificial cartilage by numerous experts in endoprosthetics. Special attention is paid to the path proposed by the authors of the present book. Its essence consists in modification of UHMWPE (the basic polymer material used in endoprosthesis friction joints) so as to impart it a microporous structure similar to a natural cartilage. The chapter dwells upon structural and process peculiarities of this material and cites investigation results of its physico-mechanical and tribological characteristics. The conclusions on biocompatibility of the developed material are considered.

6.1 BIOPHYSICAL CRITERIA OF ENDOPROSTHESIS WEAR RESISTANCE

Being a solution to a significant socio-medical problem of rehabilitation after various joint diseases, endoprostheses still remain a foreign inclusion in the human organism. Their useful operating life is rigidly restricted by the wear resistance of the friction joints generating wear debris during operation able to initiate inflammatory reactions in the ambient tissues. Even the best among manufactured at present endoprostheses can not bear comparison with the perfect self-regulating precision systems of the natural synovial joint. The key role in joint lubrication plays the cartilage, which functions besides as a damping antifrictional pad and a porous reservoir for the synovia.

If to compare endoprostheses and natural joints on the grounds of tribological criteria, we can draw the following conclusions.

1. Owing to elastoplastic deformation of the cartilage and rheological properties of contained in the pores synovia, mechanical stresses in a healthy natural cartilage are distributed always in a uniform manner. In contrast, contact areas of the endoprostheses friction joints experience peak loads. This occurs because:
   - immovable elements of endoprostheses aggravate the probability of inaccurate mating of frictional surfaces of the counterparts;
   - technological errors in shape and size of movable joints are hardly unavoidable;
   - mechanical microasperities found on the frictional surfaces of endoprostheses;
   - the force needed to deform rather tough polymer frictional counterbody and crush microasperities for balancing technological errors is commensurable to the breaking load.
2. The dynamic contacting cartilage surfaces of a natural joint experience friction via the lubricating synovial fluid. With increasing load an additional portion of the synovia is squeezed into the friction zone from the cartilage pores. As for the endoprostheses, their joints undergo friction under boundary lubrication or without any lubrication in the zones of peak contact loads. Since it is impossible to compensate fabrication errors and immovable state of fixed parts by deformation of the polymer counterbody, friction in endoprostheses gradually becomes unlubricated, which is ruinous for them.
3. Natural joints operating mainly in fluid or boundary lubrication regimes are not practically worn out. Cartilage areas experiencing wearing can be regenerated in a healthy organism. The implanted polymer friction part of endoprostheses wears at insufficient lubrication by the fatigue mechanism presupposing the formation of subsurface cracks and intensive spalling of UHMWPE particles [1].
4. The porous system of cartilages in natural joints absorbs debris in a healthy organism thanks to physiological mechanisms of dissociation and digesting of wear products. Wear debris of endoprostheses are a foreign matter leading to inflammatory processes in the surrounding tissues and at the bone-implant interface. This leads to loosening of endoprosthesis joints.

It is evident that most actual trend in refining endoprostheses designs is in approaching artificial frictional materials most closely to natural ones. In respect to cartilage this trend can be realized most fully in polymer materials whose organic origin brings them close to albumen structure. The polymer friction part of endoprostheses should be apparently made porous to simulate the cartilage structure and let the synovia into the polymer pores thus lubricating the joints. As a result, the porous friction counterbody acquires pliability to loading and will move closer to the natural cartilage structure. Besides, the polymer pores are able to absorb wear debris and lower their accumulation in the ambient tissues. This suggests that serviceability and triboengineering characteristics of endoprostheses as a whole will be raised considerably and acquire similarity to the natural joints.

6.2 NEW POLYMER FRICTIONAL MATERIALS

Experience in application of different structural polymeric materials (polyethylene, polyamides, polyethyleneterephthalate, fluoroplastics and other) in friction joints of endoprostheses has shown rather distressing results. Even the best antifrictional material PTFE yields with time debris leading to inflammation in soft tissues, osteolysis, and aseptic instability of components after endoprosthetic operations [2].

So far it is not surprising that in spite of the diversity of orthopaedic materials the weakest unit of metal-polymer and ceramic-polymer endoprostheses turn to be the polymer counterbody. Their wear rate bounds the period between the primary and revision endoprosthetics of any joints from 5 till 20 years depending on physical activity of the patient [3]. As it is been mentioned previously, the main reason of failure of metal-polymer endoprostheses is believed to be the fatigue wear of the polymer parts as a result of which they undergo cracking and crumbling [1]. The only material that proved to be reliable in endoprosthetic joints is UHMWPE. It was tested and adopted in practice in early 1960-ies by J. Charnley whose hip joints are still considered most perfect in endoprosthetics [4].

About half a million of endoprostheses of this type are implanted in the world yearly. Nevertheless, clinical practice of named endoprostheses with more than 20 years of forecasted service life evokes certain anxiety in conditions of elevated number of young patients. Wear of UHMWPE along with progressing in post-operative joint osteolysis are most dangerous for the hip at its total replacement. Just the debris, their transfer onto the bone-implant interface and the responding cellular reaction of the bone tissue are the chief cause of aseptic instability of both stem and the hip component of the metal-polymer endoprosthesis [5].

Persistent attempts in perfecting structure and properties of the polymer part of endoprostheses have lead to creation of a new material. This is a kind of cross-linked polyethylene exposed to a flow of electrons called WIAM. It displays improved resistance to wear, ability to absorb water and its oxidation is unlike that of UHMWPE [6].

An alternative to UHMWPE is carbon-reinforced polyether-etheroketone (PEEK). Wear of inserts of the total hip joint endoprosthesis made of PEEK is

30 times less than that of UHMWPE. These inserts are however highly tough, which intensifies transfer of vibration loads on the bone and, consequently loosening of endoprosthetic members.

Carbon materials of INTOST type developed in NIIGRAFIT (Russia) show high strength and porosity, self-lubricity, durability in biological media, fatigue strength, resistance to thrombi and absence of undesirable reactions of the organism to the implants and their debris. Carbon materials based on polymer binders have an advantage of adjusting implants directly during the operation using traditional surgical tools like saw, scalpel, hacksaw, and other. Unfortunately, some grades of these materials are based on the polyamide binder, which makes them unfit for endoprosthetics because of complications caused by polyamide debris [7].

In spite of a limited range of available for implantation polymer materials still new modifications are being in constant search for. Since the potentialities of mechanical properties and biocompatibility of engineering materials used in orthopedy have been in fact exhausted, it is highly actual to find the resources in lowering unfavourable affects of implants on the living tissues and bringing endoprosthetic friction joints nearer to the natural structures [8].

It often happens during operations on traumatized organs that damaged sections of the cartilage are to be removed and the hollow formed should be filled with some biological material. The implants serving to fill holes and cavities in the cartilage do not experience any substantial mechanical loading so can be made of the polymer gels. Of acute interest in this respect have become hydrogels, which are cross-linked hydrophilic polymers containing polyacids, polysaccharides, medicinal agents and other matter [9, 10]. Hydrogels of medical purposes are most often made on the base of polyvinyl alcohol (PVA). A net of spatial links formed in the PVA-water liquid system deprives it of fluidity and adds elasticity, and some other properties of a solid body [11]. The presence of a great number of hydroxyl groups in a PVA macromolecule makes it possible to immobilize biologically active compounds in the hydrogels [12].

First attempts to create an artificial cartilage using PVA-based hydrogels go as far back as 1970-ies [13, 14] and are still underway today [15, 16]. At the end of 1990-ies an artificial cartilage has been made at the University of Science and Technology, Beijing on the base of the PVA matrix incorporating a dispersed aqua phase. This cartilage has been formed as a result of multiple freezing-unfreezing of the PVA hydrogel followed by dehydration in vacuum. As a result of such treatment, elasticity modulus of the hydrogel becomes very close to that of the natural cartilage. A similar material resembling silicone by its deformation and strength characteristics has been developed in the University of Technology, Georgia, USA. It passed all mechanical, tribological, sanitary and hygienic trials since 1999 preceding clinical tests [17].

Various tissue constructions for cartilage repair using a wide range of scaffold materials have been investigated. A team from Tufts University in the

USA, Chalmers University of Technology, and Kungsbaska Hospital in Sweden has investigated the use of bacterial cellulose by *Gluconacetobacter xylinus* as a biomimetic alternative [16]. This biopolymer has the advantage of significant mechanical strength on hydration, in addition to *in situ* moldability, biocompatibility, biodegrability, and low cost production. Although further investigation is required, these materials clearly offer potential as a bio scaffold for the tissue engineering of cartilage.

Along with above frictional endoprosthetic materials, a material based on a PVA gel filled with UHMWPE powder has been proposed [18]. Its strength and density gradients can be regulated due to capability of uneven distribution of filler particles within the hydrogel matrix. The material is elastic, can keep large amounts of medicinal and biological fluids, and retain its elasticity and shape. It is possible to cut-out by scalpel implants of the geometry repeating the cartilage defect. The material composition is optimized by the criterion of loading in the place of implantation. High filling of the hydrogel matrix prevents ingrowing of the vessels into the implant, whereas low filling decreases its hardness and increases elasticity. Aqua solutions of medical agents used for the hydrogel matrix are isolating from the implant over a prolonged period. Kinetic parameters of this isolation depend on the composition and concentration of drugs as well as structure of the matrix and its filling degree.

Polarization characteristics of PVA hydrogels filled with either synthetic or natural substances were studied in [19]. The fillers were starch, medical gelatine, carboxymethyl cellulose (CMC), sodium salt of CMC and proteolytic enzyme papain (Loba-Chemie, Austria). The compositions were optimized in terms of strength criteria and colloidal stability of hydrogel systems. The fillers were differentiated by their electrophysical characteristics. Mentioned measures helped to choose biologically active compounds for immobilizing in gels.

State of the art in growing bone autotransplants [20] proves that we can hardly rely on improvement of the situation in endoprosthetics within the nearest decade if to take only the above-described path. Therefore, the research in modelling cartilage tissue by engineering materials goes ahead despite the obstacles in reproducing cartilage structure. The creation of materials similar to the cartilage can give rise to new approaches for solution of tribological problems of endoprosthetics. It has been found out by D. Dowson in the 1970-ies that endoprostheses fit with an elastic lining in the friction zone acquire a hydrodynamic lubrication regime, which is characterized by low enough wear [21]. It was however hard to implement this idea in practice because of a number of difficulties, most intricate of which were: i) the choice of elastic biocompatible and antifrictional materials, and ii) reliable fastening of the elastic lining on bearing members of endoprostheses.

In this connection, a concept has been put forward in Metal-Polymer Research Institute (Gomel, Belarus). Its essence consists in formation of a transient microporous layer on the surface of polymer parts to reproduce biophysical functions of the cartilage [22-27]. This layer carries the electret charge that improves lubrication by the synovia and biocompatibility of the endoprosthesis. Besides, it serves as a vessel for drugs and ensures their prolonged extraction into the operation wound (Fig. 6.1).

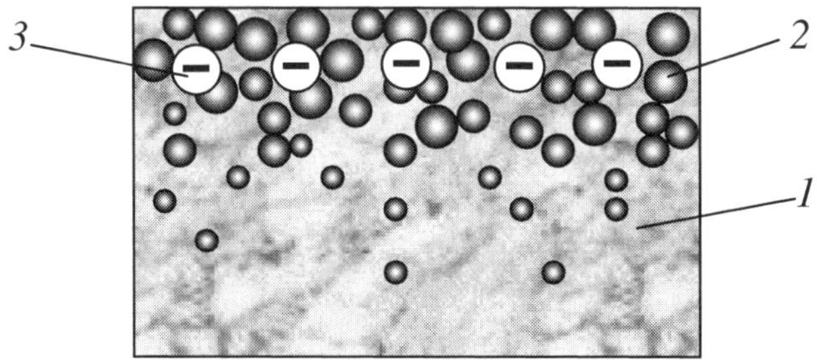

Fig. 6.1. Schematic presentation of the friction surface structure of microporous polymer counterbody of endoprosthesis: 1 – UHMWPE matrix, 2 – pores, 3 –polarizing surface charge

Investigation results of structure, physico-mechanical, triboengineering characteristics, and biocompatibility of UHMWPE-based cartilage are reported in the section to follow.

6.3 CARTILAGE-SIMULATING POLYMER MATERIAL

The technological basis of the microporous layer simulating the cartilage proceeds from representations on the gel state of polymers. Polymer gels are structurized systems formed at either solidification of liquid polymer solutes or during swelling of solid polymers. UHMWPE appeared very suitable for the gel base as it can be easily modified and imparted a needed structure or properties during processing [6, 28]. The microporous structure of gel serves as a convenient capacity which provides controlled delivery of drugs healing the operational wound. Controlled drugs delivery offers several advantages over conventional delivery. Among the benefits of UHMWPE as the endoprosthetic material are its biological inertness and possibility of radiation-induced sterilization (ionizing radiation dose about 2.5 Mrad). In addition, UHMWPE is easily deformed, so can damp impact loads on the endoprosthesis. One of the variants of producing the transient microporous layer on the friction surface of UHMWPE part is described in works [22, 23, and 30].

Specimens of block UHMWPE or its sintered powder of grades *Hostalen Gur 4120* and *Chirulen DIN 58836C* (Hoechst AG, Germany) have been used in the experiments. Medical Vaseline oil (MVO) that is a component of a great number of therapeutic medicinal preparations for curing arthritis has been chosen as the UHMWPE plasticizer. The samples based on powder UHMWPE were formed by hot pressing under T = 160–170 °C and p = 10–12 MPa. The block samples were modified by dipping in MVO at 60 °C till 170 °C temperatures and endured for a preset time. Then MVO was removed from both types of samples. With this aim, the samples were washed by hexane in a Soxhlet apparatus and underwent vacuum drying at p = 100–200 mPa.

Figure 6.2 illustrates how strength and deformation characteristics of UHMWPE change with introduction of MVO. Heating of the mixture above the polymer melting point leads to formation of the colloidal solution. This thermally unstable composition decomposes at cooling in a specific manner into phases only partially due to high viscosity of the colloidal solution. The phase with a high polymer concentration preserves its spatial continuity and forms a porous matrix, while the liquid happens to be immobilized in the matrix pores.

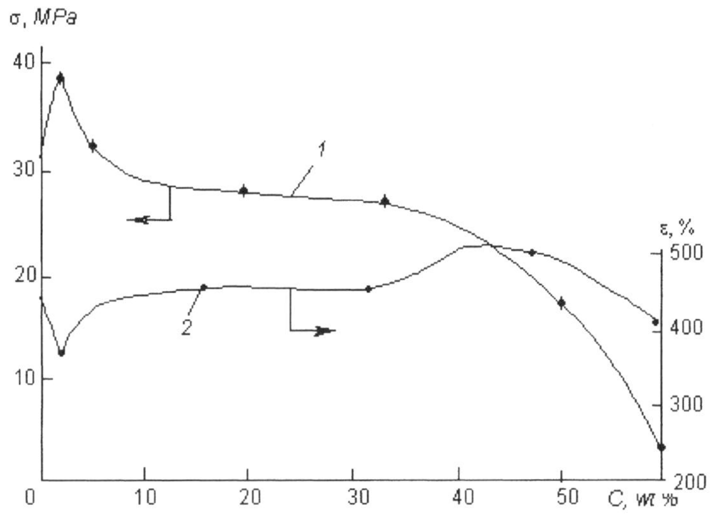

Fig. 6.2. Breaking strength at tension σ and relative elongation at rupture ε of UHMWPE-based materials versus MVO content (C)

Plasticizers raise macromolecular mobility and alleviate crystallization of the polymer. This is proved by the antiplastification effect of UHMWPE displayed in the maximums on the curves presented in Fig. 6.2 at MVO content about 2-5% [28]. With further increase of MVO content macromolecular interactions calm down, and strength of the samples monotonously diminish. Abrupt strength impairment at above 40% MVO content is attributed to intensive formation of the pores in the polymer matrix.

To avoid strength reduction of the material not participating in friction the problem of its local modification by thermal treatment in MVO has been studied. As a result, a microporous layer consisting of UHMWPE matrix incorporating liquid inclusions has been formed. The samples modified by thermal treatment in the liquid are compared in Fig. 6.3 to those made of the composite mixture of UHMWPE and MVO.

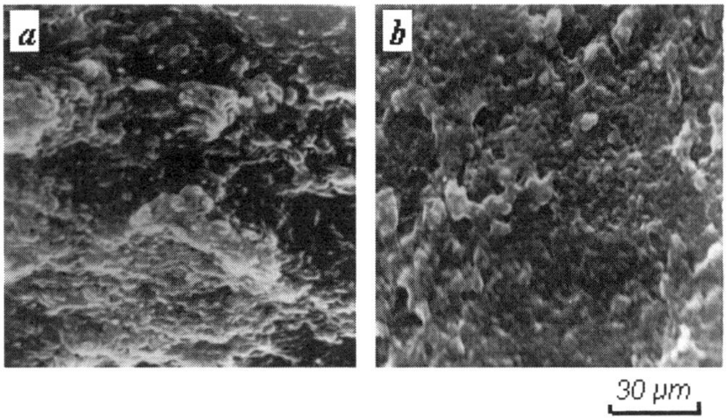

Fig. 6.3. Cross-cut photos of samples: *a* – UHMWPE thermally treated in MVO at 150 °C (the arrow shows sample surface having contacted MVO); *b* – UHMWPE + MVO (1:1 by mass)

It is evident that permolecular structure of both samples is identical although the one subjected to thermal treatment in the liquid shows an irregular porous structure. The pores are distributed in the sample with a gradient over the diameter: the maximum diameter have the pores near the sample-MVO contact, which gradually reduces till disappearance of the pores in the sample centre.

This can be explained as follows. MVO diffuses with heating into the sample surface layer that transfers in the state of a colloidal solution and the microporous structure with the gradient of pore distribution is formed during cooling [31].

The analysis of UHMWPE gels has visualized that the pores in the polymer matrix represent communicating capillaries emerging on the sample surface. The capillary length exceeds by 10-20% on the average thickness of the gel-like layer. The cross-section diameter of the pores is 1-20 μm, which corresponds to the porous system parameters of the natural cartilage [32].

Electron microscopy has shown that thickness of the microporous layer can be varied by the choice of time and temperature regimes of UHMWPE treatment in MVO from units till hundreds of μm (Fig. 6.4).

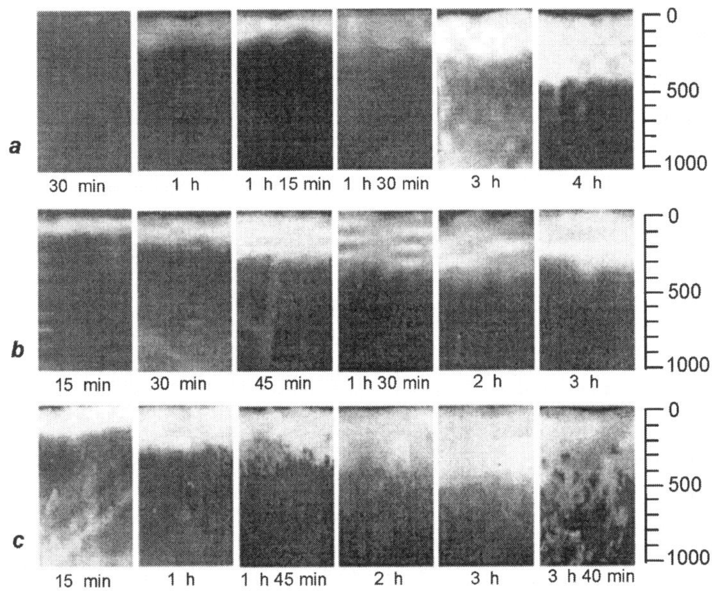

Fig. 6.4. Effect of thermal treatment duration on thickness of modified layer at temperatures: *a* – 125 °C, *b* – 130 °C, *c* – 140 °C

The time and temperature of UHMWPE modification turned to be the equivalent factors for the structural formation.

Figure 6.5 illustrates microphotographs of the modified surface of experimental samples after removal of the plasticizer. The samples obtained at different technological regimes are seen to have identical structure [33].

Table 6.1 presents data on plasticizer absorption by the polymer material at thermal treatment. The kinetic dependencies of mass variations of the samples evidence that each temperature corresponds to certain compatibility threshold of UHMWPE and MVO that can not be overcome via prolongation of the treatment.

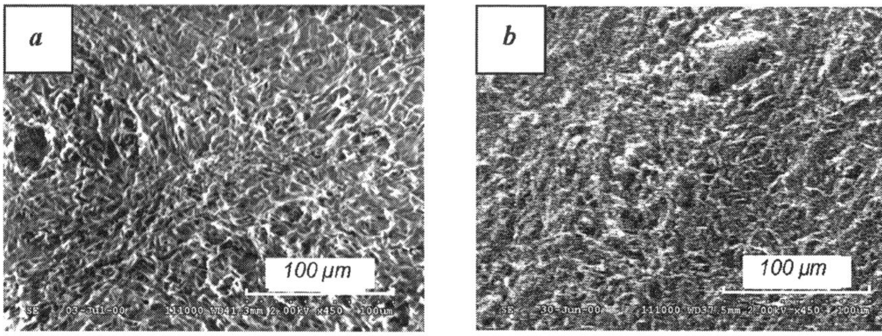

Fig. 6.5. UHMWPE sample surface after treatment in MVO and upon plasticizer removal. Time-temperature regimes: *a)* T = 130 °C, t = 45 min, *b)* T = 125 °C, t = 120 min

Table 6.1

Time and temperature effect of thermal treatment in MVO on mass increment of UHMWPE

Temperature T, °C	Weight increment (%) at treatment time, h				
	1	2	3	4	5
125	0.51	1.25	1.50	1.51	1.51
140	0.85	1.53	2.18	2.62	2.97

Absorption of the plasticizing liquid by the polymer alters geometrical dimensions of the samples (Fig. 6.6). Increment in size depends on the treatment temperature. The limiting dimensions of the samples and approach to the maximum grow with temperature elevation. The minimal changes in the sample size occur at thermal treatment temperature T = 125 °C. This fact confirms that there is a possibility to form a microporous surface layer participating in friction without any impairment in strength of the whole sample. Mentioned data corroborate with the results of the electronic microscopy (Fig. 6.4).

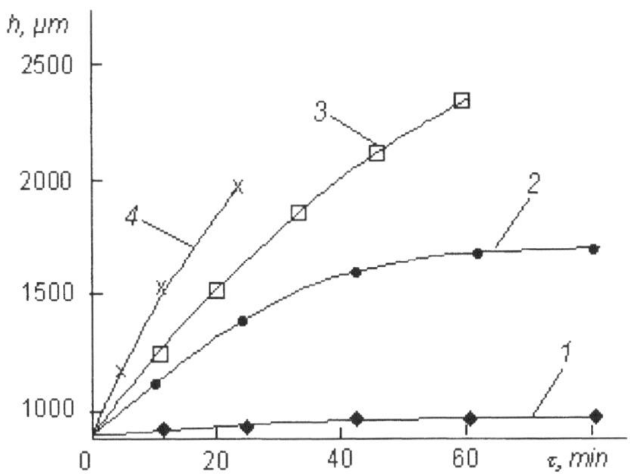

Fig. 6. 6. Time and temperature effect of treatment in MVO on thickness of UHMWPE films: ***1*** **– T=170 °C,** ***2*** **– 150,** ***3*** **– 140,** ***4*** **– T= 125 °C**

It is expedient to form the microporous layer at a temperature below 130 °C as follows: during 2–2.5 h at T = 125 °C, or not more than 30–45 min at T = 130 °C. Proceeding from the kinetic regularities of thermo oxidative destruction of UHMWPE [6], time and temperature of thermal treatment should be limited to avoid even partial decomposition of the polymer matrix.

The microporous structure presents a no equilibrium system whose specific feature is *syneresis*, i.e. spontaneous extraction of the liquid phase from the pores. The main reason of syneresis is stress relaxation appeared in the matrix at its formation. Syneresis depends nonlinearly on the plasticizer content, and parameters of the kinetic dependence are defined by the polymer and plasticizer nature as well as by the ratio of gel ingredients [28]. The maximum separation of the liquid phase occurs 100-200 h after the gel formation and reaches 1% of the sample mass. Then, the amount of the extracted liquid goes down exponentially and approaches zero in 40-45 days.

After treatment in MVO the samples are washed in solvents to extract the process liquids. The solvent evaporates practically fully from the samples during 30 days at room temperature (Fig. 6.7). Use of the vacuum drying shortens time of solvent removal to 24 h. An alternative method is drying in a thermostat at 60 °C during a week, which may however speed up aging of the polymer.

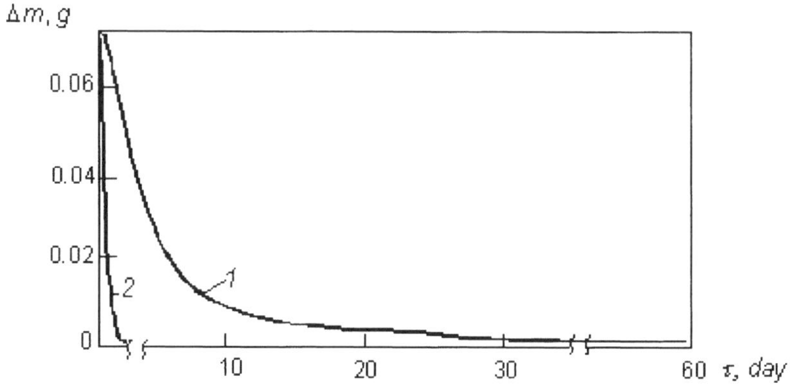

Fig. 6.7. Kinetic dependencies of sample mass variations at hexane evaporation from porous polymer matrix: *1* –in atmosphere at room temperature, *2* – in vacuum chamber

The microporous layer structure on the cartilage-simulating UHMWPE is shown in Fig. 6.8. The image of the sample has been taken by the optical microscope upon extraction of the process liquid. The sample surface layers are seen to preserve their porosity and can contain the necessary drugs.

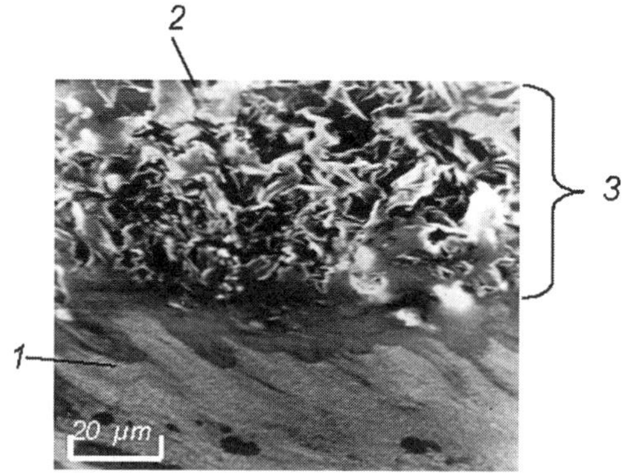

Fig. 6.8. Cross-section of UHMWPE sample with a microporous layer after plasticizer extraction. *1* – initial UHMWPE structure, *2* – sample surface, *3*– microporous layer

The crystalline structure of the main part of the sample has remained intact after the formation of the microporous layer, which is visualized by the X-ray diffraction analysis (Table 6.2). The comparison of crystallinity degree of the initial UHMWPE (49%) and the samples with the microporous layer (47%) prove obliquely that modification is not accompanied by any recombination of the crystalline structure. Slight decreasing of crystallinity in the latter case is because of the fibrillar structure of the pore walls, which is less ordered than that of the initial UHMWPE. The characteristic size Z_{ef} of crystallites of both crystalline and microcrystalline phases of UHMWPE was established to reduce after modification. It will be logical to presume that less ordered parts of the crystallites have dissolved in MVO and transferred into the liquid phase.

Table 6.2

Crystalline structural parameters of UHMWPE samples

Sample	Crystalline phase		Microcrystalline phase		Peak areas on sample spectra, rel. units		Crystallinity degree, %
	2Θ, degrees	Z_{ef}, nm	2Θ, degrees	Z_{ef}, nm	Crystalline	Micro-crystalline	
Initial	21.56	192.05	19.55	22.48	616	653	49
With micro-porous layer	21.46	173.85	19.80	17.94	815	912	47

Above-cited data are supported by the electron-microscopic images of the microporous surface layer of UHMWPE samples (Fig. 6.9).

The analysis of presented photos has lead us to the following conclusions:

- the chief elements of the permolecular structure of the microporous layer are spherulite formations shaping the polymer matrix nodes;
- the spherulites are interconnected by fibrillar bunches in the form of long filaments that contribute much to spatial continuity of the matrix, its shape stability and strength;
- the matrix occupies rather large free volume consisting of communicating micropores filled during formation of the porous layer by the process liquid.

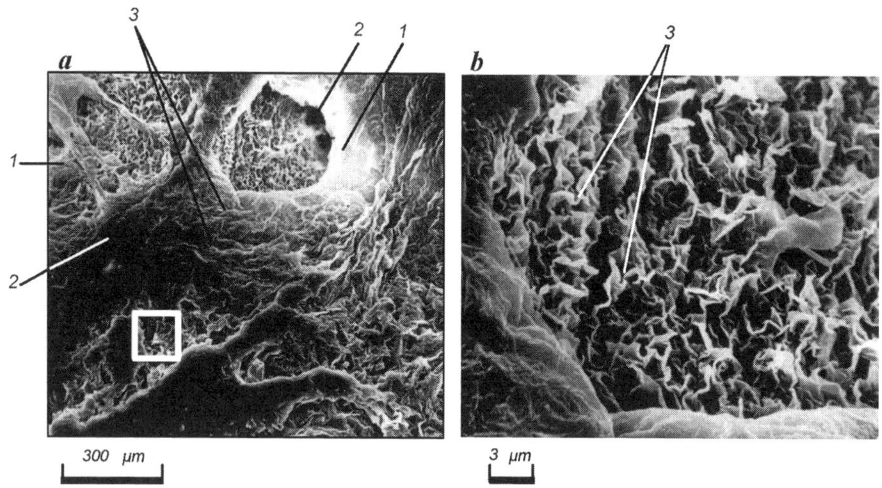

Fig. 6.9. Photomicrograph of the porous layer (*a*) and its magnified fragment in a square (*b*): *1* – spherulite formations, *2* – micropores, *3* – fibrils

The described structure looks like protein-polysaccharide formations of the natural cartilage tissue. The molecular aggregates of proteoglycanes resembling peculiar brushes constitute its major fragments. These brushes have needles representing glycosamine glycanes saturated with carboxyl and sulphate groups adding negative charge to the brushes. Owing to this, the cartilage is wetted by the polar liquids, which penetrate readily inside the free space of the cartilage matrix.

We tried to regulate wetting of the microporous layer via electrical polarization of the UHMWPE matrix. This problem turned to have two main aspects, namely:

1. Durability of joint endoprostheses *in vivo* is to a great degree specified by the lubrication regime. Under prolonged effects of high mechanical stresses the lubricating film is squeezed from the friction joint leading to seizure of metallic heads of the metal-polymer endoprostheses and formation of polymer transfer films. The hydrophobic property of UHMWPE is the main obstacle of satisfactory lubrication of endoprostheses *in vivo* [34]. So far, its elimination will result in fundamental improvement of lubrication of movable joints by the synovia, and expansion of the technological lifespan of the implanted endoprostheses.

2. Perfect lubrication of UHMWPE by water is a requisite condition for filling of the microporous polymer matrix with drugs. Oil-based drugs can easily penetrate into the UHMWPE matrix. But the medicinal products for post-operative curing are mostly water-based, and their impregnation into the micron-size pores presents a separate intricate problem. Solution of this problem is strongly actual since prolonged extraction of drugs from the endoprostheses into the operation wound is by far more beneficial than any traditional drugs use. These and some other critical problems are considered at length in Chapter 8.

The problem of lowering hydrophoby of the microporous layer of endoprosthesis friction parts was also solved by treating UHMWPE matrix in the low-temperature HF discharge plasma after removal of the process liquid from the pores [23]. This technique is extensively used in medical practice thanks to its feasibility, safety, and sterility. The plasma energy is suffice for rupture the chemical bonds in the main chain of UHMWPE macromolecule in the surface layer. The resultant radicals in the places of rupture react with air oxygen, and quickly form carboxyl and other oxygen-containing polar groups that actively participate in wetting. As the experiments have shown, the edge wetting angle of the initial UHMWPE by distilled water is 92°, whereas it is only 10° after plasma treatment. The edge wetting angle of the microporous samples is composed of, respectively 83° and 0°, which is the compete spreading of the drop. This is because the spherulite-fibrillar structure of the microporous layer (Fig. 6.9) transfers during plasma treatment into the polarized state, which is analogous to the charge state of molecular proteoglycane aggregates making up the base of the cartilage tissue.

Optical microscopic images of cross-sections of UHMWPE film samples with a microporous layer are presented in Fig. 6.10. The probability of impregnating them with medical products has been estimated by application of the brilliant green solution (alcohol solution of the main bright green dye used for disinfections) on the sample surfaces. It is seen from Figure *b* that the dye penetrates but poorly into the initial UHMWPE matrix pores, and concentrates in a thin layer in the surface vicinity. After treatment with HF discharge plasma the dye distributes over the whole volume of the pores. The intensive colouring of the sample at the boundary of its porous and pore-free parts can be attributed to elevated concentration of the dye in the solution and crystallization of the saturated solution.

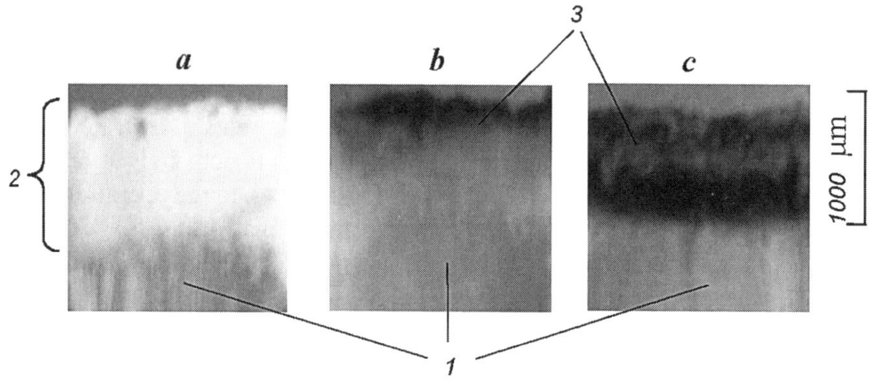

Fig. 6.10. Cross-sections of UHMWPE samples with a microporous layer: *a* – initial, *b* – the same treated by brilliant green solution, *c* – treated first by plasma and then brilliant green solution. *1* – UHMWPE with preserved structure, *2* – microporous layer, *3* – microporous layer coloured by brilliant green solution

Electron-microscopic images of the sample surface presented in Fig. 6.10, *c* are illustrated in Fig. 6.11. Thanks to colouring of micropore walls a greater number of details can be visualized on the fragments of the porous permolecular structure. One can clearly see spherulites in the polymer matrix nodes (*a*), and connecting them fibrillar formations (*b* and *c*) that pierce the free volume. Furthermore, a system of communicating micropores distributed across the modified layer thickness is distinctly observed. By its appearance the spherulite-fibrillar structure presented in Fig. 6.11 is a close analogue to that of the cartilage at corresponding magnification [32].

Polymer films and coatings treated in corona discharge were shown in [35, 36] to be wetted better than the reference once. Improved wettability is attributed to increasing surface energy at the polymer-air interface at electrical polarization of polymer. It is also proved that the edge wetting angles of polymer samples treated with corona discharge reduce with increasing surface charge density independently of the charge sign. This makes grounds to consider the corona discharge treatment of the microporous samples as an alternative to plasma treating means of reducing their hydrophobicity.

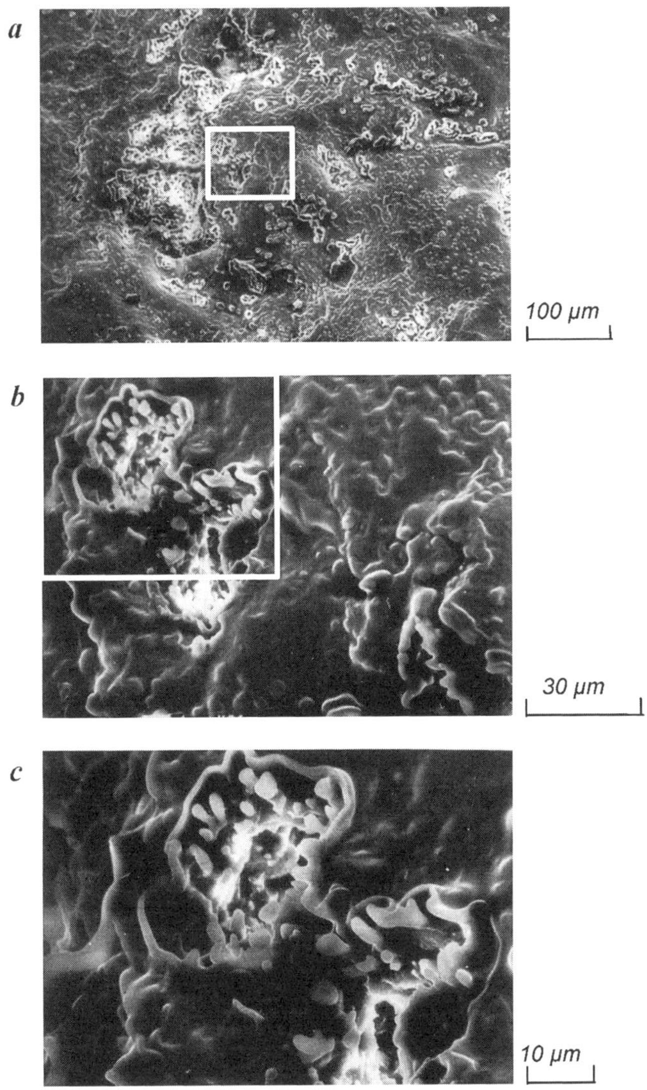

Fig. 6.11. Permolecular structure image of microporous UHMWPE sample coloured by brilliant green solution. Photos *b* and *c* show magnified fragments of *a* and *b* images found inside squares

The investigation results have been reproduced in a modification scheme of UHMWPE friction parts of endoprostheses (Fig. 6.12). This technique can be perfectly adjusted for a corresponding traditional production process of polymer parts of endoprostheses.

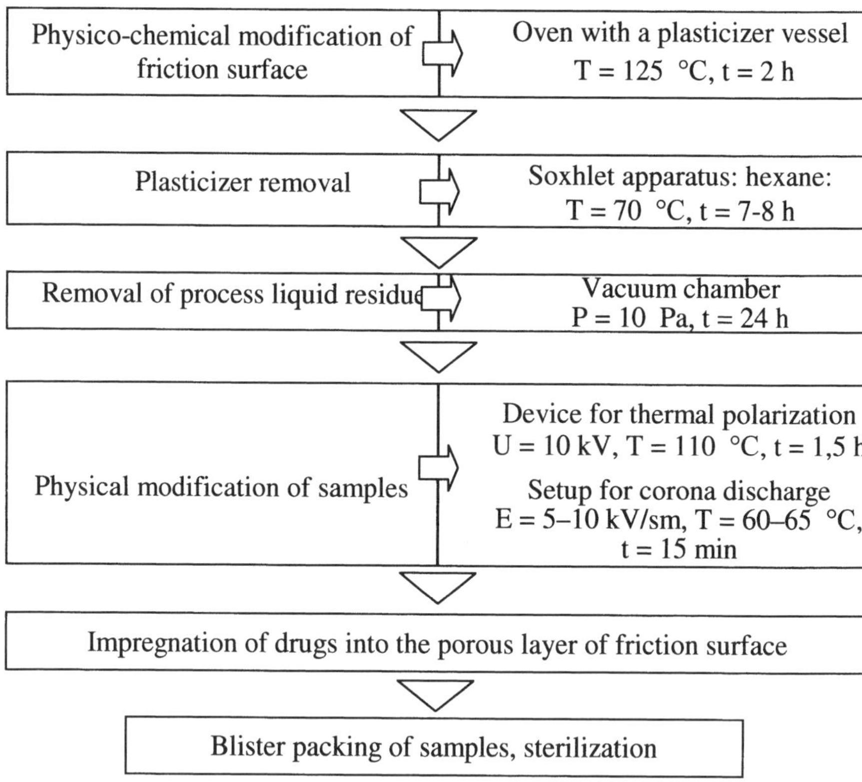

Fig. 6.12. Flow chart of modification process of polymer parts

6.4 PHYSICO-MECHANICAL AND TRIBOLOGICAL CHARACTERISTICS

Formed on the friction surface of the polymer part microporous layer, which is less strong than the initial UHMWPE slightly impairs strength of the whole part. Experimental results of cylindrical UHMWPE samples whose faces were treated in MVO are presented in Fig. 6.13.

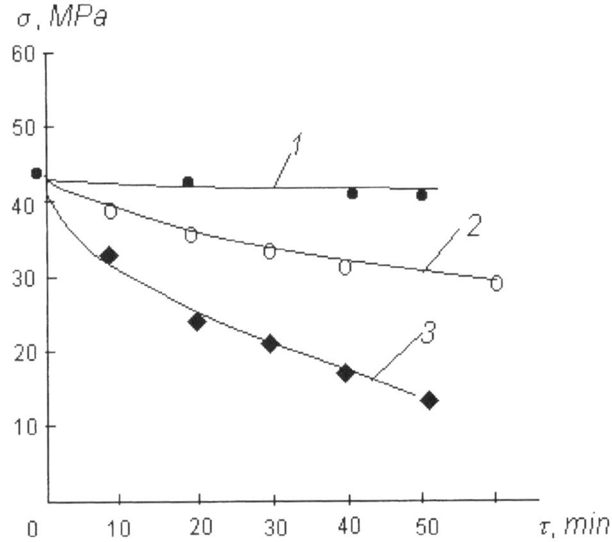

Fig. 6.13. Breaking strength at compression of cylindrical UHMWPE samples depending on time and temperature of treatment in MVO: *1* – 125 °C, *2* – 140 °C, *3* – 150 °C

The compressive breaking strength decreases exponentially with time and temperature elevation. This is probably connected with diffusive permeation of MVO into UHMWPE and its intermolecular plastification. The volume of the plasticized and less strong material increments with time and temperature growth. It has been nevertheless established that under optimum choice of treatment parameters (curve 1) strength of the samples does not worsen essentially due to even distribution of stresses in the cylindrical sample-testing machine plate contact.

Tribological characteristics of the microporous UHMWPE samples at friction against steel counter bodies (St 45, of 0.1 μm roughness) were obtained under different loading regimes on SMT-1 friction machine (rotation without lubrication) and are presented in Table 6.3. The regimes of continuous sliding simulate the conditions when the deformed microporous matrix does not almost restore. The analysis of the results obtained has proved that tribological properties of UHMWPE improve after treatment in MVO. The friction coefficient and wear rate reach the maximum depending on time and, consequently, on the microporous layer thickness.

Table 6.3

Tribological characteristics of microporous material (dry friction, p = 1 MPa, v = 0.25 m/s)

<table>
<tr><th colspan="2">Type of treatment</th><th>Friction coefficient</th><th>Wear rate I, mg/h</th></tr>
<tr><td colspan="2">Reference sample (initial UHMWPE)</td><td>0.52</td><td>0.07</td></tr>
<tr><td rowspan="4">Samples subjected to treatment in MVO under next time and temperature regimes:</td><td>T=125 °C, t = 1 h</td><td>0.13</td><td>0.012</td></tr>
<tr><td>T=125 °C, t = 1.5 h</td><td>0.26</td><td>0.014</td></tr>
<tr><td>T=125 °C, t = 2 h</td><td>0.36</td><td>0.075</td></tr>
<tr><td>T=125 °C, t = 3 h</td><td>0.12</td><td>0.025</td></tr>
<tr><td colspan="2">Samples treated in MVO at:
T = 125 °C, t = 2 h, washed in solvent and dried</td><td>0.25</td><td>0.057</td></tr>
</table>

Investigation results of the microporous UHMWPE samples at friction against steel counter bodies using USK-1 testing machine (reciprocating sliding, lubrication by physiological solution) are presented in Table 6.4. With increasing thickness of the microporous layer, lubrication of the friction pair improves initially and elasticity of micro roughness on the polymer sample increments. Further, pliability of the surface layer of UHMWPE sample becomes a more significant factor, which depends on the total volume and size of the pores. With increasing pliability, deformation processes of the sample start to grow with loading, and aggravated friction and wear of UHMWPE appear. The best tribological properties display the samples whose microporous layer thickness is about 200-400 μm. The results presented below were obtained using the samples with an optimum microporous layer thickness.

Table 6.4
Tribological characteristics of UHMWPE-steel friction pairs depending on microporous layer thickness (p = 2 MPa, v = 1 m/s)

Parameters	Modified layer thickness, μm					
	100	200	400	600	800	1000
Friction coefficient:						
- with lubrication	0.27	0.20	0.25	0.30	0.34	0.20
- without lubrication	0.20	0.12	0.10	0.15	0.17	0.15
Linear wear, mm ($n = 10^6$ cycles)	0.60	0.30	0.20	0.80	0.80	0.80

UHMWPE samples with the microporous layer were tested on a pendulum tribometer using "Diasynol" medication as a lubricant containing liquid-crystalline components and employed in medicine for treating joints by applications. The results of these tests are presented in Fig. 6.14. They prove that the friction coefficient drops with loading depending on the structure of the surface layer participating in friction.

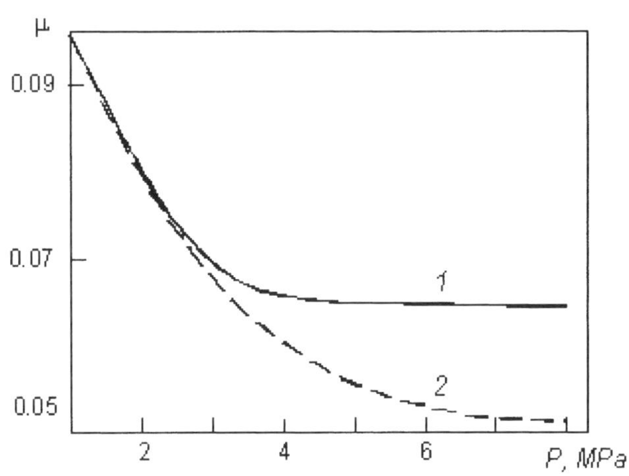

Fig. 6.14. Friction coefficient versus load for UHMWPE-CoCrMo alloy lubricated by Diasynol medication: *1* – initial UHMWPE, *2* – UHMWPE with microporous layer

The pairs with microporous UHMWPE samples display lower friction coefficients in contrast to the initial UHMWPE. Under $p > 4$ MPa the friction coefficient decreases down to the level typical of the natural joints operating at lubrication by the synovia.

The analysis of damages on UHMWPE samples with the microporous layer obtained during simulation tests of hip joints (dry friction, impact load $p =$ 12 N with swivel and return of the metal counterbody by $\alpha = \pm 50°$ angle) has shown the following regularities. Friction paths of crumbling on the initial UHMWPE samples and polymer grit appear after 10 hours of sliding (Fig. 6.15).

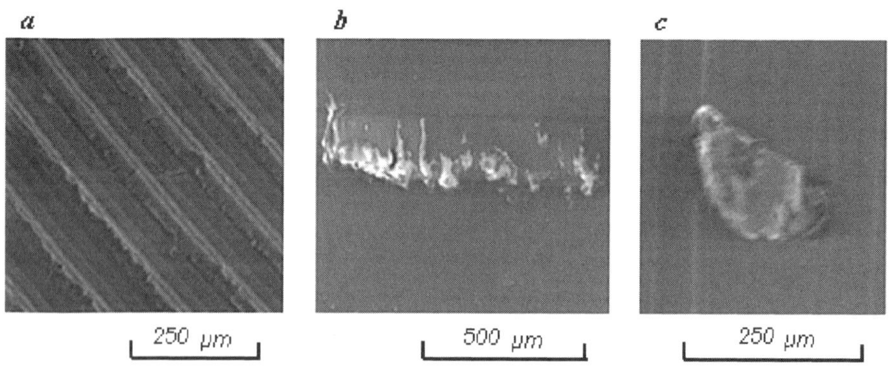

Fig. 6.15. Friction surface of reference UHMWPE sample: *a* – initial, *b* – after 10 hours of simulator operation, *c* – debris on the friction surface

One of the reasons of debris appearance in such a short period is the originally low quality of the friction surface with cutter traces (*a*) formed at turning. This is one more evidence in favour of the opinion that finishing of polymer parts of endoprostheses should be of high quality or otherwise this may result in elevated wear already at the run-in stage *in vivo* and aseptic instability. Run-in of the samples with the microporous layer takes a different in principle path, all other conditions being equal. It should be noted first of all that not a single crumbling track or traces of fatigue wear were detected on the microporous samples. Microroghness of the friction surface was found to smoothen already at the stage of run-in. It is evident (Fig. 6.16) that asperities of different levels formed at machining, smelting or warping of the surface at cutting become smoothed.

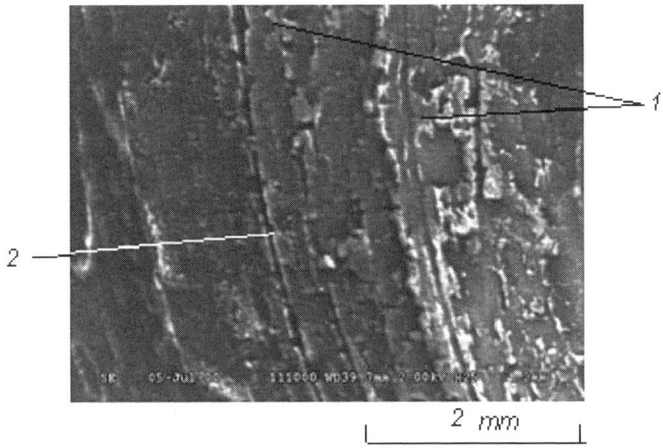

Fig. 6.16. Friction surface of the sample with microporous layer after the 100-hour of simulator operation without lubrication: *1* – smoothed asperities, *2* – cutter trace

Smoothening may presumably lead to crumpling and plastic deformation of the porous matrix walls. As a result, an elastic layer with the roughness typical of the conditions of a concrete studied pair forms on the friction surface. Owing to the gradient of porosity in the surface layer of the modified sample pressure distributes evenly over the friction surface. As a consequence, wear process of UHMWPE samples with the microporous layer is characterized by a soft loading and absence of fatigue wear of microasperities even at dry friction.

During the tests on a pendulum tribometer the friction coefficients of the pig cartilage-UHMWPE pairs lubricated by Diasynol were recorded. Similarly to UHMWPE-metal pairs (Fig. 6.14), the friction coefficient of the samples with the microporous layer was registered to lower by 30%. The optimum thickness (defined by the *min* friction coefficient criterion) of the microporous layer in pair with the natural cartilage is about 1000 μm. Note that determined earlier optimum thickness of the porous layer for the UHMWPE-metal pairs is within 200-400 μm. It can be anticipated that damages of the natural cartilage of the acetabulum in pair with the UHMWPE head of a unipolar endoprosthesis fit with an optimum in thickness microporous layer will be minimal.

To optimize the microporous layer parameters loading of the total hip endoprosthesis has been simulated based on a complex computation approach that combines the analytical contact problem solution for the spheres of close radii and the finite-element method of the stress-strain state of the endoprosthesis [37]. The computations have proved that the most loaded area of the polymer cup is the layer of about 100 μm thickness under the friction surface. This agrees with Dowson's conclusion [21] on prevailing fatigue wear mechanism of the prosthesis polymer friction parts and formation of subsurface cracks followed by polymer particles spalling from the friction surface. The porous layer created in the stressed zone filled with a liquid contributes to stress relaxation and abates fatigue wear of the polymer parts of endoprosthesis.

6.5 BIOCOMPATIBILITY

The degree of biocompatibility of implants is an integral parameter that characterizes time during which the implant may be present in the organism and function without inflicting any sickly reactions or disease. Along with requisite medico-biological and standard toxicological tests, biocompatibility of UHMWPE implants with the microporous layer has been estimated by the tissue homeostasis check, i.e. permanency of composition and properties of blood, which is the determining factor of the main physiological functions of the organism. The presented below data have been kindly furnished by Prof. Adamenko G.P. (Vitebsk State Medical University, Belarus) derived by the method and experiments on blood cells reaction to the artificial cartilage.

Thus, the object of investigations was a certified block UHMWPE of *Chirulen* grade, DIN 58836 C. The samples were modified by the method described elsewhere [28]. Their surface was plasma-treated in HF discharge, in the corona discharge (corona electrets) or electric field (thermoelectret) under elevated temperatures. Their polarity and the polarizing charge distribution were controlled. The efficient surface charge density (σ_{ef}) was estimated by the contact less compensation method using an oscillating electrode [38].

After modification and electrophysical treatment film samples in the form of a 5 mm in diameter ring were placed into a case for the immune-enzyme analysis. The structural and functional states of immune-competent cells of peripheral blood were estimated using multi channel antibodies in the reaction of indirect fluorescence [39]. With this aim, 0.2 ml of cultural cells suspension in (2–3) $\cdot$ 10^{-6} ml^{-1} concentration in a nutrient medium199 or RPMI–1640 has been used. The structural parameter of blood cells was estimated judging by the differential clusters (CD) of T-lymphocytes, the functional one – by generation of interleukins (IL-1 and IL-2), and myeloperoxidase (MPO) activity.

The structural characteristic of blood cells is defined by the degree of regulating T-lymphocytes of man's immune system. Lymphocytes are able to switch on or off the functions of other cells, or operate individually [40]. The experiments made use of T-helpers and T-suppressors with CD-3, CD-4 and CD-8 phenotypes, since just these cells govern to the utmost the reactions of transplantation immunity. Besides, the immune-regulation factor T_{CD} was estimated as ratio T-helpers / T-suppressors.

Functional activity of blood cells is characterized by the production of interleukins, among which IL-1 is the chief mediator (the substance changing cellular membrane penetrability) of the human immune system. It is produced by the cells of the macrophage-monocytic series and some other specialized cells in response to stimulation or damage brought by exogenous or endogenous agents. The acute-phase response is in its essence a protective reaction of the organism aimed at sustaining homeostasis. IL-1 raises functional activity of leucocytes and stimulates post-traumatic regeneration of the connective tissue [41]. IL-2 induces *in vitro* growing of undifferentiated cellular forms in cytotoxic natural killers and proliferation of lymphocytes of peripheral blood. Secretion and reception processes of IL-1 and IL-2 cells have been estimated by the restoration index of the blast-transformation reaction (RIBR) [39].

Metabolic activity of blood cells has been determined by the nitrosinim tetrazole test (NST-test). This reaction specifies metabolic activity of neutrophilic leucocytes connected with oxygen consumption [42]. Two test kinds were used to evaluate spontaneous (NST_{SP}) and stimulated (NST_{ST}) activity of leucocytes. The amount (%) of NST positive cells was registered by the microscope.

Myeloperoxidase is a phagocyte enzyme, which specifies bactericide property of these cells when participating in formation of microbicidal molecules. MPO activity is determined in relative units by the optic density of the cellular suspension under study [40].

The effect of UHMWPE samples on structural and functional state of immune-competent blood cells was estimated during model tests *in vitro* based on T_{CD3}, T_{CD4}, T_{CD8} as well as IL-1, IL-2, MPO, NST_{SP} and NST_{ST} indices.

The main factors of the effect of polymer samples on structural parameters and activity of blood cells are the surface layer structure and the polarizing charge value of the samples. As the laboratory check test of tissue homeostasis has shown, the samples with the microporous layer treated by HF discharge plasma possess the highest biocompatibility. The immune-regulation factor of these samples is within 1.5-2.0, which corresponds to the norm (the norm is the immune status of blood donors). RIBR index is 1.02 for IL-1 and 0.99 for IL-2, which also corresponds to the parameters of healthy people. MPO activity is 0.25 relative units (norm). NST_{SP} is 10% (the norm is 9-10%), and NST_{ST} equals to 41 (norm – 35-55%). The rest samples can be arranged in the following order according to their biocompatibility:

Block and microporous electrets ($\sigma_{ef} \geq 10^{-6}$ C/m^2)	$<$	Initial UHMWPE	$<$	Block electrets ($\sigma_{ef} < 10^{-6}$ C/m^2)	$<$	Microporous electrets ($\sigma_{ef} < 10^{-6}$ C/m^2)

In addition the following phenomena have been revealed:

- immune-competent cells preserve their intrinsic regulator and antiphlogistic activity in contact with the microporous samples, which is exhibited in a normal production of IL-1 and IL-2; neutrophilic agents elevate their bactericidal activity, T-lymphocytes preserve their receptor apparatus and regulation subpopulations;
- microporous samples do not excite any elevated activity of immune-competent cells (lymphocytes, monocytes, inducers);
- blood reactions on the natural pig cartilage and microporous UHMWPE samples treated in plasma are identical, which confirms high degree of approximation between physico-chemical structure of the man-made implants and the natural cartilage.

Besides, the results of medico-biological, chemico-toxicological, and hygienic tests of experimental samples of the polymer implants have lead to next conclusions;

- the samples do not exert any effect on organoleptic indices of aqua extractions;
- there is no methanol isolation from the samples into the model media (air, water), while formaldehyde is isolated in admissible amounts;
- haemolytic activity of substances migrating from the samples into the model medium does not surpass the admissible limit;
- the samples do not change pH of the model media;
- extracts of the samples do not produce irritating effects;
- biocompatibility of the material is supported by physiological tests based on the criteria of test culture growth and life of the protozoa;

◈ ◈ ◈

The reported results of original investigations meet modern trends in developing polymer bio absorbable materials for implants [10, 12, 15, and 27] and substantiate elaboration of the artificial UHMWPE-based cartilage. Its apparent advantage consists in the usage of UHMWPE as a reliable and biocompatible friction material for joint endoprostheses. The parameters of microporous and permolecular structures of the artificial cartilage show almost full similarity with the physico-chemical structure of the natural cartilage. In addition, the microporous polymer matrix can simulate the biophysical field of the natural cartilage by inducing the polarizing charge. The layer of the artificial cartilage can be made without impairing strength of the polymer part of the endoprosthesis. Already first variants of the artificial cartilage have proved to raise wear resistance of traditional endoprostheses designs. Viability of the new material is supported by biocompatibility tests during which the response of immune-competent blood cells to the implant has been studied. The identity of their response to the natural pig cartilage and UHMWPE-based microporous material makes grounds for most favourable clinical test results of endoprostheses with the developed artificial cartilage.

References:

1. Cooper J.R., Dowson D., and Fisher J. Macroscopic and microscopic wear mechanism in ultrahigh molecular weight polyethylene. *Wear*, 1993, V. 162–163, No. 6, p. 378–384.
2. Movshovich I.A. Instability reasons of hip joint endoprosthesis and revision endoprosthetics. *J. Orthopedics and traumatology*, 1993, No. 3, p. 5–10.
3. Morcher E.W. Endoprothetic. Berlin, Springer-Verlag, 1995, 431 pp.
4. Charnley J. Low friction arthroplasty of the hip. Berlin, Springer-Verlag, 1979, 355 pp.
5. Wroblewski B.M. 15–21 year results of the Charnley low-friction arthroplasty. *Clin. Orthop.*, 1986, V. 211, p. 30–35.
6. Kurtz S.M. The UHMWPE Handbook. Ultra-High Molecular Weight Polyethylene in Total Joint Replacement. Oxford, Elsevier, 2004, 379 pp.
7. Gavryushenko N.S., Urazguildiev Z.I., and Khoranov Yu.G. Tribological characteristics of home hip joint endoprostheses. *Proc. IV Congress of Russian traumatologists and orthopedists.* N.-Novgorod, 1997, p. 905–920.
8. Biomaterial Science. Ed. by B.D. Ratner, A.S. Hoffman, F.J. Schoen, and J.E. Lemons. Oxford, Elsevier, 2004, 864 pp.
9. Stoy V.A., and Kliment Ch.K. Hydrogels: Special Plastics for Biomedical and Pharmaceutical Applications. NY, TECHNOMIC Publ. Co., 1997, 455 pp.
10. Synthetic bioabsorbable polymers for implants. Ed. by C.M. Agraval, J.E. Parr, and S.T. Lin. Philadelphia, ASTM Int., 2000, 500 pp.
11. Ushakov S.N. Polyvinyl alcohol and its derivatives. Moscow-Leningrad, USSR AS publ., 1960, 320 pp.
12. Platai N.A., and Vasilyev A.E. Physiologically active polymers. Moscow, Khimiya, 1986, 296 pp.
13. Bray J.C., and Merrill E.W. Poly (vinil alcohol) hydrogels for synthetic articular cartilage materials. *J. Biomed. Mater.* Res., 1973, V. 16, p. 431–443.

14. Kempson G.E. Mechanical properties of articular cartilage. In: Adult Articular Cartilage, Ed. by A.R. Freeman. London, Pitman Medical, 1979, p. 333–413.
15. Oka M., et al. Development of an artificial articular cartilage. *Clin. Mater.*, 1990, V. 6, p. 361- 381.
16. Bacterial cellulose scaffolds for cartilage repair. *Materials today*, 2004, V. 7, No. 11, p. 28.
17. Sanders J. Biocompatible breakthough. Georgia Tech, 1999, V. 75, No. 4, p. 79.
18. Sementovskaya E., and Pinchuk L. A material for joint arthroplastics. *Proc. III Symp. "Inzyneria Ortopediczna i Protetyczna"*, Bialystok, Polska, 2001, p. 207–210.
19. Ukhartseva I.Yu., Shalamov I.V., Tsvetkova E.A., et al. Polarization characteristics of filled gels based on polyvinyl alcohol. *Plastic masses*, 1998, No. 4, p. 40–42.
20. Bone grafts substitutes. Ed. by C.T. Laurencin. Philadelphia, ASTM Int., 2003, 250 pp.
21. Auger D.D., Dowson D., Fisher J., and Jin Z.M. Friction and lubrication in cushion form bearing for artificial hip joint. *Proc. Inst. Mech. Eng*. Pt. H, 1993, V. 207, p. 25–33.
22. Belarus patent 2673, A 61 F 2/30, A 61 L 27/00. Joint endoprosthesis. L.S. Pinchuk, E.A. Tsvetkova, V.I. Nikolaev, and V.A. Goldade, 1999.
23. Tsvetkova E.N., Kadolich Zh.V., Goldade V.A., and Pinchuk L.S. Structural changes in the surface friction layer of a polymeric endoprosthesis cup. *Mechanics of Composite Mater.*, 2000, V. 36, No. 5, p. 365–372.
24. Pinchuk L.S., Goldade V.A., and Tsvetkova E.A. Polymer endoprosthesis friction materials with cartilage-simulating structure. *Abstracts of World Ttibology Congress,* London, 1997, p. 745.
25. Tsvetkova E., Kadolich Zh., and Pinchuk L. Polymer insert of hip joint cup endoprosthesis with modified friction surface. *Applied Mechanics and Engineering*, 1999, V. 4, Special issue: NCBS'99, p. 247–250.
26. Kadolich Zh.V. Physical modification of metal-polymer joints for raising wear resistance via simulation of biophysical properties of natural cartilages. Ph.D. Thesis, Gomel, 2002.
27. Pinchuk L.S. Polymers in joint endoprostheses. In: Problems in modern materials science. *Proc. III session of Scientific Council MAAN*, Kiev, 1998, p. 35–42.

28. Pinchuk L.S., Tsvetkova E.A., and Nikolaev V.I. Frictional material with the cartilage-simulating structure. *Friction and Wear*, 1995, V. 16, No. 3, p. 505–510.

29. Wu X.Sh., and Dong L.Ch. Controlled drug delivery systems. NY, TECHNOMIC Publ. Co., 1996, 287 pp.

30. Tsvetkova E., Pinchuk L., and Kadolich Zh. Cartilage simulation surface of frictional polymer material intended for endoprostheses. *Proc. II Sympozjuma «Inżynieria ortopedyczna i protetyczna»*, Białystok, Polska, 1999, p. 335–338.

31. Papkov S.P. Physico-chemical bases of processing polymer solutions, Moscow, Khimiya, 1971, 372 pp.

32. Pavlova V.N., Kopyeva T.N., Slutsky L.I., and Pavlov G.G. The cartilage. Moscow, Medicine, 1988, 320 pp.

33. Kadolich Zh.V., Gradzka-Dalke M., Pinchuk L.S., and Anisov A.P. Investigation of friction surface of polymer implants of the hip. *Friction and Wear*, 2001, V. 22, No. 1, p. 78–83.

34. Tsvetkova E.A. Effect of UHMWPE hydrophilic properties upon friction in artificial joints. *Int .J. of Applied Mechanics and Engineering*, 2002, V. 7, p. 51–54.

35. Blitshteyn M. Wetting tension measurement on corona-treated polymer films. *Tappi*, 1995, V. 78, No. 3, p. 138–143.

36. Mironov V.S. Electrophysical activation of polymer materials at frictional and electric effects, Dr. Sci. Eng. Thesis, Gomel, 1998.

37. Shilko S.V., and Nikolaev V.I. Simulation of adaptive reactions at endoprosthetics. *Proc. Int. Conf. Polycom'98*, Gomel, MPRI NASB, 1998, p. 321–225.

38. Russian State Standard 25209 - 82. Plastics and polymer films. Methods of determining surface charges of electrets.

39. Adamenko G.P. Receptor and mediator mechanisms of interactions between neurophils and monocytes in human blood at norm and disease with immunological component. Dr. Sci. Med. Thesis, 1993.

40. Sachek M.G., Kosinets A.N., and Adamenko G.P. Immunological aspects of surgical infections, Vitebsk, 1994, 140 pp.

41. Nasonov E.L. Interleikin 1 and its role in human pathology. *Therapeutic archives*, 1987, V. 59, No. 2, p. 112–115.

42. Roitt I.M. Essential immunology. 6th ed. Oxford, Blackwell Sci. Publ., 1989, 300 pp.

Chapter 7. SIMULATION OF BIOPOTENTIALS IN JOINTS

As an ancient craft studying living objects, medicine has been absorbing the achievements of the progressive scientific thought since its onset. Its present successes are unimaginable without the fundamental knowledge accumulated by other natural sciences, physics and chemistry above all. Modern representations on metabolism disclose all types of transformations of matter and energy occurring in the organism. Habitual functioning of joints, their adaptation to external conditions and curing of joint diseases are at present interlinked with the regulating effect of physical fields. The dominating place in achieving positive effects on the osseous tissue, cartilage, ligament, tendon, and muscles occupy the phenomena of the electric and magnetic origin. The overall fundamental legacy left by the outstanding physiologists I.M. Sechenov, S.P. Botkin, I.P. Pavlov and their disciples has visualized that the regulation of physiological processes provides coordinated activities of the organism as an integral whole and its separate systems under varying conditions of the environment.

A basic property of the living tissue is its ability to be in the state of nonequilibrium electrical polarization whose measure is a bioelectrical potential. The laws of its nucleation and disappearance in muscles and cartilages, ligaments and bones are interrelated with the results of joint endoprosthetics. The distribution of bioelectret potentials intrinsic for a healthy organism in bony and soft tissues alters during sickness. An efficient means of restoration of joints subjected to endoprosthetics is believed to consist in imparting a property to the prosthesis to be a source of electromagnetic field. Its aim is to compensate the disturbed natural distribution of bio potentials in a joint resulted from a surgical operation. It would be ideally to ensure the distribution of bio potentials in the tissues surrounding the endoprosthesis similar to that of a healthy limb.

Chapter 7 is devoted to electrical effects employed at endoprosthetics, including electrically stimulated growth and rehabilitation of joints; perfection of biocompatibility of endoprostheses carrying the polarizing charge; regulation of the tribological parameters of implanted endoprostheses. Furthermore, new data on electrophysical properties of the biological fluids like blood and synovia are cited, that disclose their response to the effect of electromagnetic fields. This chapter also presents information on novel approaches aimed at improving functions of artificial joints by exposure to electric and magnetic fields.

7.1 BIOPOTENTIALS AS A PROPERTY OF LIVING MATTER

Bioelectrical potentials or bio potentials represent electrical potentials generated in the tissues or individual cells of living organisms. The bio potentials play the most important role in the process of excitation and inhibition of cells.

The prehistory of studying bio potentials dates back to the 18th century when the scientists attempted to analyze the nature of a shock inflicted by the fish with electricity generating organs. This is how the first scientific substantiation has been achieved about the existence of animal electricity. This was the time when an Italian anatomist and physiologist L. Galvani started investigations of bio potentials on other objects. A scientific dispute between him and a physicist A. Volta on the origin of animal electricity has brought to a discovery of a new principle of obtaining electrical current using a galvanic element. In 1837 an Italian physicist C. Matteucci was the first to measure bio potentials using a galvanometer during his experiment in animals. E.H. Du Bois-Reymond, a German physiologist, pioneered a systemic examination of bioelectrical potentials and proved in 1848 that there existed a standard potential difference between the innards of a cell and surrounding it liquid, which he called a *membrane potential*. The value of the membrane potential in rest varies regularly when the cell is excited. J. Bernstein is ascribed to be the first (1868) to analyze bio potential oscillations occurred during spreading of the excitement over a nerve fibre lasting a few thousandth portion of a second (action potential). In 1883 a Russian biologist N.E. Wedensky used a telephone to record the charges that accompany nerve impulses [1].

The use of electron-amplifying machinery and inertia less oscillographs in physiological experiments (1930-1940-ies) is associated with the names of American physiologists G.H. Bishop, and Nobel Prize winners J. Erlanger and H.S. Gasser. Investigations of bio potentials of separate fibres and cells have become possible thanks to the methods of introducing measuring microelectrodes into the cells. The mechanism of bio potential regeneration has been studied on the gigantic nerve fibres of squids by estimating their penetrability for Na^+ and K^+ ions depending on the membrane potential. For deciphering the ionic mechanism of generating the potential of action and elaboration of the membrane theory of bio potentials (1947–1952), English physiologists A.L. Hodgkin, A.F. Huxley and B. Katz were awarded the Nobel Prize [2, 3]. The results of their research have formed groundwork for employed at present electrophysiological methods of diagnostics such as electrocardiography, electroencephalography, electromyography and other.

The phase theory of bioelectrical potentials developed by D.N. Nasonov is based on the understanding of protoplasm as a phase relative to the surrounding aqueous medium. According to the theory, charge carriers are spreading between the cells and medium depending on solubility of the substance in protoplasm, its adsorptive capacity on micelles and chemical binding with the protein substrate [4].

The coarse cells of algae turned to be a classical object for studying the ionic nature of bio potentials. Using these cells, Du Bois-Reymond detected in 1882 the cognate features between bioelectrical phenomena in animals and plants. The plants possessing the growth motion were the first research objects. Thereafter, an Indian biophysicist J.Ch. Bose has established that bio potentials and electrical responses on excitements are intrinsic for all plants. Designed by him sensitive self-recording galvanometers were able to record the electrical reactions of plants to physical and chemical effects.

In early 1990ies, scientists of the Moscow State University attempted to study electrical properties on the example of a simplest of higher plants, i.e. a tissue in the form of merged cells without cell walls. The investigations of electrical responses of the plants on mechanical and heat damages, regeneration of the root, stalk and leaves have shown that the mechanism of electrical signals spreading over the nerve fibre is similar to the one generated in the simplest [5].

Electrical polarization has been detected in the majority of biopolymers like proteins, enzymes, polysaccharides, polynucleotides, and etc. The polarization is induced by a directed orientation and uneven distribution of the dipoles, impurities and defects. To these are also added the ions that provide the conduction of their own and of the impurities as well as electrons, and structured water bound with macromolecules [6].

As far back as the 1960ies the origination of quantum biochemistry began, which sprung up at the junction of molecular biology and quantum chemistry. Quantum computations have made it possible to transfer from the simplest notions on the composition, distribution and spatial localization of the atoms in organic molecules to evaluation of distributions of electron density in them, energy characteristics, and afterwards to interrelations between charged particles and the biological function of organic compounds [7].

The terms and main regularities of variations in bioelectrical potentials as a fundamental characteristic of the living matter are set forth below.

The rest potential (U_r) is the difference of potentials between cytoplasm and extra cellular liquid, which exists in the alive cells in the state of physiological rest. It appears due to unequal concentrations of K^+, Na^+ and Cl^- ions on the sides of the cellular membrane and it is unlike penetrability for these ions. U_r of most of cells is generated by the diffusion of K^+ ions from the cytoplasm out. In the skeletal muscle fibres U_r is maintained by the diffusion of Cl^- ions from the outside medium into the cytoplasm. The range of U_r measurements in nerve and muscle fibres is within 60-90 mV. The inner side of the membrane is commonly positively charged relative to the external one. The electrical current passed through the membrane and instability of its ionic penetrability brings about changes in U_r.

The reduction in U_r is called *depolarization.* The passive depolarization appears when a weak current is passed through the membrane (from the cell) leaving its ionic penetrability intact. The active depolarization is stimulated by increased penetrability of the membrane for Na^+ ions or reduced for K^+ ions. A prolonged depolarization of the membrane leads to inactivation (inertness, passivity) of sodium channels and raised potassic permeability. This results in the reduction or full loss of the cell excitement.

Increase in U_r is called *hyper polarization.* The passive hyper polarization occurs when the electric current passes from the outside into the cell. The active hyper polarization takes place at elevated penetrability of the membrane for K^+ and Cl^- ions. Local hyper polarization of the membrane sets during activation of the ionic channels by a physiologically active substance isolated from the nerve-ending at its excitement.

The potential of action (U_a) represents swift oscillations (spike) of the membrane potential generated by the excitement of the nerve and muscle cells. It appears as the signal-irritant reaches some threshold, which if exceeded, can effect neither the amplitude nor U_a duration. The U_a is caused by the activation of electrically excited ionic channels. The ascending U_a phase in the nerve and skeletal muscle fibres is attributed to raise penetrability of the membranes for Na^+ ions. Their flow inside the cells over open channels leads to fast recharging of the cellular membrane. Its inner side is charged negatively at U_r, and acquires a positive charge when U_a is in the peak. Inactivation of Na^+ channels and activation of K^+ ones following the U_a peak results in a drop of U_a. Its restoration till some initial value is preceded by the track depolarization or hyper polarization of the membrane. The duration of U_a is 0.1-3.0 ms in the nervous cells and 10-100 ms in the myocardium cell.

Refract ability of the cells means the reduction of their excitability that accompanies the appearance of U_a. As soon as the U_a peak is reached, the excitement disappears fully, which is called the absolute refract ability. Drop in U_a leads to the restoration of excitability of the cells till the initial value within a few ms (relative refract ability). Refract ability is one of the factors determining the maximum pulsation rhythm of the cells. Various drugs that prolong the period of the relative refract ability (antiarrythmic) lower the frequency of heart contractions and obviate violations in its rhythm.

The theoretical notions on the mechanisms of the emergence and transformations of the membrane potential based on a simple model of the double electrical layer undergo a continuous evolution. Their refined models cover in homogeneity of dielectric properties across the membrane thickness [8], intracellular gradients of electrical fields, the effect of dipole-induced field of peptide links in membrane proteins [9], the presence of a cation-exchange layer on the cell surface [10], the membrane induction parameters [11]. Certain approaches for calculations of bio potentials that take into account cellular metabolism, ion inflows and outflows in the cells and outside them, structural changes in the membrane and cytoplasm, and so on are in the process of elaboration. The bioelectrical potentials spontaneously arising in living organisms underlie the bioelectret effect [12].

7.2 ELECTRICAL FIELDS IN MEDICINE

Bio potentials characterize two opposing functions performed by the cellular membranes, i.e. the barrier that protects the cells from the foreign matter, and the transporting function that supplies substances necessary for the cell vitality. The experience in utilizing electrical fields in medicine aimed at regulation of named functions has visualized their valuable assistance in diagnostics of diseases, delivery of drugs into sick organs, physiotherapeutic curing, and so on.

Diagnostic methods of diseases based on recording bio potentials are available presently for various organs generating bio potentials. Human organs viewed from the theory of the bio field are modelled as 3D anisotropic conductors with a specific bio potential distribution [13]. Each organ is characterized by its proper resistance, zero potential areas, and high (1–10 mV) absolute negative and positive bio potential points.

A method of bio field diagnostics most widely applicable today is *electrocardiography*. It is used for functional investigations of heart diseases and is based on recording kinetic dependencies of heart bio potentials. *Electrocardiogram* represents a kinetic curve of the potential difference of the electric field recorded by special devices during the heart systole. The *electrocardioscope* takes the parameters of heart activity using an electron-beam tube. The *electrocardiophone* transforms bio potentials of the heart into sound signals. For diagnostics of rhythm disorders, heart conductivity, interventricular and interatrial septum defects, more intricate methods are used. *Intracardialogical electrography* registers the time dependence of heart bio potentials using electrodes placed on the heart cavity walls. The *intracavitary electrocardiogram* of the heart presents a kinetic curve of its electric field recorded by the electrodes introduced in the cavities over the blood vessels.

A vast group of methods has been developed for diagnostics of the functional state of brain. Using the *electroencephalography* the cerebric bio potentials are recorded in time. By means of *electroencephaloscopy* the bio potentials of separate brain zones are displayed as a set of dots with varying brightness and size. *Electroencephalophony* is based on converting infrasonic oscillations of bio potentials into sonic ones. *Electrocorticography* is often applied during surgical operations. This is a method of studying the cortex of cerebral hemispheres by recording its bio potentials with electrodes placed in the needed areas. The method of *electrosubcorticography* is aimed at recording the bio potentials of sub cortical structures using introduced electrodes.

The nervous system condition can be analyzed by the method of *electroneurography* during which time-dependent changes of the action potentials (U_a) of the peripheral nerves are recorded.

For the diagnostics of gastrointestinal tract one can use the following methods. *Electrogastrography* is intended to study the motor activity of the stomach via recording its bio potentials in time. *Electrointestinography* is a similar method for studying intestines. The bio potentials of various sections of gastrointestinal tract are identified by the frequency of their oscillations (small intestine 0.1–0.3, and 0.015–0.30 Hz large intestine).

In ophthalmology the bio potentials are recorded to study nystagmus, i.e. involuntary rhythmic biphase (fast and slow phases) movements of the eyeball. *Electrooculography* investigates the bio potentials of the oculomotor muscles and outer layers of the retina. *Electroretinography* assists in recording bio potentials of the retina appearing at light-excitement of the eye.

Functional diagnostics of the muscular system is unimaginable today without *electromyography* – a method of graphical registration of bio potentials of skeletal muscles. Using the kinetic dependencies of bio potentials (myograms) it is possible to record time and oscillation amplitudes of muscle bio potentials to estimate their activity. *Electromyomasticatiography* is a simultaneous graphical registration of bio potentials of masticatory muscles and movements of the lower jaw.

Computer electrostructurography and *computer electrotopography* are the methods of recording electric images of the structural and functional states of man's innards [14]. Needle-like (non-invasive and painless) electrodes envelope man's body in a studied anatomic cross-section. Asymmetry of the active (defined by bio potentials) and reactive (current-induced response) electric characteristics are recorded by passing the high-frequency (120 kHz) electric current (field intensity – 10 V/cm). The degree of asymmetry is estimated by a diagnostic computer program intended for evaluation of pathologic changes and sick organs

Galvanodermic reactions (GDR) of a man on the electric signal depend on the distribution of his bio potentials, state of epidermis and sweat glands. The amplitude, polarity and configuration of GDR are responsive to disorders of the nervous system and internal organs. Analogous information can be obtained by the method of *electro acupuncture* based on the measurements of electrical resistance of biologically active coetaneous points.

The method of electromagnetic HF-radiation is utilized for non-invasive *in situ* determination of the vitality of soft tissues. The difference in amplitude and phase of the radiation reflected from the dying off tissue areas (U_a=0) in contrast to the healthy ones, complies with the severity of necrosis [1].

The regulated drug delivery within the organism is an urgent medical problem. Medical preparations usually consist of two main parts, namely, the active medicinal substance (MS) and components of the medicinal form (MF).

The latter makes the drug convenient for usage and provides favourable conditions for its effect in the organism. The traditional MF are tablets, capsules, ointments, or solutions for injections. They, however, are not considered an optimum means for this purpose and often fail to carry the MS in the needed part of the organism. Modern directions in medicine have assigned the MF a function of a prolonged and continuous carriage of MS into a targeted organ according to a set program. Many elements of these systems consist of polymers, therefore the new generation of MF is called a macromolecular therapeutic system (MTS). They were first created by A. Zaffaroni and his associates in the USA, Alza company, in the 1970ies as antianginal, antiseasick and antihypertensive preparations [15]. After this, MTS for curing diabetes and chemotherapy in oncology were developed with a long-term action.

In Fig. 7.1 one can see the kinetic dependencies of MS migration *in vivo* from different MF.

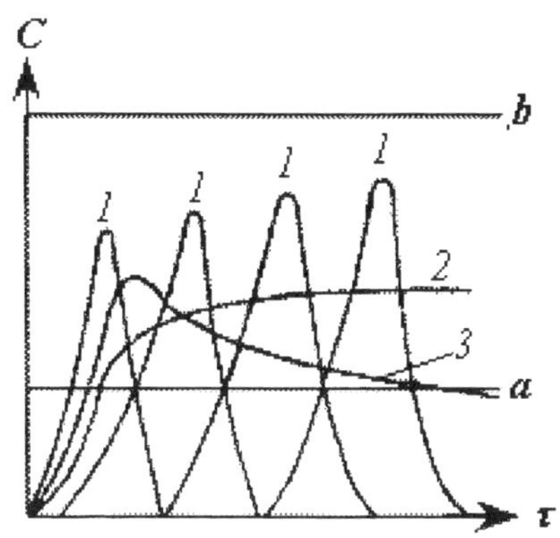

Fig. 7.1. Kinetic dependencies of medicinal form concentration (*C*) in blood plasma during: 1 – periodical injection of traditional MF; 2 and 3 – continuous delivery from MTS; *a* and *b* – lower limits of the efficient and toxic concentrations of MS

A traditional MF is unable to provide a uniform delivery of MS into the organism. The majority of advanced MS are actively participating in metabolism and their concentration in blood drops rapidly and demands an additional dose. A serrated scheme $C = f(\tau)$ (curve 1) is nonoptimal from the standpoint of therapy of chronic diseases. A uniform isolation of MS from the MTS within a required time (2) or ejection of some elevated dose of MS followed by a constant or attenuating isolation from the MTS (3) are more preferable.

The task of regulated delivery of MS into the organism following a set program should account for or employ the bioelectric potentials. Two trends can be traced in this aspect.

Firstly, the bio potentials condition the regulated isolation of MS from MTS due to either dissolving or bio destruction of the polymer shell of MTS. This may occur when the system of retaining MS in the polymer carrier is altered. MS ions can be kept in MTS via the electrostatic attraction forces to the polymer shell possessing an unlike charge. MS of the hydrogel MTS are kept in a spatial macromolecular lattice of the hydrophilic polymers with the parameters varying in response to swelling. The membrane MTS form a reservoir for MS to govern its diffusion rate through the walls. The membrane penetrability is regulated by changing its charge state or by some other mechanisms. Microcapsules and liposomes are the representatives of the membrane MTS. The microcapsules represent a miniature MTS (0.1–100 μm) whose shells are made of polyelectrolytes. Liposomes with a mono- or polylamellar phospholipid shell present micro reservoirs for MS. A lamella is a two-layered membrane structure consisting of oriented, viz. polarized, molecules of phospholipids. The structure-regulating processes of the polymer components in MTS resulting in MS liberation take place in the organism bio field and, hence, are dependent on the distribution of bioelectrical potentials. These regulation mechanisms are individual for each kind of MTS. Whatsoever interesting this topic be, it goes beyond the frames of our discussion and will be omitted.

Secondly, micro-MTS circulating in the blood or other system carry a polarizing charge and can be sorbed by the target cells whose bio potential carries an opposite sign. The polymer shell of micro particles is afterwards absorbed by the cells (phagocytosis) and the MS is liberated. The bio field of the cells stimulates sorption of the polar molecules of MS on the cellular membranes.

Physiotherapy is a section of applied medicine that studies the therapeutic effect of natural and artificial physical factors on the human organism. The aim of the physiotherapeutic effect is either rehabilitation or compensation of the membrane potential of cells lost during disease. The electrical fields are known to be among most efficient physiotherapeutic factors.

Electrical treatment is a kind of physiotherapy that uses a dosed effect of electrical current and electric or magnetic fields on patients [1].

The low-voltage electrotherapy is wide-spread today. The current is passed via the terminals and wires to the electrodes placed on a patient. The current can be passed continuously (*galvanization*) or as pulses alternating with pauses. This stimulates ionic motion and initiates physico-chemical processes at the cellular level, which leads to restoration of bio potentials. The response reactions of the human organs and systems improve blood circulation, sleep, sooth pain and inflammation, recover injured tissues, and so on. The effect of currents is intensified during a *medicinal electrophoresis* when the electrode is placed on a lining moistened in MS. The MS penetrates into the tissues and effects the area jointly with the current. The medicine that spreads poorly over the affected area can be impregnated in a gel-like polyelectrolyte composition. The *electrophoretic liberation* of MS from the gel exerts a strong curing effect [16]. Both methods induce a synergetic complex effect of the current and MS whose efficiency exceeds the sum of the factors taken separately.

The effect of the high-voltage pulse current has been called *darsonvalization* by the name of its founder, a French doctor G.A. D'Arsonval. A weak spark discharge appearing between the skin and glass electrodes filled by a luminous gas irritates the nerve endings thus intensifying blood circulation, nutrition of the tissues and relieving pains.

Franklinization is a therapeutic method employing a constant high-voltage electrical field. The needle electrodes are placed overhead or over the wound, ionize air molecules in a strong electric field that are adsorbed by the tissues and polarize them. Breathing of the ions is also useful for treating diseases of the respiratory tract.

UHF therapy means the use of the alternating electrical field of ultra-high frequency. The condenser plates installed over the injured area bring about a feeling of warmth, relieves pains, renders an antiphlogistic effect, dilates vessels and improves nutrition of tissues.

Microwave therapy consists in the effect of super-high frequency fields (about a centimeter-decimeter range) on the injured tissues. The radiation energy penetrates deep into the tissues (a few cm) leading to dilation of the vessels, pain alleviation and improved nutrition.

Electro aerosol therapy presupposes the use of aerosols whose particles bear an electric charge for curing aims. Electro aerosol inhalation means inhalation of a medical aerosol. Colloidal particles of MS are attracted electrostatically to mucous shells of respiratory organs where the charge of the particles dissipates on the cellular membranes and the substance is quickly absorbed in blood. This procedure exerts a synergetic effect.

Electro analgesia is aimed at relieving painful sensation during the effect of the electric field or current on the central nervous system or directly on the painful area. *Electro puncture* is a variant of this procedure, during which the biologically active points are effected by the electric current using needle electrodes. The authors have proposed a device for the electro puncture using needles of electrically polarized dielectrics [17]. The electric current applied in acupuncture points exerts a general stimulating effect on the patient. This is accompanied by the appearance of oxidation products in blood, which lowers the amount of free radicals and paramagnetic centres in it [18].

The aforementioned methods of electric treatment are among most frequently used techniques, but their range is by far broader. Once considered a novelty the physiotherapeutic procedures intended for stimulation and restoration of the natural distribution of bioelectrical potentials in the organism are now a reality and continue developing [19].

The potentialities of using the theory of bioelectrical potentials in traumatology and orthopaedics are set forth in the chapter to follow.

7.3 ELECTRICAL EFFECTS IN TRAUMATOLOGY AND ORTHOPEDICS

Positive results of the usage of electrical fields in endoprosthetics of joints are related to, first of all, electric polarization of implants made of dielectric materials. An *electret* is a dielectric able to preserve its electrical polarization within a prolonged time. It is an electrical analogue to a magnet [20]. Two main aspects of using electrets in endoprosthetics, namely electrical stimulation of osteoreparation and biocompatibility of implants are presented below.

Electrical stimulation of osteoreparation, i.e. growth and rehabilitation of bones, became possible after the detection of electrical signals in the osseous tissue. These signals are generated without the application of any external electric field or mechanical loads.

The bioelectrical potentials recorded *in vivo* on long bones using insulated from each other electrodes contacting the bone are of the order of an mV [21]. The osseous tissue is more electrically positive than the marrow. A significant negative charge is formed in the vicinity of fractures and bony neoplasms [22]. The clenched surface of a bone charges negatively at flexion and is positively at extension [23]. The bio potentials are supposed to be generated in the bones by piezoelectric and electro kinetic mechanisms. The ability of bones for exhibition a piezoelectric effect (generating electric charges at straining) has been first recorded as far back as the 1950ies [24]. The electro kinetic potentials difference arises in the direction of the flow of the biological fluids containing ions (blood, synovia), in the capillaries of the osseous tissue, cartilage and connectivum [25].

The potentials recorded in bones are grouped into the static and dynamic ones [26]. To the first group we relate the potentials of rest, stress, growth and regeneration. The dynamic potentials involve piezoelectric, pyroelectric and electro kinetic ones. The rest potential of the osseous tissue measured by a potentiostat shows that the epiphysis possesses the most positive potential, while the diaphysis is charged negatively and the metaphysis is neutral or weakly negative (Fig. 7.2).

The potentials distribute unevenly also over the bone cross-section [27]. Consequently, the metal implants wetted by biological fluids and contacting unlikely charged bone areas are inevitably involved into the electrode processes.

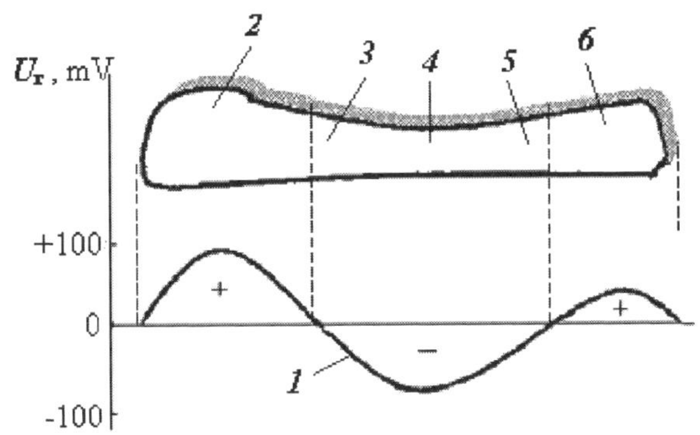

Fig. 7.2. Distribution of the rest potential (*1*) along rabbit is femur: *2, 6* – epiphysis; *4* – diaphysis; *3, 5* – metaphysis

We distinguish between three stages of the osseous tissue healing after traumas from the standpoints of electro genesis [26]. The first stage (10-14 days) corresponds to the amplitude of the action potential U_a= 450–500 mV. During the second stage the amplitude U_a lowers down to 300 mV. Then, the potential drops abruptly to U_a=50 mV together with other electrical parameters of the bone till a background level. The rest potential U_r of the regenerating bone tissue was found to be always more negative than that of a sound bone. Once violated under some outer or inner factor this regularity may hamper or restrain the process of the bone tissue reparation. Stimulation of the osseous tissue growth is accompanied by the activation of the marrowy blood circulation.

The appearance of U_r in the osseous tissue is attributed to the vital activities of its cells [27], whereas that of U_a with generation of piezoelectric, pyroelectric and other mechanoelectric effects in response to straining of the collagen and hydroxyapatite components of the osseous tissue. This is the reason why the bio potentials redistribute in case of bone deformations.

The electric stimulation of bones is used to cure fractures, osteoporosis and shortened limbs. The electric field is generated by the electrodes applied on skin or introduced surgically inside a patient's body [28]. Besides, induction coils can be used with this aim when placed on a sick extremity [29]. The experiments in white rats have confirmed the dependence of osteoreparation on the electrode potential difference of the implant metal contacting the bone [30].

The investigation results reported in [31] are concerned with the effect of electret implant fields on the compensation of injured bony and cartilage tissues. The electrets based on tantalum oxide (Ta_2O_5) and polytetrafluoroethylene

(PTFE) were formed as coatings on the implants of highly pure tantalum. The Ta_2O_5 coatings of 0.15–0.45 μm thickness were obtained by the electrolytic anodizing. The polymer coatings of 30–40 μm thickness were formed of a PTFE aqua suspension followed by drying and fusing. The coatings were electrified using a corona discharge or a constant field applied to the coatings with the help of auxiliary liquid electrodes.

When endured in a physiological solution, the electrified samples showed a reduced surface charge density, which restored partially after the removal from the solution. The experiments with cell structures in animals and under clinical conditions have, however, proved that the electrified samples were able to stimulate physiological processes even after a protracted endurance in conducting liquids.

The growth of fibroplasts was studied in a pure cellular structure. In the reference experiments the fibroplasts were grown under preset conditions in common flasks, and in flasks having electret PTFE films on the bottom. The microscopic analysis has recorded a greater amount of fibroplast colonies in the flasks with the electrets in contrast to the reference ones. Two to three weeks later the fibroplast cells got oriented along the force lines of the electret fields. An optimum magnitude of the electric field voltage has been noticed for the fibroplasts. The cells were less in size while their density was the maximal at a certain distance from the electret film that corresponded to a highest regenerative capability of the cells [31].

Above results conform well to the experiments in animals. Radial bones of rabbits subjected to a transverse osteotomy were recovered using special pins with Ta_2O_5 coatings. The pins with electret coatings induced an accelerated mineralization and recombination of the callus in contrast to non-electret ones. The greater mineralization of the cartilage tissue was observed in the contact with a negatively charged surface of the implant. The positively charged implants exert less effect on mineralization, like non-electret specimens. In the experiments in dogs the electret PTFE fixatives were placed in the region of the tibial osteotomy. As compared to the reference fixatives (non-electret) the electret ones hastened noticeably osteoparation, shortened hardening time and reconstruction of the callus, and promoted early restoration of the mechanical strength of the bones.

Clinical experiments on treating closed, open and gunshot fractures were carried out with the use of metallic locks fit with electret coats. The mean rehabilitation time and functional recovery of the extremities was proved to shorten 1.5–2 times, and the number of bone non-union reduced noticeably [31].

The mechanism of the electret field effect on the osteoreparation process can be viewed as follows. Within the conducting and screening medium of the organism the electret field spreads to a slight distance and effects therefore only the neighbouring cells. More distant cells are activated, presumably, by the nervous conductors and via a token passing of the excitement. This activation may spread to considerable distances and stimulate osteoreparation.

Some examples of using electret implants are cited below.

The osteoreparation stimulator [32] is made as a polymer film. Its surface contacting the bone bears a negative charge, whereas its reverse side has a metallic coating (Au, Ar, Al, or other) to prolong the electret service life in the organism. The charge surface density is maximal in the centre and decays with approaching the stimulator's edges. The stimulators of a needed configuration and size are cut from the electret film samples and attached cutanuously or to bone fragments for curing fractures.

A device for osteosynthesis [33] is made in the form of a metallic perforated plate coated with a dielectric material carrying a polarized charge. The maximal value of the surface charge density coincides with the mass centre of the plate and diminishes to the plate edges at a given gradient.

The carbon plastic endoprostheses of joints were furnished with piezoelectric transducers [34]. Their signal value is a measure of the mechanical stresses in the bone-endoprosthesis contact, which reflect the callus formation rate. The described system in which the current signal accelerates osteogenesis may form the base for smart endoprostheses able to regulate the state of osseous tissues in the zone of endoprosthesis fixing.

The effect of electrical fields on biocompatibility of implants is an intensely explored today direction of medicinal techniques. From the standpoints of physical chemistry, biocompatibility of solid bodies is conditioned by the adsorptive processes. Hydrogen atoms and hydroxyl groups OH^- are known to adsorb on the implant found in a biological fluid. The former creates positive charges on the implant surface during adsorption, while the latter generate negative charges. The content of H^+ ions in the solution imposes an effect on the solution pH. There exists a threshold value of pH for each implant, under which the concentration of adsorbed OH^- and H^+ ions is equal. This pH value is called a zero charge point (ZCP). For TiO_2 implants pH_{ZCP} = 6.2, for Al_2O_3 it is pH_{ZCP} = 2.3. In most cases the implants are not electrically neutral but carry either a negative or positive adsorptive charge that initiates the formation of a double electrical layer (DEL) in the biological fluid [27].

Biocompatibility is believed to be dependent on the surface charge density ratio of the implant to the bioelectric potentials of the surrounding osseous cells. The surface charge density of the implant seems to shift pH_{ZCP} from the value corresponding to the electrically neutral implant and define the DEL structure round the implant. These notions present a physico-chemical basis for regulating biocompatibility of implants using electrical polarization.

The experience in endoprosthetics supports an opinion that biocompatibility depends much on the surface charge density of the implant. The proofs are the next [35]:

- cellular cultures grow well on the surface of polystyrene films treated with plasma or corona discharge and poorly on untreated ones;
- in the initial reaction between the implant and living tissue, leucocytes respond to the charge state of the implant surface by liberating signal bands of cytokines that promote healing of the operation wound;
- most favourable biological reactions from the point of biocompatibility are characteristic for bioactive polymers with high concentrations of ionogenic groups and radicals, i.e. the materials with surface-immobilized bio molecules, biodegradable materials and some other.

Plasma treatment of artificial PTFE-based blood vessels results in the generation of surface polarized charges on the implants [36]. This leads to improved adhesion of epithelium to vessels and speeds up their growing into the tissues. Negative charging of the inner surface of artificial vessels to avoid thrombosis has become a classical example of the usage of electrets for controlling biocompatibility of the implants and blood [6].

Tyrosine-substituted polycarbonates are employed for preparation of biodegradable fixatives of small bones of the wrist and hand. Their surface layer contains a considerable quantity of radicals and perfect biocompatibility almost similar to that of hydroxyapatites. The pins of a modified polycarbonate carrying a surface charge were implanted into a rabbit is hollow bone. A 48-week observation has proved a tough adhesion to the bone. The histological analysis has visualized that the bony needle-like formations (spicules) penetrate into the cracks and slits formed on the pin surface in the organism [37]. This reaction of the organism to the tyrosine-substituted polycarbonate differs them drastically from other polymers of orthopaedic purposes.

Physico-chemical methods of surface modification of implants have proved to be a promising means of regulating biocompatibility [38]. So-called polymer brushes represent a monolayer of macromolecules adsorbed on a solid body surface. They are oriented normally to the surface, densely packed and have one end of their chain linked in adsorptive way with the solid body. This imparts a charge to the solid body, which is similar in sign to that of the free end of the chain. It defines to a great degree biocompatibility of the implants. The surface of any implant covered with a polyethylene glycol brush is hostile towards protein adsorption but is highly compatible with blood [39].

The surface modification schemes of solid bodies by the polymer brushes are illustrated in Fig. 7.3 [40]. The implant surface potential can be varied substantially depending on the nature of functional groups on the free ends of the brushes (*a–c*). The polysilamine-based brushes (a polymer with a vinyl silyl group on one end of the macromolecule and a diamine one on the other) are sensitive to variations in the ambient medium. The surface properties of implants can be changed by regulating either pH (2 till 10) or temperature (20–40 °C) thus making it positively charged and hydrophilic or negatively charged and hydrophobic. In the left side of Fig. 7.3, *d* the polysilamine chains are curled up into a ball (pH > 6, amonigroups are deproteinized), in the right corner the chains are strictly oriented (pH > 6, the amonigroups are proteinized). This helps to control the implant surface bonding with the components of the biological medium (proteins, amino acids, cells, etc.).

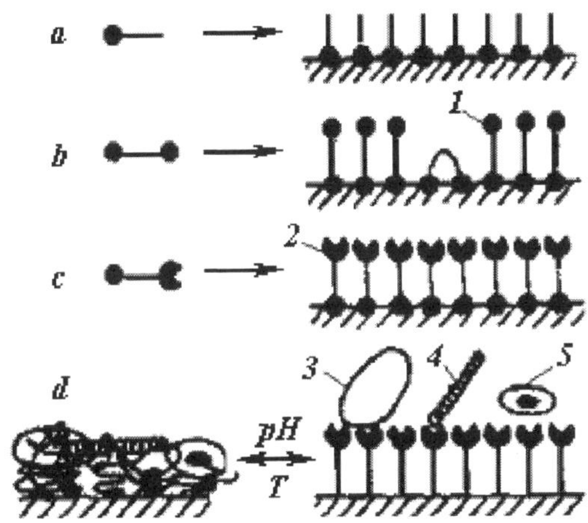

Fig. 7.3. Diagrams of solid surface modification (*a-c*) by oligomers with functional groups on one (*a*) or both ends (*b, c*) of molecules, and a scheme of regulating activity of implants (*d*): 1 – hydroxyl, silanol or thiol; 2 – amino group; 3, 4 and 5 – proteins, DNA and micro organism cells

An optico-electronic technology developed lately for regulating properties of solid bodies [41] can be also used to improve biocompatibility of implants. The original techniques of manufacturing thin-film structures carrying a surface charge of one or another sign expand the potentialities of the Langmuir-Blodgett method consisting in the transfer of condensed films from the liquid surface onto the implant as they cross the liquid-gas interface.

7.4 ELECTROPHYSICAL PROPERTIES OF BIOLOGICAL FLUIDS

Just as natural, so artificial human joints are operating *in vivo* at lubrication by a synovial fluid. It is shown in [42] that its quality and delivery properties to the articular cavity impose an effect on intrastructural changes that take place in the transitional zone between the articular cartilage and synovial membrane preceding the joint disease. It is been reported earlier that the mechanism of synovia transport to the joints is realized in a bio field created by the living tissues. This is why the synovial fluid displays a specific complex of electrophysical properties, which predetermines its sensitivity to electromagnetic fields. A review of traditional methods of laboratory analysis of the synovia applied today in medicine and a method proposed by the present authors to estimate its structural state using the electret-thermal analysis is given below.

The methods of laboratory analyses are acknowledged as most informative and simple means of studying synovia. For the diagnostics of joint diseases we usually evaluate four functions of the synovial fluid:

- the metabolic function consists in removal of decomposition products of the synovial shell cells and wear debris of the cartilage via the vascular system;
- the tribological function presupposes the provision of joint lubrication and a uniform pressure distribution over the cartilage surface during loading of the joint;
- the trophic mechanism is intended for transporting energy-generating substances to the vessel-free cartilage;
- the barrier function is to protect the joint from foreign protein compounds and its own denatured proteins by means of phagocytosis.

A healthy synovia is sterile, transparent and viscous. It can be extracted by the puncture and estimated by the colour, viscosity, transparency, the character of a mucin clot and cellular composition.

The synovial colour depends on the amount of foreign inclusions, which are also a sign of joint disease.

Viscosity is a function of hyaluronic acid (HUA) content in the synovia. An expressed reduction in viscosity at rheumatoid arthritis is because of low concentration of hyaluronic acid HUA.

The state of mucins (glycoproteins, complex compounds of HUA with proteins) is evaluated by impregnation of a glacial acetic acid into synovia (Ropes test) [43]. The resultant mucin clot is homogeneous, elastic and viscous in the case of a normal synovia, and is diluted and turbid if the synovia is pathological.

Physico-chemical methods of analysis are used to define the content of proteins, glycogen and other polysaccharides as well as lipids in the synovia.

The concentration of proteins of a healthy synovia makes up 1/3 of their concentration in blood serum, which rises during rheumatoid arthritis and may reach that in blood [44]. A spectrophotometric method elaborated in recent decades assists in determining proteins in the synovia based on the changes in optical density of their absorption [45], and is able to register some structural changes in protein complexes. The data on varying content of enzymes with age and governing the exchange processes in horse blood and synovia are presented in [46]. According to this analysis, age dynamics of enzymatic activity in the synovia and serum is identical and reflects the violation of exchange processes in the joints and organism as a whole. The normal content of polysaccharides in the synovia is by 10% higher than in blood, and drops during rheumatoid polyarthritis. Lipids present a fatty fraction of the synovial fluid and contain cholesterol derivatives allied to liquid crystals [47]. Their concentration is about 1/3–1/2 of that in blood serum [44].

Judging by the changes in quantitative correlation of synovial cells it is possible to distinguish between the inflammatory processes and their severity. The signs of inflammation in the synovia are the increased content of neutrofils (50–90%) and lowered concentration of lymphocytes (0–8%). A ragocyte means the appearance of leucocytes in the synovia whose cytoplasm contains granuloma inclusions called vacuoles that look like grape seeds and are an indication of rheumatoid arthritis. The disclosure of the ragocyte phenomenon has made a profound contribution in pathophysiological understanding of inflammations in the synovial fluid [44]. The synovia is subjected to a bacteriological analysis in the event of infectious onset of synovitis [43, 48].

In spite of indisputable value of studying synovial properties for diagnostics of joint diseases, its biochemical analyses are still applied rather rarely in CIS countries. For instance, one of the latest reference books on medical laboratory analyses [49] does not even recommend biochemical investigation of the synovia at rheumatoid lesion of joints.

It is hard to single out a most informative and sensitive method among numerous laboratory procedures of studying synovia. Although the attempts of PC-aided generalized estimates are met in literature a unified complex parameter for defining the severity of synovial pathology has not been devised yet. In our viewpoint, a complex criterion of the functional state of the synovia can be the structural parameters of a 3D molecular complex consisting of a protein-polysaccharide base and other synovial components.

The method of electret-thermal analysis has emerged in the physics of dielectrics to estimate charge carriers distribution in a substance and register different from zero total electric torque in this substance. Its usage for studying the polarizing charge formed in dielectrics under electrical treatment is restricted in many countries by standards [50, 51]. The essence of the method consists in recording the current generated in the sample as a result of heat-induced disordering of dipoles, liberation of charge carriers from the traps and their transport. In the recent years the electret-thermal analysis is more frequently used for investigation of such biological and medical objects as collagen, haemoglobin, polysaccharides and other [6, 12, 52, 53]. The objects of these investigations are traditionally the solid phase bodies. For example, the hydrated state of blood cells was studied using frozen blood specimens under negative temperatures [52].

The present authors attempted to use the electret thermal analysis for biological fluids, namely, blood and synovia. Blood was chosen as a research object because, first of all, synovia and blood plasma are very close in cellular composition [47], and second, blood analysis is requisite for diagnostics of any illness since it carries information on the state of all human systems.

Blood is known to be a dynamically balanced multi component system whose structural parameters are its paramount property. Notice that the efficacy of blood circulation is conditioned by the configuration, extension of the capillary network, from the one hand, and dimensions of blood cells, their propensity to deformation, aggregation and deposition on capillary walls, from the other [54]. Moreover, the heaviness of ischemic heart disease is said to depend on the deficiency of the coronary bed, which results from the pathology of erythrocytes, their abruptly increased aggregative ability and elevated viscosity of blood plasma [55]. There exist numerous evidences to the metabolic changes in blood cells during ischemia that are attributed to their surface charge, shape and volume transformations [56]. Sensitivity of blood cells to the electric field is supported by their electrophoretic mobility [57] and ability to retain the polarizing charge [58]. The pathological changes are not, however, among the signs that may classify the functional state of ischemic patients [59], whereas cardiologic practice in analyzing blood cells [60] is unable to define their structural parameters.

In the experiments carried out at MPRI NASB we used periphery blood of groups O, A, B, and AB with the positive rhesus factor donated by young volunteers (20–35 years old) of both sexes. A blood sample was placed onto an aluminium electrode cleaned by ethyl alcohol. The electrode was covered with a polymer (PTFE) 100 μm thick film on which another electrode was placed, after which the sample was installed in the device for thermally stimulated depolarization. During the sample preparation the blood does not manage to coagulate (10-15 s), so anticoagulants were not used. The blood sample placed into a closed space between the electrode and PTFE film could not dry up during the experiment, and was subjected to depolarization under $T<100$ °C in the liquid phase.

In another experiment the blood sample was dry out at room temperature on the lower electrode and was overlaid with the second electrode without a PTFE film.

The current generated in the circuit between the electrodes was recorded during heating of the samples at a rate 5 °C/min. A measuring cell scheme is illustrated in Fig. 7.4 [61]. The dependence of current I on temperature T can be presented by a spectrum of thermally stimulated currents (TSC) that carries information on the mechanisms stimulating the appearance of the current. The check experiments have shown that the PTFE film is devoid of the polarizing charge, and TSC are not recorded during its thermal depolarization under the studied temperature range 20–150 °C. To determine precisely the activation energies of the polarizing charge relaxation with corresponding peaks on TSC spectra we used the method of thermal cleaning [62]. A sample was heated at a constant velocity and exhibited the first peak, after which was quickly cooled down to room temperature. Upon a repeated heating the next peak was recorded under a higher temperature, and so on. Judging by the initial section of curve *I* (*T*) elevation to the maximum, the activation energy which corresponded to the peak under study was determined by the Garlick-Gibson method.

The results cited below are the averaged values of not less than five replicate measurements of each blood group conducted in identical conditions using the samples taken from different patients.

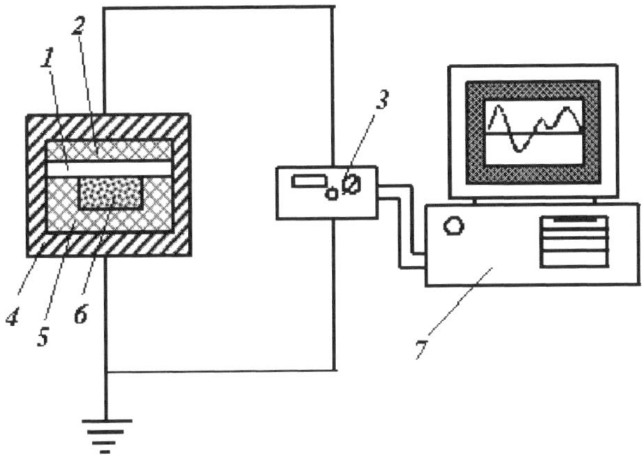

Fig. 7.4. Diagram of a device for recording thermally stimulated currents: 1 – PTFE film; 2 and 5 – upper and lower electrodes; 3 – transducer-amplifier; 4 – screen; 6 – blood sample; 7 – PC with special software

A remarkable phenomenon was revealed during the experiments. The spectra of thermally stimulated currents evolved during heating of blood samples without any preliminary electrical treatment. This can be explained as follows. Blood is a multi component system containing plasma (aqua solution of mineral salts, amino acids, proteins, enzymes, steroid compounds including liquid-crystalline derivatives of cholesterol, and etc.) in which blood cells – erythrocytes, leucocytes and thrombocytes are found in a suspended state [63]. The presence of salts presumes some amount of free charge carriers in the liquid blood. The chemical structure of organic compounds of plasma and lipid blood cells contain numerous polar groups (NH, CO, OH) that may impart a clear-cut dipolar treats to blood components under certain circumstances. Along with above named, blood contains the compounds with the properties of liquid crystals (cholesterol derivatives, some proteins, etc). It is believed that separation of charges and orientation of dipoles in liquid samples are because of the difference in the work function value of the electrodes and PTFE film. Heating of this system is accompanied by the disordering of spatial charges and dipolar structures in blood. The appearance of a low-temperature peak on TSC spectra of the simplest biopolymers – collagen and gelatine is commonly associated with dipolar disorientation, while the high-temperature one with relaxation of the spatial charge [6].

All blood samples studied have shown three main peaks on TSC spectra (Fig. 7.5).

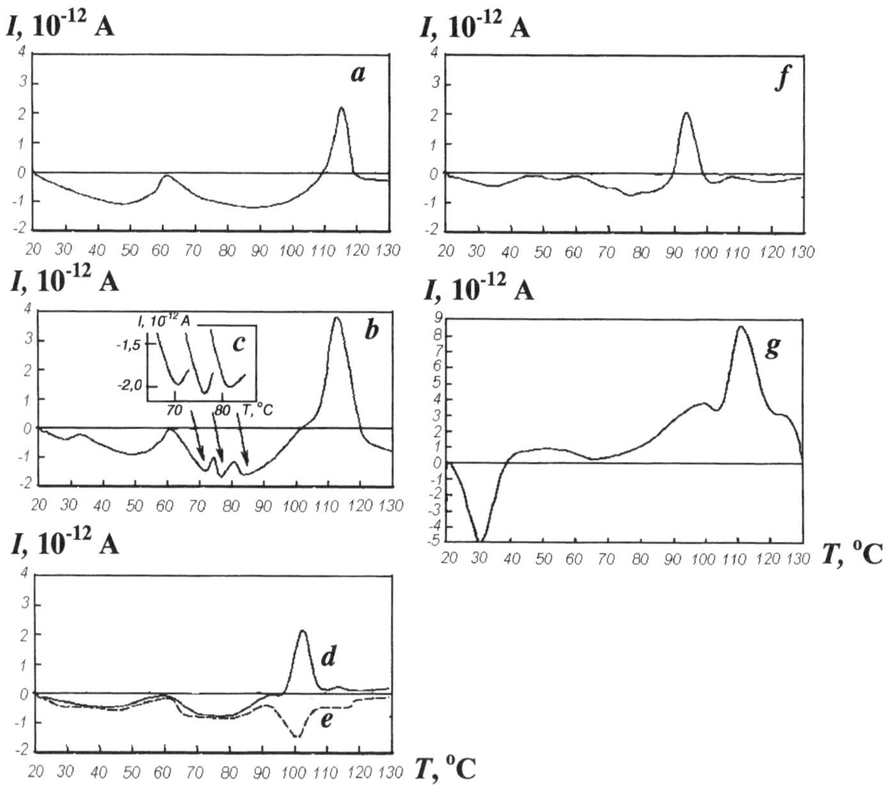

Fig. 7.5. TSC spectra of different human blood groups: *a* – O; *b* and *c* – A; *d* and *e*– B with high-temperature peaks of different polarity; *f* – AB; *g* – A, dried up on the electrode

The low-temperature sloping peak of the negative charge relaxation corresponds to T = 30–50 °C. Its appearance is attributed to the following factors. The polar groups found in the blood structure are the reason of hydrophilic properties of its components and the presence of hydrate shells on the majority of them. The bound state (structurized) of a part of blood water resulted from the formation of hydrogen bonds with polar fragments of organic molecules may bring about an electret effect [52]. The presence of some areas on TSC spectra responsible for the bound water in biopolymers has been discussed elsewhere [6, 12, and 53]. The low-temperature peak on the TSC spectra is apparently a response to the thermally stimulated destruction of coordination structures containing bound water. The charge relaxation activation energy that corresponds to this peak equals to W_1 = 0.45 eV. The mean hydrogen bond energy alters within 0.10 till 0.25 eV [64]. In case the organic compounds with a complex spatial structure and polar groups are present in blood, water molecules might join into coordination structures having high linkage energies. Presumably, charge liberation going hand in hand with structural recombination of thermo tropic liquid crystals based on cholesterol derivatives makes its contribution into the low-temperature peak as well.

The medium-temperature peak (T = 70–90 °C) is more intensive than the first one and consists, as a rule, of a few peaks. Consequently, the breakage of several coordination structures takes place in the indicated depolarization temperature range. In our view, this peak reflects denaturing of protein compounds in blood, i.e. irreversible changes of the tertiary protein structure without disturbing polypeptide chains. The spatial structure of proteins is known to be stable if above 90% of the intramolecular hydrogen bonds have been preserved in the protein globule [65]. This suggests that a great number of bonds get broken under 70–90 °C and force protein structures to a cooperative transfer into the denatured state. Heat-induced charge relaxation of ionized and protonated groups responsible for the electrostatic attraction of the protein molecule fragments do their share in the denaturing of protein blood compounds. Van der Waals interactions conditioned by the attraction of the constant, induced or virtual dipoles are also involved in the process [66]. The undulatory configuration of the medium-temperature peak in figure (*b*) supported by thermal cleaning (*c*) conforms to three energies of the bonds (within W_2 = 0.5–0.7 eV) participating in polarization of protein structures of the A blood group.

The most intensive high-temperature peak at 105-120 °C responds to the phase transition of blood during a solid film formation of coagulated blood. Thermo oxidative destruction of organic compounds in blood is accompanied by sticking of lipid shells of erythrocytes, denatured proteins, enzymes of peptide chains and other. Thermal decay of the newly formed structures leads to the relaxation of the spatial charge formed with probable participation of Maxwell-Wagner's polarization.

The TSC spectrum of dried out on the electrode blood (*g*) belongs to the sample without a PTFE film. All three peaks of this sample are extremely intensified. The reason can be that dried in air bound water not only remains in the blood structure but its links with complexing agents become even more intensive. What is more, the organic compounds undergo oxidation, which adds to their dipolar moments.

Being close in the spectrum outline, TSC of different blood groups (Figs. a-f) have dissimilar high-temperature peak locations on T-axis, and differ in both intensity and polarity. The high-temperature peak maximum shifts to the lower temperatures with different blood groups: 115–118 °C (group O), 108–110 °C (group A), 98–100 °C (group B), 95–97 °C (group AB) [67]. Observed in the experiments variation of the high-temperature peak polarity on TSC spectra of blood samples taken within a day from one and the same donor (*d, e*) does not fall for some unequivocal interpretation. These effects are, presumably, connected with certain life cycles of the human organism. The reversible biochemical processes taking place in blood give out positive and negative potentials in the double electric layer encircling blood cells [1]. Depending on the physiological factors (blood pressure, physical activity, presence of stimulators in the organism, and so on) certain oscillations in the distribution of these potentials may vary and lead to alterations in the TSC peak intensity and polarity [68].

In virtue of its origination from blood plasma, *the synovial fluid* has a cognate base, contains water, lipids, proteins, urea and other biological liquids. Its pseudo-elastic properties the synovia, however, owes to a high content of HUA and hyaluronates, while uncommon lubricating characteristics to cholesterol ether derivatives, viz. thermo tropic liquid-crystalline compounds [47]. The described system is likely to have uncommon electrophysical properties that specify its lubrication mechanisms of joints and sensitivity to bio potentials.

A TSC spectrum of the synovia taken from an arthritic knee joint (initial stage) is illustrated in Fig. 7.6 [69].

The peak of positive currents at 28–31 °C on the spectrum of the initial synovia apparently corresponds to the heat-induced breakage of the dipolar ordered mesophase structure of the synovia. The hydrate shells of biopolymers [6] undergo failure under the same temperatures, which effects both configuration and size of the peak.

The peak of negative currents at 65 °C arises from the breakage of coordination links within the spatial structure of the protein-polysaccharide synovial complex and is comparable by its intensity with the low-temperature one. Both structures undergo destruction during friction, which is reflected in diminished TSC peaks of the synovia taken from the friction zone of the pendulum tribometer after 2 hours of operation (p = 2 MPa, v = 0.1 m/s). The protein-polysaccharide coordination compounds turn to be most damaged, whereas a specific layered structure of the synovial liquid-crystalline components though undergoes some tribochemical variations but remains largely intact.

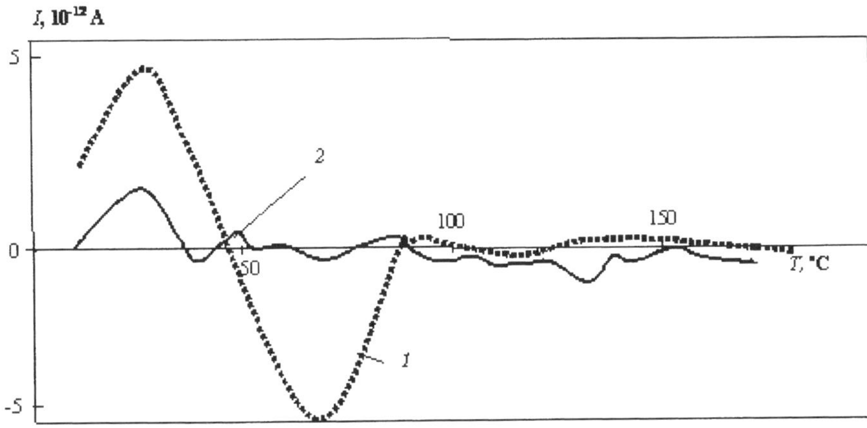

Fig. 7.6. TSC spectra of synovia: *1* – initial; *2* – after friction in UHMWPE–CoCrMo pair [69]

Based on investigation results of temperature transformations of the synovia presented elsewhere [39, 47, 63, 69] we have drawn a scheme of its phase transitions (Fig. 7.7). It conforms well to TSC spectra shown in Fig. 7.6.

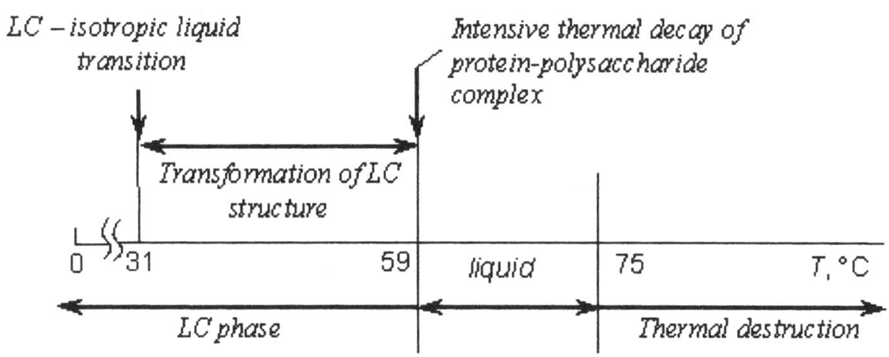

Fig. 7.7. Temperature transitions in synovial fluid

Above results prove that both blood and synovia display a spontaneous quasi-electret effect induced by the coordination character of the links in their permolecular structure. The parameters of named effect *in vivo* depend upon the nature of biophysical and biochemical processes in the organism. This makes grounds to model the synovia as a 3D molecular complex formed of coordinated structurally and interrelated functionally components. In this connection, structural characteristics of this complex seem to be a suitable criterion for estimation of the functional state of the synovia, while the electret-thermal analysis may serve as a simple and informative tool for diagnostics of joint diseases. Thanks to minor volumes of synovial samples (~ 0.1 ml) necessary for this analysis, thermal depolarization can be referred to as a microanalysis, representing a modern trend in clinical laboratory research methods [70].

Presently, a series of investigations of blood components (plasma, regular elements) and synovial fluid (collagen, HUA, phospholipids) are being carried out at MPRI NASB jointly with Belarussian clinics dealing with the curing process of joint diseases by the electret-thermal analysis. The results obtained are checked up by the traditional methods of biochemical clinical analyses.

7.5 ELECTRET PARTS FOR ENDOPROSTHESES

The formation of the electret charge in endoprostheses whose electric field is able to promote an optimum distribution of bio potentials in the operative wound has been recognized as an efficient means of eliminating after-effects of surgical intervention. The accumulated experience in various fields of engineering sciences justifies the usage of electrets with the aim of improving performances of joints with fine clearances between conjugated parts. A relatively weak field of electrets is targeted to regulate lubrication, capillary permeation, liquid spreading and other processes in which surface phenomena play a crucial role [20].

The properties of an electret as a source of electric field are subjected to variations due to natural relaxation of the polarizing charge. This is why the problem of a long-term serviceability of endoprostheses containing electret elements remains actual in spite of the fact that the weak and super-weak electrical fields exert a most efficient effect on the bio field of living tissues [1-6].

During the experiments a joint endoprosthesis was modelled by a pendulum tribometer resting on a cylindrical UHMWPE sample of Hostalen GUR 4120 grade (Hoechst AG, Germany). A counterbody was a prism made of a CoCrMo alloy used by Protek Co. (Switzerland) as a metal friction material in metal-polymer hip and knee endoprostheses. The friction coefficient was identified by attenuation of the pendulum oscillations displayed on a computer.

The lubricating media were a physiological solution (0.9% solution of NaCl), synovial fluid taken from a human knee joint and a lubricating liquid based on Vaseline oil containing liquid-crystalline compounds.

The polymer samples were electrically polarized following a common technique of producing corona and thermoelectrets [20]. Their charge state was monitored using the electret-thermal analysis. A support and pendulum prism structures are illustrated in Fig. 7.8. An UHMWPE sample is fixed in the support frame by screws (*a*). The friction zone of the sample looks like a cylindrical groove into which the lubricant is poured. The metallic prism (*b*) with a suspended pendulum is fit with a cylindrical indenter installed in the groove.

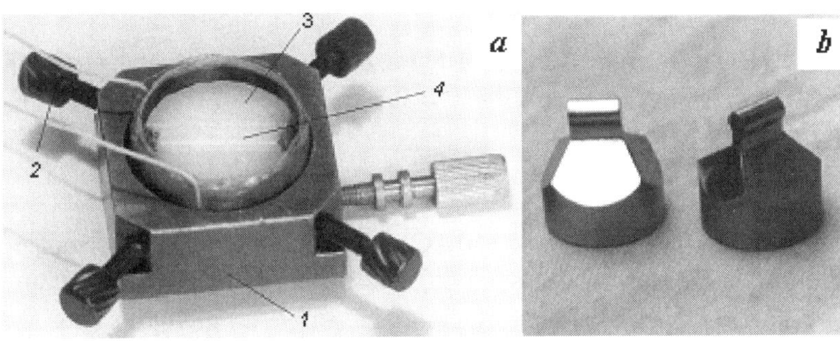

Fig. 7.8. A support unit (*a*) and a prism (*b*) of the pendulum tribometer: *1* – case; *2* – screw; *3* – polymer sample; *4* – groove

It is evident from Fig. 7.9 that the described structure may form a stable polarizing charge in the UHMWPE sample, which is preserved during friction.

The TSC spectra of UHMWPE-based corona- and thermoelectrets are seen to coincide in principle. They differ only in the peak value and stability in the course of friction (the polarizing charge of thermoelectrets is more stable). The peaks on TSC spectra conform to temperature transitions in UHMWPE [71]. During friction in the physiological solution the polarizing charge of corona electrets drops rapidly down to zero, while that of thermoelectrets remains stable and undergoes less changes in contrast to dry friction.

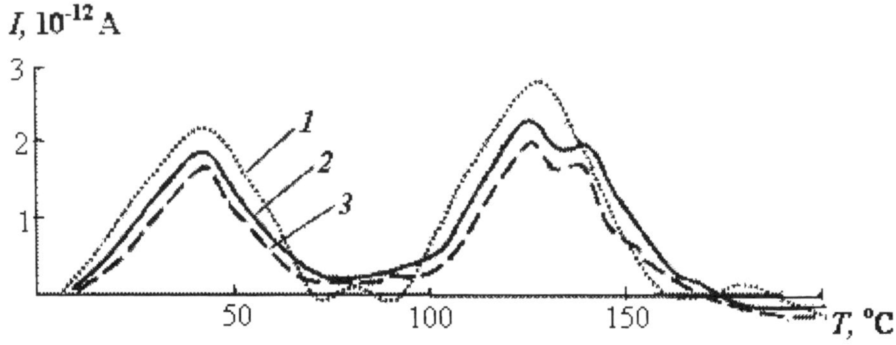

Fig. 7.9. TSC spectra of UHMWPE-based thermoelectrets experiencing unlubricated friction. (p = 2 MPa, v = 0.1 m/s). *1* – initial; *2* and *3* – after friction during 0.5 and 1.5 h

The low-temperature peak (40 °C) on TSC spectra of thermoelectrets disappears after friction in the synovia, while the intensity of the main peak corresponding to UHMWPE melting point (135–137 °C) increases (Fig. 7.10).

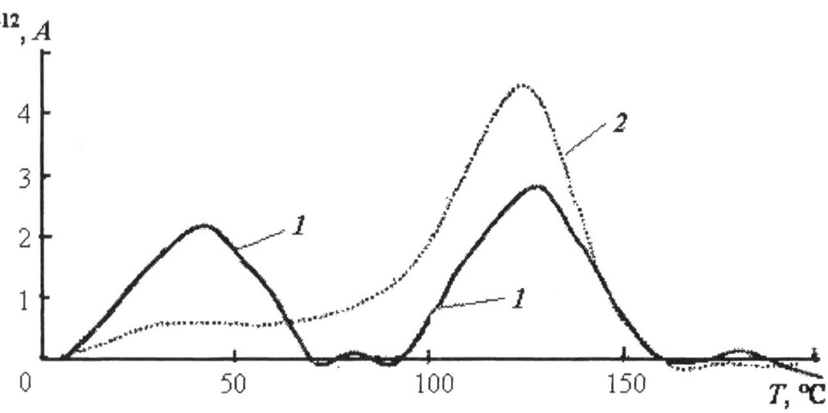

Fig. 7.10. TSC spectra of thermoelectrets: *1* – initial; *2* – after friction in synovia (p = 2 MPa, v= 0.1 m/s)

The low-temperature peak appears during the surface charge relaxation of the mechanoelectret cut of a sample exposed to friction in the course of studying TSC spectra. The relaxation λ-transitions of UHMWPE occur within the same temperature range. The peak corresponding to the melting point of UHMWPE is formed due to heat-induced decomposition of the polymer crystalline structure subjected to a relaxation polarization in the course of forming the thermoelectret [20].

The obtained results were explained in terms of the following hypothesis [72]. The structure of cholesterol liquid crystals and coordinated protein-polysaccharide compounds incorporated in the synovia alters during friction. Shear stresses and frictional heating disturb the spontaneous ordering of the liquid crystals, which results in formation of polar molecules unlinked in layers. The coordination compounds dissociate and result in counterions. These fragments adsorb on the friction surface and neutralize the surface charge of the mechanoelectret. The field of the thermoelectret accelerates diffusion of the fragments in the lubricating layer and their adsorption acquires the features of chemisorption (Fig. 7.11).

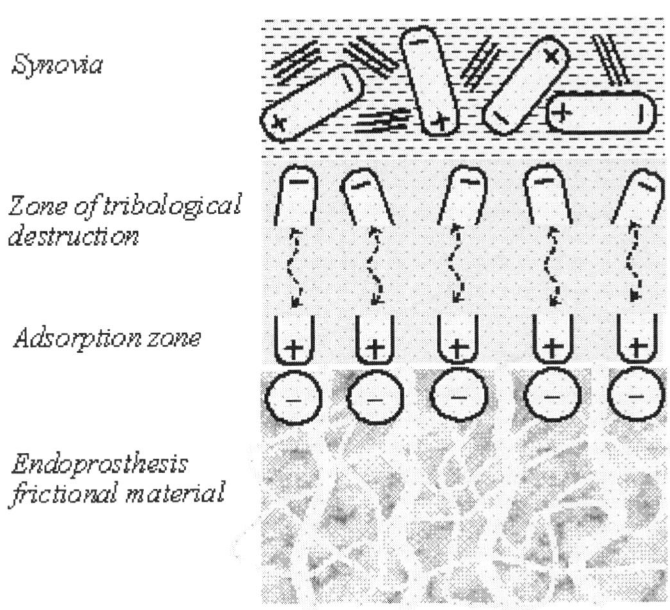

Fig. 7.11. An adsorptive model of synovial fragments on the electret friction surface

The presence of a single peak at 135–137 °C is a proof that the adsorption of the synovial fragments takes place on the crystalline elements of UHMWPE. Intensification of this peak after friction in the synovia visualizes that thermally stimulated depolarization adds desorbed from the friction surface charged particles to the charge carriers liberated during the thermoelectret charge relaxation. This fact agrees with the intensification effect of the synovial adsorption on the cartilage surface being in dynamic contact described in [73]. This hypothesis also conforms to the results of the electret-thermal analysis of the synovia cited above.

When rubbing in a physiological solution the friction coefficient of the pair under study, all other conditions being equal, remains constant with a high degree of reproducibility independently of whether the polymer support has been electrically polarized or not. Lubrication with the synovia makes the pair sensitive to the presence of the polarizing charge in the support. As it is shown in Fig. 7.12, the friction coefficient diminishes as much as twofold in case of the polarized support. This effect takes place independently of the manner of the polarizing charge formation in the support, be it a thermoelectret method or corona discharge. Notice that in the pair lubricated by the synovia the friction coefficient is less when the friction surface is charged negatively as opposed to a positively charged surface.

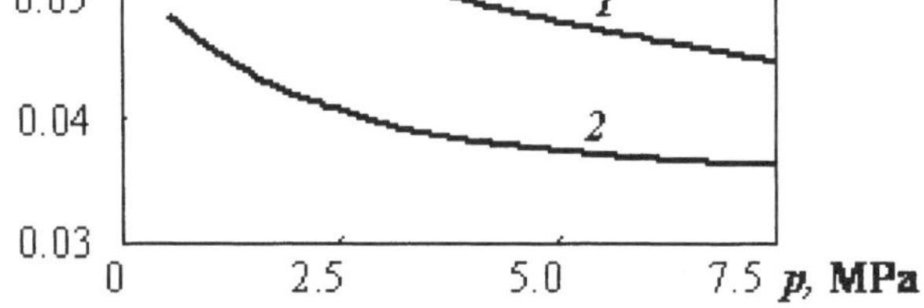

Fig. 7.12. Friction coefficient versus load in the pair lubricated by synovia: *1* – non-polarized UHMWPE support; *2* – support with a thermoelectret charge

Named results comply well with the experiments based on a reciprocating friction machine (p = 2 MPa, v = 0,1 m/s, τ = 4 h). In these experiments a

"Diasynol" preparation containing liquid-crystalline compounds was used as a lubricating medium [47]. The initial UHMWPE in pair with a CoCrMo counterbody has shown μ = 0.072–0.090. The corona electrets from the same UHMWPE displayed μ = 0.024–0.026 in the case of a negatively charged surface and μ = 0.024–0.030 with a positive charge [69].

7.6 MAGNETIC FIELDS IN MEDICINE

The permanent magnetic field is a global physical factor influencing all terrestrial biological objects. A.L. Chizhevsky was the first to discover in 1920 that the geomagnetic field exerts an effect on the biosphere. However, as far back as the extreme antiquity the physicians tried to use magnets in therapeutic aims. The interest to curing with the help of magnetic fields has revived by the mid 19th century in connection with the development of strong electromagnets. At the beginning of the 20th century when the active physiotherapeutic methods were invented including radiotherapy, UHF, ultraviolet radiation, etc., the magneto therapy was abandoned again. In the 1960-70ies the scientists started to explore the therapeutic effect of magnetic fields at the junction of biology, medicine, physics, chemistry and engineering, which gave rise to a new branch of medicine – medical magnetology. Unfortunately, by the initiative of the President of USSR AS Alexandrov A.P. in 1982, the works in this sphere were ceased due to, first of all, falsification of the results based on subjective perceptions of the patients and distrust in the therapeutic effect of magnetic fields. Nevertheless, *magneto therapy*, i.e. the use of magnetic field with therapeutic and prophylactic aims, has justifiably occupied a fitting place among other medicinal procedures. Although magneto therapy has received a great deal of attention in our times, it is still related to non-traditional methods of curing. Most often, people address to it when other medical treatments turn to be powerless (chronic diseases) or injurious in respect to side effects. The major aspects of application of magnetic fields in medicine are considered below.

The magnetic field is a specific type of matter in which the moving electrical charges are brought into interactions. The existence of a living matter is accompanied by the transfer of substances and, consequently of electric charge carriers. Therefore, the living matter is a source of the magnetic field. In Table 7.1, we name the variants of magnetic fields that present most interest from the standpoints of biology and medicine [74].

Table 7.1
Classification of magnetic fields [74]

Origin	Variation in time	Spreading in space	Intensity
Natural fields (geomagnetic field, fields of magnets)	Permanent Variable	Homogeneous Inhomogeneous	Weak Medium
Artificial fields Fields of bio objects	Fluctuating Pulsating Noise-type		Strong Super-strong

Actually, all above-named fields differed in their physical parameters are used today with therapeutic and prophylactic aims.

The mechanisms of biological effect of the magnetic fields are based on their interaction with the structural elements of the living tissues having their own magnetic fields. The types of interactions depend on the peculiarities of the living tissue [75], namely

- the presence of free radicals in the biological object;
- diffusion in the cells, particularly, motion of ions through the membranes;
- anisotropy of magnetic susceptibility of DNA molecules and protein macromolecules;
- Brownian movement of protein molecules possessing anisotropy of magnetic properties and a dipolar moment different from zero;
- the probability of altering valent angles in paramagnetic molecules.

Organic compounds, biological tissues included, are related by the level of their magnetic properties to the dia- and paramagnetic substances [76]. It is hard to explain their appreciable sensitivity to the weak magnetic fields used with curing aims. The interaction energy of permanent and variable magnetic fields with bio molecules is by several orders of magnitude lower as opposed to the energy of heat-induced molecular motion under physiological temperatures. A fundamental contribution into understanding of the mechanisms of the biological effect of magnetic fields was made by A.L. Buchachenko. He formulated the notions about the field effect on the probability of initiating reactions that run with different spin multiplicity (multiplicity is a number of probable orientations of the molecular spin in space), in particular the free-radical reactions [77]. Initiation of the processes that amplify a weak outer excitement is considered as a resonance after-effect, i.e. consistency of the physical and time-spatial characteristics of the magnetic field with the properties of the target bio molecule [76]. It is believed that the magnetic field plays a role of a peculiar trigger actuating certain biological mechanisms. It is able, for instance, to alter the structure of liquid-crystalline substrates, to which membranes also belong, and thus change their penetrability, and consequently, their exchange processes [78].

It was attempted to explain the biological effects of the magnetic field in terms of its influence on the water incorporated in living organisms. Indeed, the water present in the cells acquires specific electrical properties. Its interaction with the external magnetic field results in so-called magnetization [79]. This alters the quasi-crystalline structure of the intracellular water and metabolism in the cells.

Two hypotheses have been put forward, first that the magnetic field stimulates breakage of ionic bonds with protein molecules, which augments ionic activity [74] and, second, on the magneto hydrodynamic inhibition of the circulation of salt water solutions in living tissues [76]. Besides, the variable magnetic field is said to generate electrical currents in the tissues containing free charge carriers thus exerting a diversified biological influence.

Numerous structures at the submolecular and permolecular levels existing in the organism realize above-mentioned phenomena. Their changes are transformed in the reactions of cells, tissues and the reactions of the organism as a whole.

Physiological and therapeutic effects of the magnetic field on living organisms are accompanied by the next phenomena [74]:

- reactions of organisms on the curing effect of magnetic fields are characterized by their diversity and instability due to individual features of some tissues and the integral organism;
- therapeutic magnetic fields exert a normalizing effect on the organism, i.e. the effect against a background of the elevated function leads to its decreasing, while the oppressed function is intensified;
- the organism reaction on the effect of magnetic fields is often a change for an opposite one;
- some organism reactions have a threshold (the field intensity should not be lower the threshold value) or a resonance character (the reaction appears only if the field and organism parameters are conforming);
- after a one-time exposure of the field, the organism reactions are preserved during 1–6 days; after a course or prolonged procedures, the reactions keep running for 30–40 days;
- the pulse or variable fields effect commonly result in more stable and expressed variations in contrast to the permanent magnetic fields.

The nervous system is most sensitive to the effect of magnetic fields. The conditional-reflexive activity is directed under their action mainly to the inhibition processes that effect favourably sleep and emotional strain. Along with this, improved tone of cerebral vessels and blood supply has been recorded.

The gray matter of the brain responds to the effect of magnetic fields by the activation of regenerative processes in the cells (bio stimulating effect). The magnetotherapy is known to lower sensitivity of peripheral receptors (anaesthetic effect) and ameliorate the conductance function (rehabilitation of traumatic and peripheral nerves).

The humoral regulation, i.e. coordination of physiological and biochemical processes in the organism through the liquid media (blood, lymph, tissue liquids) is disposed to magnetic fields owing to stimulation of the neuroendocrine system. Excitation of some cerebric areas evokes a chain reaction during which the targets (peripheral endocrine glands) get activated and branch out in the form of numerous metabolic reactions. As opposed to the depressive effect of the majority of irritants, the magnetic field renders a stimulating effect on the functions of thyroid gland.

The cardiovascular system is also sensitive enough to the magnetic field influence: the pulse slows down, heart contractions become more efficient, endocardial haemodynamics improves. The contraction function of the myocardium is intensified due to a shortened stress phase and lengthened relaxation. The arterial pressure shows a clear tendency to lowering. Under the effect of the field of up to 30 mTl induction the blood flow velocity increases after 10–30 min in all chains of microcirculation, blood thinning and opening of the reserve capillaries are observed. In this manner, the antiedematous function of magnetic fields is realized. They also effect epithelial penetrability, which promotes better supply of drugs through the epithelial barrier and the efficiency of medicinal electrophoresis. The permanent and variable magnetic fields intensify regeneration processes in osseous tissues. The fibrous and osteoblasts are formed in the regeneration zone much faster and the bony matter grows more intensively [74].

A hypo coagulation effect of the therapeutic magnetic fields has been discovered, which is attributed to the activation of the anticoagulation system and hampered thrombus formation. Notice that over dosage of the fields or their prolonged effect may activate the blood coagulation system [80]. In addition, the fields influence the electrical and magnetic properties of blood elements participating in hem coagulation [81].

The erythrocyte sedimentation rate (ESR) is usually retarded under the effect of magnetic fields. In addition, the intensified phagocytic activity of leucocytes, and normalization or stimulation of metabolic processes is observed [74]. In particular, metabolism of carbohydrates and lipids is activated leading to the reduction of cholesterol concentration and elevated content of fatty acids, and phospholipids in blood.

The lymphatic system is also sensitive to variations in the magnetic field intensity. The activation of lymphatic tissues results in improved blood supply to lymphatic nodes, growing number of lymphoid cells in the peripheral blood and their raised resistance [82].

The neuromuscle system of the man responds to magnetic fields by intensified muscular work capacity at both general and local tiredness. Small therapeutic doses of magnetic stimulation ameliorate the activity of genitals and sexual functions.

The magnetotherapy procedure and facilities are connected with the usage of variable, low-frequency and less often permanent magnetic fields.

The permanent fields are generated by magneto plastics (polymer materials filled with magneto-solid particles), metallic or ceramic permanent magnets having surface magnetic induction from 60 till 100–130 mTl [83].

The variable low-frequency fields are generated by the inductors that transform the electric energy into magnetic one. Commonly, the magnetic fields are used whose induction varies sinusoidally with time (variable field), or spreads along the injured area as square pulses of a single (travelling) or both polarities in turn (revolving field). Their frequency is usually 0.125–1000 pulse/s and magnetic induction is below 100 mTl. Numerous modifications of devices are available for feeding the inductors to create the fields of different configurations [84].

Less often the pulse magnetotherapy is employed whose therapeutic action is based on the high-intensity magnetic pulses (1.2–1.7 Tl) generated at 10–40 pulse/min frequency. The *high-frequency magnetotherapy* envisages curing by the high and ultra-high frequency fields: 13.56 MHz with $\lambda = 22.13$ m wavelength, 27.12 MHz with $\lambda = 11.05$ m, and 40.68 MHz with $\lambda = 7.37$ m. The magnetic pulses of named characteristics are generated at 50 pulse/s frequency. This rather intensive effect promotes heat generation in the tissues whose quantity is a function of the radiation frequency and intensity, electrical parameters and thermoregulation properties of the tissue [85].

Speaking about the technical control over the magnetic field parameters, we would like to draw your attention to a device invented in mid 1960ies and named Squis (superconducting quantum interferometer sensor). It is capable of measuring weak magnetic fields with a record fine sensitivity. Squis-magnetometry has become an actual methodological trend in studying bio magnetism [86]. Its significance for joint endoprosthetics is connected with possibility of recording small quantities of ferromagnetic debris in the tissues surrounding the endoprosthesis.

Electromagnetic contamination of the environment is a term introduced by the World Health Organization to denote ecological conditions characteristic of the recent five decades and exerting pathological effect on the man and biosphere [87]. The danger of the high-frequency radio noise was recognized in the process of studying the effect of electromagnetic field on living organisms. Harmless in small dosage the magnetic fields of $B_r \geq 70$ mTl intensity might be a source of stress factor and evoke discoordination of endocrinal organs, reduce the intensity of energy exchange, infringe penetrability of cellular membranes, result in hypoxia and dystrophic phenomena [74]. As the physical fields of anthropogenic origin intensify, the total intensity of the electromagnetic field in some points of the Earth augment as much as by 2–3 orders of magnitude in contrast to the natural background. It raises abruptly specifically in the vicinity of power transmission lines, radio, power and TV stations, radar and radio communication facilities, city electric transport. This sharp growth in the intensity of electromagnetic fields has acquired a property of a constant factor whose grave consequences for the evolution of life on the Earth are hard to predict. Most vulnerable to this factor are the nervous, immune, endocrinal and sexual systems of the man. In this connection, all developed countries of the world have established limiting levels of electromagnetic radiation and organized National Committees summoned to protect the population against non-ionizing radiations. They coordinate the respective activities with the International Commission on Non-ionizing Radiation Protection (ICNIRP).

7.7 LUBRICATION OF ENDOPROSTHESES IN MAGNETIC FIELD

All living tissues are known to belong to diamagnetics (the sum of magnetic moments of bio molecules is close to zero and relative permeability is μ_r = 0.99995), and only some of their constituents (oxygen, iron salts, hydro peroxides, etc.) have a proper magnetic moment, i.e. are referred to as paramagnetics (μ_r = 1.00005). Their μ_r values are of the same order and the magnetic field effects therefore only weakly the conduction current in the tissues. The greatest effect the magnetic field renders the permolecular liquid-crystalline structures of the tissues that are oriented relative to the magnetic induction vector of the external field. In this regard, from the viewpoint of endoprosthetics it would be helpful to consider the effect of magnetic fields on the lubricating layer of the synovial fluid in the artificial limb. As far as the synovia contains liquid-crystalline compounds [73], the lubricating layer is anticipated to interact with the magnetic field whose source can be the joint endoprosthesis itself.

The magnetic field has been generated during the experiments by a solenoid built into a support of the pendulum tribometer (see Fig. 7.8, *a*). Analogously to the electret field, that of the solenoid does not affect friction at lubrication with a physiological solution. The presence of the synovia in the friction zone promotes gradual reduction of the friction coefficient upon actuating the solenoid. The dependence of the friction coefficient on the pre-treatment time of the lubricating layer in the electromagnetic field is of an exponential character (Fig. 7.13). Without the field application the kinetic dependence of μ at lubrication with the synovia is similar to curve *1*. It is evident that lubricity of the synovia is higher than that of Diasynol preparation containing the liquid-crystalline components.

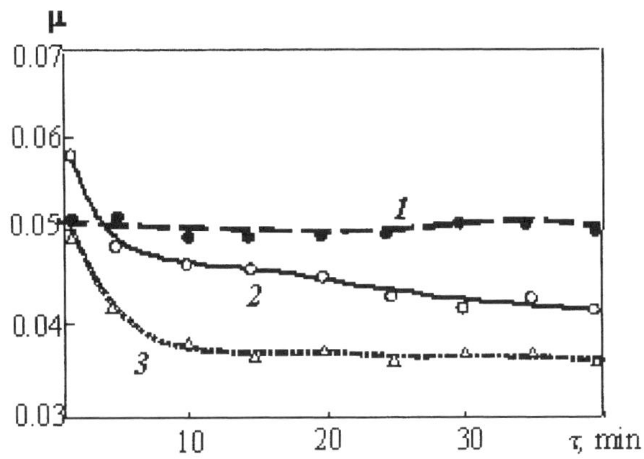

Fig. 7.13. Friction coefficient versus pre-treatment time of lubricating layer in magnetic field for different lubricants: *1* – physiological solution, *2* – "Diasynol", *3* – synovia

With increasing direct current in the solenoid coil, reaching of a stable friction coefficient in the pair is accelerated (Table 7.2). Above data conform to the vertical direction of the solenoid magnetic field. In case a solenoid installed in the support whose field is directed horizontally, the friction coefficient, all other conditions being equal, reduces less.

Table 7.2
Friction coefficient (μ) versus solenoid field value at lubrication with "Diasynol"

Current strength, A	Field intensity, kA/m	μ versus endurance of lubricating layer in the field, min					
		0	5	10	20	30	40
0.03	0.6	0.060	0.055	0.045	0.043	0.041	0.042
0.06	1.2	0.059	0.057	0.053	0.054	0.049	0.045

Above-presented results can be explained in terms of a specific structure of liquid crystals. The stable thermodynamic state of cholesteric liquid crystals complies with a layered arrangement of molecules with the maximum intensity of molecular interactions in each layer and the least between them. The molecules of each layer are oriented in a certain direction and turned to some angle in relation to the direction of molecular orientation of the neighbouring layer. The treatment of the lubricating layer in the solenoid field assists in an optimum orientation of the liquid crystal layers towards the friction direction from the standpoints of tribology. In the absence of the magnetic field the lubricating layer structure is not optimum and in case of lubrication by the physiological solution it will not undergo optimization at all. The structure of the liquid-crystalline layers under study is to some extent inert, so to rearrange it into an optimum tribological state one should apply an external magnetic field. The rearrangement time depends, all other conditions being equal, on the lubricating medium nature.

An optimum tribological orientation of the liquid-crystalline components of the synovia *in vivo* is believed to be induced by the bio field of the joint. Different lubricating media are said to possess various magnetic susceptibility [69]. Evidently, magnetic susceptibility of a natural synovia is a function of a number of factors, primarily of the presence of joint pathology, age and individual features of a patient.

◈ ◈ ◈

The concept of designing artificial limbs as a source of electric or magnetic fields set forth in the present chapter can be considered as a means of realizing a modern trend in endoprosthetics of creating biomechanical systems simulating natural joints. Justifiability of this approach is consistent with intrinsic for the synovia and blood ability to manifest a quasi-electret effect. A specific kind of adsorption of tribodestruction products of the synovia on the friction surface occurring in the proper field of the endoprosthesis results in a layer that protects the friction surface from mechanical damages. Above conclusions have lead to a hypothesis that the electromagnetic field (either natural or induced) stimulates the formation of lubricating synovial layers whose structure approaches an optimum one from the tribological point of view. This lends impetus to a targeted application of electric and magnetic fields in orthopaedics. Their energy is transformed by the living organism not only in membranes and cells, but also at the level of such refined friction units as the synovial joints.

References:

1. Makarevich A.V., Pinchuk L.S., and Goldade V.A. Electrical fields and electroactive materials in biotechnology and medicine. Gomel, MPRI NASB. 1998, 106 pp.
2. Hodgkin A.L., and Katz B. The effect of sodium ions on the electrical activity of the giant axon of squid. *J. Physiol.*, 1949, V. 108, p. 37–77.
3. Hodgkin A.L., and Huxley A.F. A quantitative of membrane destruction current and its application to conductance and excitation in nerve, *J. Physiol.*, 1952, V. 117, p. 5–44.
4. Nasonov D.N. Local reaction of the protoplasm and spreading excitation. 2nd ed. Moscow–Leningrad, USSR AS Publ. 1962, 426 pp.
5. Zatsepina G.N., and Kolomiets A.A. Permanent electric field and active characteristics distributing by symplast of higher plants excitation signals. *Biophysics*, 1990, V. 35, No. 2, p. 331–335.
6. Mascarenhas S. Bioelectrets: Electrets in biomaterials and biopolymers. In: Electrets, Ed. by G.M. Sessler. Berlin, Springer-Verlag, 1987, p. 321–346.
7. Kulbi-Garfias C. Biological applications of the quantum chemistry theory. NY, Research Signpost, 2002, 150 pp.
8. Cherepanov D.A. Electric field in a three-layered membrane. *Biophysics*, 1991, V. 36, No. 4, p. 584–588.
9. Sizonenko V.L. Effect of multicharge ions on membrane potential. *Biophysics*, 1991, V. 36, No. 5, p. 822–825.
10. Golovanov M.V. A new concept of the structure of double electric layer on the cell surface in high-concentration electrolyte conditions. *Biophysics*, 1991, V. 36, No. 3, p. 463–466.
11. Volobuev A.N., Zhukov B.N., Ovchinnikov E.L., et al. Nonlinear modeling of spreading of the action potential. *Biophysics*, 1991, V. 36, No. 3, p. 546–551.
12. Kulin E.T. Bioelectret effect. Minsk, Nauka i Tekhnika, 1980, 216 pp.
13. Shlyguin V.V. Peculiarities of potentials distribution in organism tissues as an inhomogeneous and anisotropic conductor. *Biophysics*, 1991, V. 36, No. 2, p. 348–351.
14. Romanchuk P.I., and Kryukov N.N. Electric location of the functioning of internal organs. *Int. Congr. "Weak and super-weak fields and radiation in biology and medicine"*. St. Petersburg, 1997, p. 22–23.
15. Zaffaroni L. Macromolecular therapeutic systems. In: Biomedical and Dental Applications in Polymers. New York, Plenum Press, 1981, p. 293–314.

16. Kemenova V.A., Mustafin R.I., Alexeev K.V., et al. Application of interpolymer complexes in pharmacy. *Pharmacy*, 1991, No. 1, p. 67–72.
17. USSR Patent 1834655, A 61 H 11/00. A device for segmental reflexotherapy. L.S. Pinchuk, V.A. Goldade, E.A. Tsvetkova, et al, 1993.
18. Orlova M.V. Kinetics of properties of coeruloplasmin of human blood under the effect of microcurrents on acupuncture points. *Biophysics*, 1991, V. 36, No. 4, p. 732–733.
19. Bogolyubov V.M., and Ponomarenko G.N. General physiotherapy, Textbook, 3rd ed. Moscow, Medicine, 1999, 432 pp.
20. Kestelman V., Pinchuk L., and Goldade V. Electrets in engineering: Fundamentals and applications. Boston, Kluwer Academic Publ., 2000, 281 pp.
21. Friedenberg Z.B., and Bringtion C.T. Bioelectric potentials in bone. *J. Bone Joint Surg. Am.*, 1966, V. 48, p. 915–923.
22. Friedenberg Z.B., and Smith H.G. Electrical potentials in intact and fractured tibia. *Clin. Orthop.*, 1969, V. 63, p. 222–225.
23. Shamos M.H., Lavine L.S., and Shamos M.I. Piezoelectric effects in bone. *Nature*, 1963, V. 197, p. 81–83.
24. Fukada E., and Yasuda I. On the piezoelectric effect of bone. *J. Phys. Soc. Jpn.*, 1957, No. 10, p. 1158–1169.
25. Pollack S.R. Bioelectrical properties of bone. Endogeneous electrical signals. *Orthop. Clin. North. Am.*, 1984, V. 15, p. 3–14.
26. Tkachenko S.S., and Rutsky V.V. Electrostimulation of osteoreparation. Moscow, Medicine, 1989, 208 pp.
27. Karlov A.V., and Shakhov V.P. Systems of external fixation and regulation mechanisms of optimum biomechanics. Tomsk, STT, 2001, 480 pp.
28. Revell P.A. Pathology of Bone. Berlin, Springer-Verlag, 1986, 390 pp.
29. Black J. Tissue response to exogenous electromagnetic signals. *Orthop. Clin. North Am.*, 1994, V. 15, p. 15–31.
30. Kornilov N.V., Anisimov A.I., and Kanykin A.Yu. Electrochemical aspects of implantation of metal structures in traumatology and orthopedy. *Int. Congr. "Weak and super-weak fields and radiation in biology and medicine"*. St. Petersburg, 1997, p. 72–73.
31. Morgunov M.S., Homutov V.P., and Sokolova I.M. Application of electrets in traumatology and orthopedy. *Proc. 8th Int. Symp. on Electrets*. Paris, 1994, p. 863–868.
32. USSR Patent 1454482, A 61 N 1/00. Osteoreparation stimulator. M.S. Morgunov, V.V. Rutsky, and V.P. Khomutov, 1989.

33. USSR Patent 1122328, A 61 N 1/00. A device for osteosynthesis. V.V. Rutsky, V.P. Khomutov, M.S. Morgunov, and V.N. Svidler, 1984.
34. Hastings G.W. New advanced materials in orthopaedics. *Proc. 6th Int. Conf. Polym. Med. and Surg.*, 1989, p. 1/1–1/6.
35. Ratner B.D. Biocompatibility: the convoluted path to a working definition. *Trends in Polym. Sci.*, 1994, V. 2, No. 12, p. 402–403.
36. Simon F., Hermel G., Lunkwitz D. et al. Surface modification of expanded poly (tetrafluoroethylene) by means of microwave plasma treatment for improvement of adhesion and growth of human endothelial cells. *Proc. 5th Dresden Discuss. Polym. and Med., Macromol. Symp.*, Konigstein, 1996, V. 103, p. 243–257
37. James K., and Kohn J. Applications of pseudo-poly (amino acid) biomaterials. *Trends in Polym. Sci.*, 1996, V. 4, No. 12, p. 394–397.
38. Polymer surfaces and interfaces. Ed. by M.J. Feast, S. Munro, and R.W. Richards. New York, John Willey & Sons, 1993, 450 pp.
39. Polyethyleneglycol chemistry, biotechnical and biomechanical applications. Ed. by J.M. Hurris. New York, Plenum Press, 1992, 388 pp.
40. Nagasaki Yu., and Kataoka K. An intelligent polymer brush. *Trends in Polym. Sci.*, 1996, V. 4, No. 2, p. 59–64.
41. Special polymers for electronics and optoelectronics. Ed. by J.A. Chilton, and M.T. Goosey. New York, Chapman and Hall, 1995, 470 pp.
42. Vidinov N., and Djerov A. Ultrastructural changes in the transitional zone between articular cartilage and synovial membrane during the development of experimental osteoarthritis. *Int. Orthop.*, 2002, V. 26, p. 137–140.
43. Ronin V.S., Starobinets G.M., and Utevsky N.L. Textbook for classes on the methods of clinical laboratory investigations. Moscow, Medicine, 1977, 335 pp.
44. Shutsyanu Sh., Ionesku-Blazha V., and Moange M. Clinics and curing of rheumatoid diseases. Bucharest, Medical Publishers, 1983, 628 pp.
45. Beloenko E.D., Slobozhanina E.I., Kozlova N.M., et al. Spectrophotometric investigation of synovial effusion in differential diagnostics of villous-nodal synovitis. *Orthop. and Traumat.*, 1990, No. 5, p. 32–34.
46. Zamazyi A., and Peredera V. Changes in cellular composition of some enzymes in synovial fluid of horses with age. *Horse-breeding and Equestrian Sport*, 2002, No. 3, p. 25–26.
47. Ermakov S.F., Rodnenkov V.G., Beloenko E.D., and Kupchinov B.I. Liquid crystals in engineering and medicine. Minsk, JS Co. Asar; Moscow, CheRo, 2002, 412 pp.
48. Gerasimov A.M., and Furtseva L.N. Biochemical diagnostics in traumatology and orthopedy. Moscow, Medicine, 1986, 234 pp.

49. Lifshits V.M., and Sidelnikova V.I. Medical laboratory analyses. Manual. Moscow, Triada-X, 2000, 312 pp.

50. Russian State Standard 25209-82. Plastics and polymer films. Methods of estimating surface charge of electrets.

51. ASTM Standard D 4470-95. Standard test method for static electrification.

52. Bridelli M.G., Capeletti R., Losi S., et al. Water induced TSCD in hemoglobin and myoglobin. *Proc. 8th Int. Symp. on Electrets*, Paris, 1994, p. 869–874.

53. Bridelli M.G., Capeletti R., Maraia F., et al. Bioelectret state induced by water in lipases. *Proc. 10th Int. Symp. on Electrets*, Delphi-Athens, 1999, p. 213–216.

54. Shafranova E.I., and Snegireva N.S. Blood as a structured medium. Blood rheology and deformability of erythrocyte membranes. *Mechanics of Comp. Mater. and Structures*, 1999, V. 5, No. 4, p. 42–50.

55. Tsapaeva N.L. Diagnostics of functional and organic failure of the distal coronary bed and patients with ischemic heart disease. *Public Health*, 1999, No. 4, p. 49–53.

56. Shepotinovsky V.I., and Mikashinovich Z.I. Metabolic variations in blood cells at various forms of ischemic heart diseases. *Problems of Medicinal Chemistry*, 1984, V. 30, No. 1, p. 25–28.

57. Matyushichev V.B., Shamratova V.G., Savrasova I.V., and Gutsaeva D.R. Interrelation of electrokinetic properties of erythrocytes and their corpuscular volume and content in blood. *2nd Int. Congr. "Weak and super-weak fields and radiation in biology and medicine"*. St. Petersburg, 2000, p. 43.

58. Soroka N.F., Gubkin S.V., Kapralov N.V., and Shalamov I.V. Electrophysiological investigation of blood serum of the sick with rheumatoid diseases in combination with infectious hepatitis C. *Russian Gastroenterological J.*, 1998, No. 4, p. 20–25.

59. Aronov D.M., Sidorenko B.A., Lupanov V.P., et al. Actual problems of classification of the functional state of the sick with ischemic heart disease. *Cardiology*, 1982, V. 22, No. 1, p. 5–10.

60. Mrochek A.G., and Tyabut T.D. Methods of studying thrombocytes and erythrocytes in clinical cardiology (methodical recommendations). Minsk, Minzdrav, 1985, 23 pp.

61. Pinchuk L.S., Kravtsov A.G., ad Zotov S.V. Thermally stimulated depolarization of human blood. *Techn. Phys.*, 2001, V. 71, No. 5, p. 115–118.

62. Van Turnhout J. Thermally stimulated discharge of polymer electrets. Amsterdam, Elsevier, 1975, 240 pp.

63. Tortora G.J., Grabowski S.R., and Schmidt-Prezbindowski K. Principles of anatomy and physiology, 10th ed. Chichester, Wiley, 2002, 1248 pp.

64. Loakes D. Recent research developments in organic and bioorganic chemistry. New York, Research Signpost, 2002, 150 pp.

65. Stepanov V.M. Molecular biology. Structure and function of proteins. Ed. by A.A. Spirin, Moscow, Vysshaya Shkola, 1996, 335 pp.

66. Knorre D.G., and Myzina S.D. Biological chemistry, 2nd ed. Moscow, Vysshaya Shkola, 1986, 479 pp.

67. Belarus Patent 5720. G 01 N 33/49. Method of determining blood groups. L.S. Pinchuk, A.G. Kravtsov, S.V. Zotov, et al., 2004.

68. Pinchuk L.S., Goldade V.A., Sessler G.M., et al. Electret-thermal analysis of blood. *Medical Engineering and Physics*, 2002, V. 24, p. 361–364.

69. Kadolich Zh.V. Physical modification of polymer–metal pairs to improve their wear resistance based on modeling biophysical properties of natural joints. Ph.D. Thesis, Gomel, 2002.

70. Devyatova O.N., Afonina G.A., Krylova N.S., and Chimitova T.A. Biochemical microanalysis in practical laboratory, usage of capillary blood. *Laboratory Practice*, 1989, No. 8, p. 73–74.

71. Bartenev G.M., and Barteneva A.G. Relaxation properties of polymers. Moscow, Chemistry, 1992, 384 pp.

72. Pinchuk L.S., Tsvetkova E.A., and Kadolich Zh.V. Effect of electromagnetic fields on friction in joint implants. *J. Friction and wear*, 2001, V. 22, No. 5, p. 69-73.

73. Ermakov S.F. Tribophysics of liquid-crystalline materials in metal and biopolymer conjunctions. Dr. Sci. Thesis, Gomel, MPRI NASB, 2001.

74. Ulashchik V.S. New methods and techniques of physical therapy. Minsk, Belarus, 1986, 176 pp.

75. Aristarkhov V.M., Piruzyan L.A., and Tsybishev V.P. Physico-chemical bases of primary mechanisms of the biological effect of magnetic fields. In: Reactions of biological systems to magnetic fields. Moscow, Nauka, 1978, p. 6–25.

76. Dorfman Ya.G. Physical phenomena in living objects under the action of permanent magnetic fields. In: Effect of magnetic fields on biological objects. Moscow, Nauka, 1971, p. 15–23.

77. Buchachenko A.L. Chemical polarization of electrons and nuclei. Moscow, Nauka, 1974, 341 pp.

78. Vonsovsky S.V. Magnetism. Moscow, Nauka, 1971, 1032 pp.

79. Klassen V.I. Magnetization of aqua systems, 2nd ed. Moscow, Khimiya, 1982, 296 pp.

80. Zabrodina L.V. Effect of permanent magnetic fields on coagulation system of blood in experiments. Ph.D. Thesis, Moscow, 1972.

81. Maloletkina L.A., and Ulashchik V.S. Curing physical factors and hemocoagulation. Minsk, Belarus, 1983, 118 pp.

82. Demetsky A.M., and Alexeev A.G. Artificial magnetic fields in medicine. Minsk, Belarus, 1981, 94 pp.

83. Tsvetkova E.A. The development of materials based on polymers for permanent magnets used in medical techniques. Ph.D. Thesis, Gomel, 1999.

84. Bogolyubov V.M., and Ponomarenko G.M. General physiotherapy. Textbook, 3rd ed., Moscow, Medicine, 1999, 432 pp.

85. Remizov A.A. Medical and biological physics. Moscow, Vysshaya Shkola, 1999, 616 pp.

86. Wedensky V.L., and Ozhoguin V.I. Supersensitive magnetometry and biomagnetism. Moscow, Nauka, 1986, 200 pp.

87. Grigoryev Yu.G., Stepanov Yu.S., Grigoryev O.A., and Merkulov A.V. Electromagnetic safety of the man. Ref. and Inform. ed. Moscow, Russian Nat. Com. on protection against non-ionizing radiation, 1999, 146 pp.

Chapter 8. ADVANCES IN JOINTS ENDOPROSTHETICS

Advanced trends in endoprosthetics lie in between such sciences as medicine, engineering and biology. Therefore, like any other forecast all challenges in this field bear the treats of subjectivity and are not always apprehended unambiguously. Nevertheless, the present authors attempted to make an unbiased overview of modern endeavours in refining current generations of endoprostheses and the nearest prospects in their development.

The main conclusion drawn from this analysis suggests that the traditional designs of endoprostheses have mostly exhausted technical resources of modification and little space is left for further progress. The attempts to improve biocompatibility of implants have lead to the use of hydroxyapatite coatings on metal parts. However, the achievements in surface and genetic engineering have not been so far assimilated in endoprosthetics. The requirement of raising wear resistance of movable joints has confined to hardening of the rubbing pairs and has almost depleted the resources. UHMWPE as a polymer frictional material has not had an alternative in endoprosthetics for at least 50 years.

The analysis of engineering ideas in respect to improving existing endoprostheses and understanding mechanisms of functioning of natural limbs has brought us to a conclusion that the nearest prospects in refining endoprostheses consist in simulation of physiological functions intrinsic for natural joints. This is referred to, first of all, the development of the artificial cartilage and regulation of bioelectrical potentials in joints. The need of refusal from the direct contact of the bone with metal elements and their substitution by the composite ones is also quite apparent. The surface charge of endoprosthesis elements contacting living tissues and the magnitude of the electromagnetic field generated by them should be a subject of monitoring. Although some potentialities of ameliorating wear resistance of metal-polymer friction pairs still exist, it seems timely to change cardinally the designs of movable junctions. The friction materials whose wear debris is friendly to surrounding tissues are becoming popular. There also arose a necessity of designing such friction pairs that are capable of insulating their debris and lubricating fluids from the living tissues. It is quite probable that combined endoprostheses will soon appear in which artificial materials will be harmonized with bone transplants. This trend in transplantation clears the way to joining the research potential of genetic engineering.

The strategy of endoprosthetic operations on joints expects reconsideration as well. The chief commandment of a surgeon not to do harm is in contradiction with the necessity of bone resection for preparing endoprosthesis bed during operation. This problem is also connected with creation of a novel generation of endoprostheses.

Above cited and related problems are to be considered in this chapter.

8.1 MODIFICATION OF ENDOPROSTHESES

The main problems that restrict service life of joint endoprostheses *in vivo* are related to insufficient strength, biocompatibility and wear resistance of their elements.

Mechanical strength of endoprostheses is laid down at designing with a significant margin of safety. The criteria of correct design solutions are a complex of laboratory, bench and simulation tests to which endoprostheses are subjected prior to clinical approbation. This is why endoprostheses may break *in vivo* as a result of accidents or in case of inadequate biocompatibility and wear resistance. In the biological environment of the organism strength of endoprosthesis materials may impair due to insufficient biocompatibility. This leads to either corrosive damage or worsened durability of the materials followed by fracture.

Since endoprostheses are rigidly fixed to the joint-forming bones, the problem of their mechanical strength can not be treated separately without their interrelation with that of the bones and bone-implant coupling. Judet's saying: "Experimental means learning from failures" – is applied to joint endoprosthetics, where it is precisely from all sorts of failures that a great deal have been learned in the last 40 years. At least 10-15 years of results in a uniform group of patients is required in order to achieve an honest statement on the performance of a procedure [1].

The endoprosthesis design to a great degree defines stress state of conjugated bones. Using the finite elements analysis the authors of [2] studied the pelvis bony socket and compared it with radiological imaging using threaded acetabular cups of three different shapes (parabolic, conical, and hemispherical). In all three cups the stress in the bony socket increased from lateral towards medial. Compressive stress was found on the superior and inferior parts of the cup, but mainly on the superior aspect, seen radiologically as a new trabecular bone formation. The maximum compressive stresses were seen in the cranial curvature of the conical cup, with less in the parabolic form and least in the hemispheric form. The tensile stress at the bottom of the socket increased from the hemispheric to the conical shape. This model uses computer simulation to predict bony changes with different designs of implants. The ability to simulate biological conditions is a valuable addition to the testing of mechanical strength.

This suggests that the problem of inadequate mechanical strength of endoprostheses can not exist in a pure form. It arises as soon as the endoprostheses mismatch biocompatibility and wear resistance demands.

Biocompatibility is a major factor that governs integration of the bone tissue and implant. The analysis of endoprostheses materials by the criterion of biocompatibility is presented below. The main structural materials for endoprostheses are metals, polymers and ceramics. They are referred to as bioinert materials, which can not form a firm link with the bone. In spite of successes in endoprosthetics, approximately 1 % of total hip arthroplasties fail and are revised each year, and a further 0.5–1.0 % fail clinically or radiographically [3]. Data from Swedish National Hip Arthroplasty Registry, a national audit of the outcome of total hip arthroplasties, has indicated that the principal reason for implant failure is aseptic loosening [4]. When the mechanical load causes shift of the bioinert implant relative to the bone bed, it undergoes fast loosening. This may result in endoprosthesis failure or bone fracture [5].

To perfect biocompatibility, the bioinert elements are overlaid by porous coatings. The implants of hydroxyapatite, corals or produced by metal powder sintering have a porous surface structure. As the bone grows into the pores the implant obtains a so-called secondary or biological fixation. To allow for a capillary blood supply into the ingrown bony tissue, the pore diameter should not be below 100 μm. It is believed that microshifts at the bone-implant interface may cause rupture of such capillaries. This brings about tissue necrosis, bone inflammation and instability at the interface [6]. As it can be seen, the porous surface on the implants can not guarantee fixation stability in the bone.

As for the bioactive glass and glass ceramics as well as for dense synthetic hydroxyapatite, they interact with the bone via formation of chemical bonds. The mechanisms of their formation depend on the nature of bioactive materials [7]. However, bioactive materials are insufficiently strong and therefore are unfit for the load-bearing elements of endoprostheses.

Biodegradable (resorbed) tribasic calcium phosphate and bioactive glass are dissolved *in vivo* and are substituted by the bony tissue. The resorption mechanisms are practically uncontrollable, depending on the state of tissues and the organism itself. This is the reason of unstable relationship between the bone and material [8] and restricted usage of biodegradable materials in endoprosthetics mainly as coatings.

Allotransplants are characterized by higher biocompatibility as compared to engineering materials. Nonetheless, the bony allotransplant can not be fully engrafted to a host bone. The biointegration mechanism is realized via lysis of the allotransplant and partial ingrowth of the osseous tissue. An ideal implant material most readily integrating with the host osseous tissue has proved to be autotransplants.

As can be seen, the means of refining biocompatibility of structural elements of endoprostheses consist mainly of application of bioactive or biodegradable coatings on the elements made of inert materials. We now can imagine a perfect joint endoprosthesis as an autotransplant created by the methods of genetic engineering and furnished with an epiphysis covered by a cartilage participating in reconstruction of the joint.

Wear resistance of prostheses gives rise to the majority of complications in endoprosthetics of joints. This problem is in the centre of orthopaedists' attention not only because the prostheses reach their limiting wear too quickly. Ingress of the wear particles into the bone-implant contact zone followed by the reaction of cells and leading to aseptic instability of endoprostheses is by far more dangerous. Let's try to analyze the ways of improving wear resistance available today for different types of endoprostheses.

Endoprostheses with the metal-metal friction pairs have a long-standing history and have reached a state of relative stability in which they are found for already ten years. The friction material is mainly stainless steel or CoCrMo alloy. Attempts have been undertaken in surface hardening of titanium friction pairs using diffusion oxygen hardening or titanium nitride coatings. The use of zirconia friction pairs is unlikely to have any promises [9]. Further refinement of surface finish of the friction parts can not improve anyhow wear resistance of metal endoprostheses since a so-called equilibrium roughness of surfaces occurs during run-in *in vivo*, which corresponds to individual features of the biological medium of the organism and friction regimes. The height of microasperities of the initial friction surface is as a rule less than the parameters of the equilibrium roughness.

Most joint endoprostheses that incorporate the ceramic-ceramic pair employ Al_2O_3 [10]. Modernization of technologies has made the cost of ceramic heads of high finish quality comparable to that of the heads made of CoCrMo alloys and the probability of their brittle fracture is close to zero. The reserves of improving wear resistance of endoprostheses with ceramic-ceramic friction pairs are likely to approach depletion. Their modification has taken the path of reducing stiffness of the friction pair. With this aim, pelvic components of hip endoprostheses are made three-layered with a polymer liner to absorb impact loads (Fig. 3.8,*f*).

It is evident that vistas in raising wear resistance of endoprosthesis friction components via their hardening are restricted by mechanical properties of technical materials. Nevertheless, even superhard materials used in friction pairs can not eliminate the problem of wear debris formation during run-in and their contamination of the bone-implant contact zone.

We believe that the resources of wear resistance of the UHMWPE–metal and UHMWPE–ceramic joints are connected with the probability of forming cartilage simulating polymer structures. The results cited in chapter 6 suggest that such pairs are capable of abating contamination of the contact zone by wear debris as opposed to the pairs made of more wear resistant materials. In addition, the microporous UHMWPE inflicts a minimum damage to the cartilage during friction and can be used as a friction material in unipolar endoprostheses. Engineering solutions unfolding this idea are set forth in the chapter to follow.

8.2 ENDOPROSTHESES WITH ARTIFICIAL CARTILAGE

The idea of creating joint endoprostheses containing a polymer element simulating cartilage has been a subject of animated discussion for already three decades. The English tribologist, D. Dowson, is ascribed to be an ideologist of this trend. He considers that the artificial cartilage can be an efficient means of reducing the amount of wear debris and perfect lubrication of the prosthesis [11]. As a model of endoprostheses incorporating an artificial cartilage he suggested to take a cushion-form bearing. Along with this, there were attempts to create endoprostheses containing thin layers (1–3 mm) of polyurethane [12], silicone rubber [13] and hydrogel [14] with elasticity modulus similar to a natural cartilage, which layer is strongly bonded with a rigid substrate. A larger portion of the shear load in such structures is born just by these layers and rather low stresses are operating at the interface with the substrate. In this manner the fatigue wear of the liner is impeded [15]. When lubricated with a fluid, a lubricating 0.5 μm thick film is formed in the contact sites, which means that microasperities of a low-modular elastic layer undergo deformation and elastic restoration thus favouring the elastohydrodynamic regime of the film recovery [16].

A model of cartilage simulating microporous layer has been described elsewhere [16]. Based on the methods of the nonlinear contact mechanics this work explains regularities of the lubricating liquid circulation in the friction pair clearance and communicating pores of the cartilage. It is underlined that the model realizes the lubricating mechanisms by sweating and booster lubrication.

The described models and attempts of their adoption have not, however, lead to the development of new endoprostheses fit for clinical testing. An artificial cartilage Salubria [18] based on a hydrogel has been elaborated in the Georgia Technological University (USA). It perfectly absorbs water, surpasses silicon in biocompatibility, and is sufficiently strong and pliable. The Salubria is at present subjected to a complex of medico-biological testing including in animals. The developers are sure to use the material in orthopaedics for "repairing" damaged cartilage.

Advanced designs of endoprostheses containing UHMWPE-based cartilage have been created whose structure and properties were described in chapter 6.

Endoprostheses of the knee [19] include a horseshoe plate furnished with three wedge-like elements for fixation in the tibial bone. The base is made by the powder metallurgy method from titanium. A microporous UHMWPE liner is press-fit into the plate pores. The pores of the liner are filled by a natural or artificial synovia (Fig. 8.1).

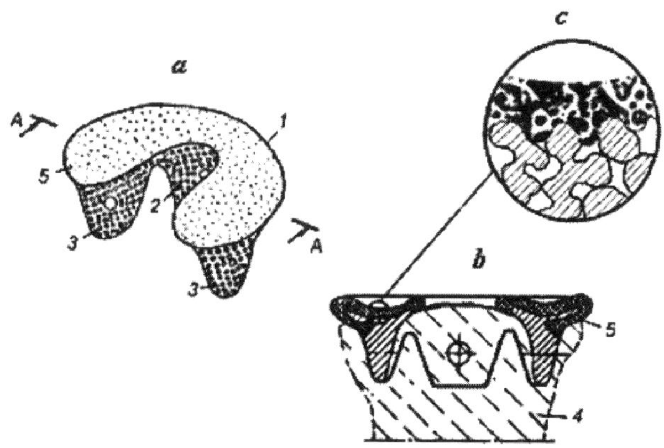

Fig. 8.1. Endoprosthesis of the knee: *a* – general view, *b* – cross-sectional view A-A, *c* – polymer-metal interface structure: 1 – plate, 2 and 3 – fixatives, 4 – tibia, 5 – UHMWPE layer

The endoprosthesis is fixed to the tibial plateau resected so as to preserve ligaments. With this aim, wedge-like elements are impacted into the bone and screws are used. Secondary fixation proceeds during ingrowth of the cancellous bone osteons into the titanium base pores that are more than 100 μm in size. A layer of microporous UHMWPE covering the base functions as a cartilage. During static loading of the knee the synovia is squeezed from the polymer micropores and separates friction surfaces in the artificial joint. In the course of rotation or sliding the microporous layer undergoes local deformation and the synovia is thus supplied into the cartilage-counterbody contact zone. This structure damps impact loads and protects the prosthesis friction surface from damaging. When the joint is unloaded the synovia returns under the action of capillary forces into the micropores of the polymer layer. The lubricating fluid is replenished via generation of the synovia by the retained portions of the synovial shell.

The formation of microporous layer simulating cartilage on the polymer part of the endoprosthesis seems to be an optimum method of modification of the traditional endoprostheses designs. The method reported in [20] is aimed at formation of a microporous layer of a given thickness (about 1 mm) on the UHMWPE prosthesis parts (ball-and-socket configuration of a unipolar hip endoprosthesis). The pore distribution over the layer thickness is uneven with pore diameter increasing closer to the friction surface. The layer is formed as follows. The polymer part is brought into contact with Vaseline oil and endured under a specific temperature about that of UHMWPE melting point. After this, the part is immersed into a solution of polyvinyl pyrrolidone in the organic solvent thermodynamically compatible with Vaseline oil, e.g. hexane. The treatment is exercised under the softening temperature of polyvinyl pyrrolidone. Then, the semi-finished polymer is cooled down and the process liquids are extracted from it.

The surface layer of the part transforms after the treatment in plasticizing liquids into a state of colloidal solution. When cooled to below UHMWPE melting point, it decomposes into the phases: i) a microporous UHMWPE-based matrix with 1–10 μm pore diameter; ii) a layer of polyvinyl pyrrolidone deposited on pore walls; iii) oil solution in hexane concentrated in the pores. After extraction of hexane a microporous cartilage-simulating layer is formed on the friction surface of the part. Hydrophilic coatings consisting of polyvinyl pyrrolidone are perfectly wetted on the pore walls of UHMWPE matrix by the synovia. It penetrates from the implanted part into the pores from the joint cavity under the capillary pressure effect. The porous layer thickness can be regulated by varying thermal treatment regimes. Among the advantages of the method are bioinertness of the substances used and absence of inflammations in the joint.

The artificial hip joint liner made of UHMWPE [20] acquires after treatment a structure simulating that of a natural acetabulum. The microporous layer operating *in vivo* like a natural cartilage is formed on the spherical surface of the cavity. The rest part of the liner is made of the initial pore-free UHMWPE. Thus, the liner [21] is formed as an integral polymer block without a coating and, consequently without adhesive seams able to impair reliability of the structure. What is more, the strength of the basic part of the liner is not lowered by the presence of pores.

The liner [21] has been studied using imitator tests whose scheme is illustrated in Fig. 5.22. The friction pair consisting of an UHMWPE liner and CoCrMo head was subjected to a cyclic loading without lubrication in a 2% water solution of carboxymethyl cellulose. During the loading cycle (about 1 s) the impact load N = 12 N was applied along the liner axis and swing-return of the head to $\pm$ 50° angle. The friction torque M was registered in the friction pair under study.

A characteristic view of kinetic dependencies of M in different moments of testing for 4 h without lubrication is shown in Fig. 8.2. Initially, during 20 min (Fig. 8.2, *a*) M values remain practically invariable. Then the friction torque reduces noticeably (*b*) due to, probably, run-in. simultaneously, a phenomenon unusual for the pairs with the initial (non-modified) liner is observed. The values of M become unstable and change with time following the law close to a sinusoidal. The vibration amplitude within the 100–200 min (*c*) time interval reaches the highest magnitude, whereas M in the reference pair diminishes monotonously. The sinusoidal dependence is preserved during the whole experiment.

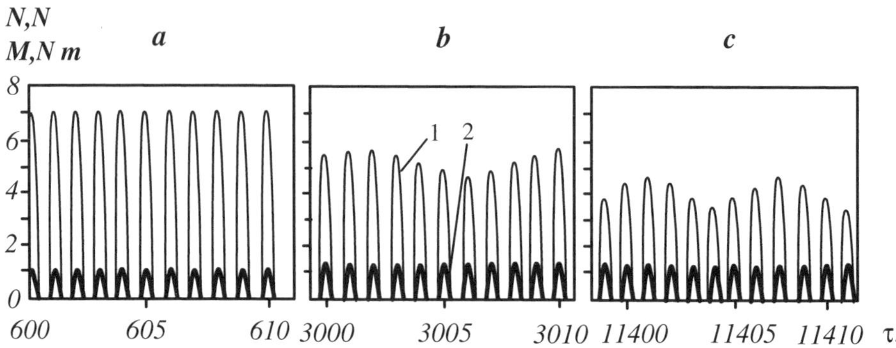

Fig. 8.2. Friction torque (*M*, *1*) versus test time *τ* under normal load variation (*N*, *2*): *a* – during first 10 min, *b* – 50 min later, *c* – 190 min after test start

This phenomenon is attributed to the participation of the microporous layer in friction [22]. Due to its elasticity the actual contact area enlarges substantially and imposes an effect on its deformation mode and heat exchange between the two components. The sinusoidal character of changes in maximum values of *M* is evidently owing to superposition of two competing processes, namely: 1) deformation of pore walls under cyclic compressing and frictional loads, and 2) reversible restoration of their shape during unloading and under the frictional heat effect generated also in cycles. This process is complicated by the wearing of microasperities of the liner and removal of wear debris at cyclic loading.

Lubrication alleviates significantly (as much as 20 times) *M* at the initial test stage in contrast to dry friction (Fig. 8.3). There are the sections of lowering and stabilizing of *M* on its kinetic curves for the pairs with reference and modified liners. Note that run-in period of the reference liner made up 40 min, while that of the modified – 5 h. This is probably connected with a specific mechanism of restoration of the elastic deformation of the microporous layer [23]. After the run-in the friction torque of the modified liner is lower than the reference one has.

The tests consisting of 1 mln cycles (corresponds to 1 year of operation *in vivo*) have shown linear wear of the initial UHMWPE liners 0.20–0.23 mm, and of the modified ones 0.07–0.11 mm [24].

The sinusoidal behaviour of the tissues in response to mechanical loads is typical for the cartilage, bone and other biological structures [25]. Apparently, the analogous character of the friction torque variations in the artificial joint will effect favourably its useful operating life.

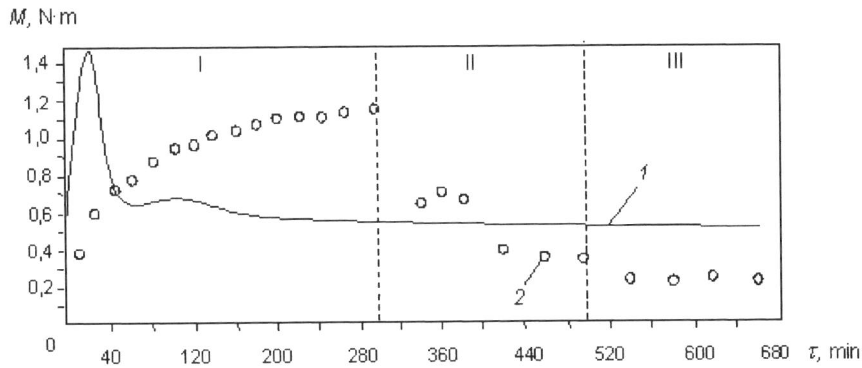

Fig. 8.3. Kinetic dependencies of *max* friction torque values for the head–liner friction pairs at lubricated friction for cycles duration: I – 5 h, II – 3.5 h, III – 2.5 h with stops between cycles for 12 h. *1* and *2* – liners from initial and modified UHMWPE

Other advantages of the polymer cartilage model are connected with the possibility of using micropores for medicinal and technological aims.

Target-oriented drug delivery into the operative wound can be performed using the joint endoprosthesis [26]. Such an endoprosthesis contains a polymer part fit with a microporous layer, which carries a medical product able to prevent complications in the operating wound.

Introduction of drugs in the endoprosthesis structure furnishes next benefits:

- the drug isolates directly in the operating wound where pathological processes may develop but is not supplied with blood as in traditional designs;
- protracted isolation of drugs from the polymer matrix micropores provides for a long-term action during healing of the wound;
- the amount of isolated drug depends upon the microporous layer deformation, i.e. moving activity of the patient.

The drugs on aqua base that are perfectly compatible with biological fluids of the organism are used to fill the pores. They are:

- antibiotics exerting bactericidal effect on micro organisms in the growth stage, such as penicillin products, cephalosporins or aminoglycozoids, lyncomicin hydrochloride, sodium fuzidin;
- their antiseptic and anti-inflammatory substitutes: dioxin, chlorohexidine, indometacin;

- enzymatic products preventing formation of ossificates from hematoma: lidase, arteparon.

As tribological tests have shown, along with their major purpose these substances function as perfect lubricating media. Being injected into the artificial cartilage pores they are able to perform two functions: 1) prevention of pathological processes in the zone of operative wound, 2) lubrication of the friction zone of endoprostheses during the initial period after implantation.

The drugs injected into the UHMWPE matrix are extracted from it following an exponential dependence on time. The greater amount is isolated in the first 3–4 days, and then the extraction reduces approaching zero 7–9 days later [26]. This regime is considered to be optimum for prevention of complications. During isolation of drugs the vacant volume of the pores is filled by the synovia.

Reliability of fixation of the head of hip endoprostheses on a tapered stem of standard dimensions presents a problem for the unipolar endoprostheses whose heads are made of UHMWPE. During operation *in vivo* complications may appear in connection with loosening of the head fit. Walking, standing and other types of loading of the artificial limb cause the stiff metallic wedge to supersede the low-modulus polymer. As a result, the conical hole in the head becomes oval in cross-section. To avoid this, a threaded recess is made in the head into which a metallic plug is screwed with a standard conic hole for fixation on the stem. Reliability of such structure depends on the strength of the plug-head junction.

A head design of a unipolar hip endoprosthesis is presented in Fig. 8.4 [27].

A cartilage-simulating microporous layer is formed on the spherical surface of the UHMWPE head. A layer of the polymer of *h* thickness on the thread profile in the head hollow is subjected to thermal treatment in the plasticizing liquid together with the cartilage-simulating one. The plasticizer in the thread profile pores is substituted by bone cement using special technological processes. After this a metal plug is screwed into the threaded hole. Upon curing the bone cement appears to be mechanically fixed in the polymer matrix and forms a strong adhesive joint with the thread surface of the plug. This limits the probability of spontaneous unscrewing and elevates reliability of plugging. Notice that the detachable conical junction of the metal plug and endoprosthesis stem correspond to adopted in orthopaedics standards. The microporous layer on the frictional surface of the head ensures a mild wear mode of the cartilage in the acetabulum as compared to other engineering materials.

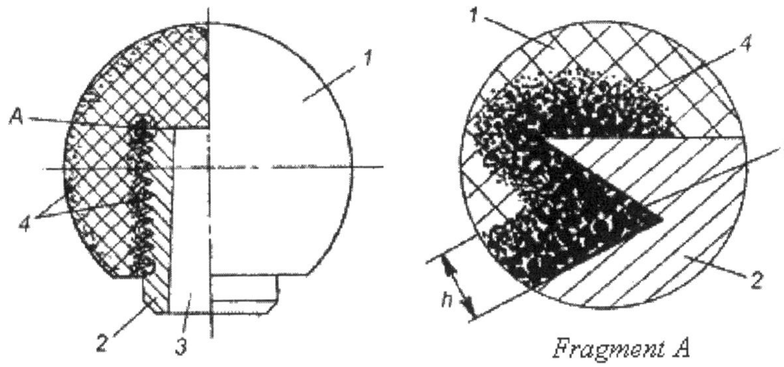

Fig. 8.4. A head of unipolar hip endoprosthesis: 1 – spherical UHMWPE head, 2 – metal plug, 3 – standard conical hole, 4 – modified surface layer, 5 – communicating micropores filled by bone cement

Testing of endoprostheses of the hip joint with described artificial cartilage preceding clinical tests was performed in Belarus in 2000-2002.

By the courtesy of Prof. Dabrowski J.R., the service life of the endoprostheses was estimated in Bialystok Polytechnic University (Poland). A hip joint simulator was used with this aim that operated for a period corresponding to one year of operation *in vivo*.

A pilot batch of UHMWPE liners was manufactured with in accordance to Technical Specifications TU RB 03535279.078-99 "Manufacture of reinforced elements of artificial joints with a surface layer simulating the cartilage structure" and a Technical Instruction TI 8-1-99 "Manufacture of reinforced elements of artificial joints with a cartilage-simulating surface layer". The experimental liners were evaluated in hygienic respect in animals at the Belarussian Research Sanitary and Hygienic Institute. Biocompatibility of the microporous samples was estimated in terms of the structural and functional state of the immune-competent cells of the peripheral human blood by Prof. Adamenko G.A. at Vitebsk State Medical University. The identical reactions of the cells to the natural cartilage and microporous samples treated in HF discharge (σ_{ef} ~10^{-7} C/m^2) are the evidence of a high-degree similarity of immunological parameters of the samples and cartilage tissue (see Ch. 6.5).

An experimental batch of UHMWPE heads furnished with a microporous layer was manufactured according to Technical Specifications TU RB 400084698.112-2000 "A polymer head of the unipolar hip joint endoprosthesis with modified surface" and Technological Instruction TI 9-2000. These heads have successfully passed physico-mechanical and tribological tests as components of endoprostheses corresponding to TU RB 500576133.001-2001 "Unipolar metal-polymer endoprostheses of the hip joint" that have been developed at Grodno Affiliation of Belarussian Engineering Academy. Their hygienic safeness was confirmed by the Belarussian Research Sanitary and Hygienic Institute. After named tests the Ministry of Public Health of Belarus has drawn up a decision on clinical testing of endoprostheses. They are being carried out at present at Belarussian Research Institute of Traumatology and Orthopedy, Grodno Regional Traumatology and Orthopaedic Centre and Gomel State Medical University. The first test results initiated in Grodno are reported in [28].

8.3 METAL-POLYMER FRICTION JOINTS

To prolong service life of endoprostheses means, first of all to improve insufficient wear resistance of polymer friction parts by their combining with more stiff design elements. In a specific biological environment of the organism this task may be solved by the friction couples of unusual structures.

The pelvic component of the hip prosthesis [29] permits to join in the artificial joint the benefits of a wear resistant metal-metal junction and those of a softer artificial self-lubricating cartilage.

A classical shock-absorbing structure of the pelvic component of the metal hip endoprosthesis illustrated in Fig. 3.8, *b* consists of a metal cup and a liner insulated by a polymer interlayer. The layer is not involved in friction and pressure distribution over the liner. This structure is unable to hamper the removal of metal wear debris from the friction zone and contamination of the surrounding tissues.

Named drawbacks are overcome by a design of the pelvic component shown in Fig. 8.5.

This component incorporates of a metallic (titanium, stainless steel) cup 1, placed in it polymer (UHMWPE) liner 2 and press-fit in the liner thin-walled hemispherical insert 3 (CoCrMo alloy). The structure is fixed at implantation in the pelvic bone 4 using teeth 5 over the outer surface of the cup. The femoral component consists of a metal stem 6 whose neck 7 carries a spherical head 8 (CoCrMo alloy). The stem is immovably fixed in the medullary canal of femur 9. The head forms a movable junction with insert 3. The holes made in the insert are filled with UHMWPE so that the inner surface of the insert and surfaces of projections 10 of UHMWPE liner emerging through the holes form a continuous hollow sphere. The surface layer of projections 10 is transformed into a gel-like state. Depending on the solubility of UHMWPE in a process liquid and time-temperature regimes of treatment, thickness *h* of the gel layer can be 0.1–1.0 mm. Further on the process liquid is removed from the polymer matrix and substituted for the synovia or medical products.

The structure operates as follows. During loading of the endoprosthesis the spherical head 8 is found in a movable contact with the spherical friction surface of the acetabular component formed of the patches of metal insert 3 and polymer projections 10. During friction the self-lubricating UHMWPE and synovia found in the UHMWPE matrix undergo frictional transfer from the surface of projections 10 onto the head and contact surface of insert 3. This is why the friction pair of the endoprosthesis is all time operating under the boundary or liquid friction regimes under which any seizure or welding bridges leading to scuffing are improbable in principle.

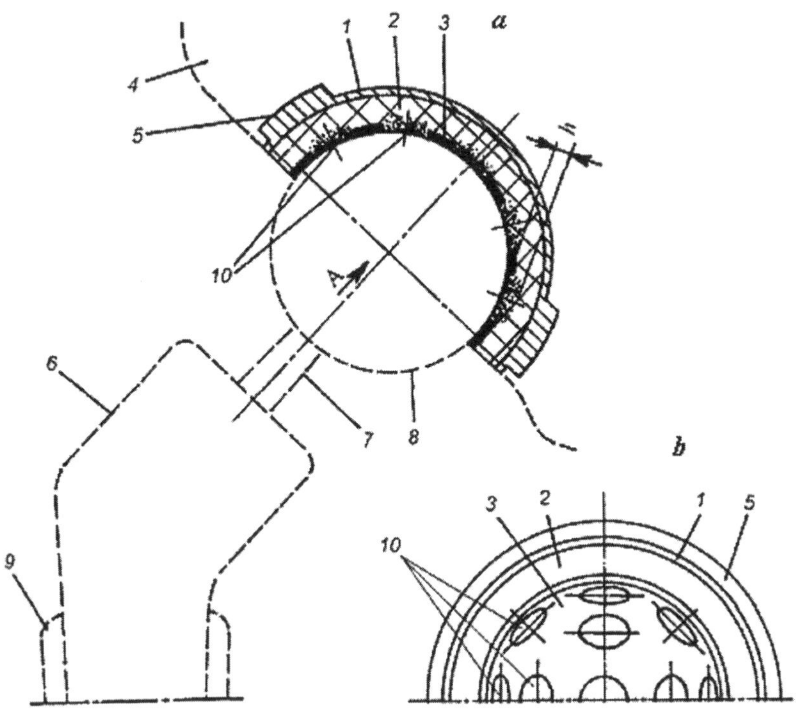

Fig. 8.5. Pelvic component of the hip joint endoprosthesis [29]: *a* – cross-section of implanted endoprosthesis in frontal plane, *b* – view of pelvic component along arrow A: 1 – cup, 2 – liner, 3 – insert, 4 – pelvic bone, 5 – tooth, 6 – stem, 7 – neck, 8 – head, 9 – femoral bone, 10 – insert segment with microporous structure

Under peak loading easily deformed polymer material of liner 2 is squeezed through the holes in the insert. This compensates wear of projections 10, and second, intensifies with load increment lubrication of the endoprosthesis friction pair. Squeezing of projections 10 through the holes in the insert and separation of the synovia is alleviated if liner 2 is made of the UHMWPE gel. Simultaneously, a stiff contact between the metal insert and head prevents polymer projections 10 from accelerated wear, whereas elasticity of the polymer material of liner 2 provides damping of impact loads and adds resilience to the artificial limb intrinsic for a natural joint.

The presence of a system of micropores on the friction surface of projections 10 promotes absorption of the wear debris (both metal and polymeric) by the porous matrix of the gel. This is strongly important since accumulation of wear particles in the surrounding tissues may initiate organism reaction to a foreign body thus aggravating the situation by the probability of osteolysis, metallosis and chemical abscess.

The distribution of contact pressures over the acetabulum friction surface is a major factor defining the frictional transfer parameters, isolation velocity of the lubricating liquids in the friction zone and drugs in the operation wound, and wear rate of the endoprosthesis as a whole. The contact pressure distribution is determined by the ratio between the metal and polymer element areas on the friction surface of the pelvic component. This ratio is easily regulated by varying the number, position and diameter of the holes in insert 3.

A combined structure [30] of the pelvic component of the hip endoprosthesis differs from the previous one in unloading mode of the UHMWPE liner from the peak loads, which is exercised with the help of a rigid antifrictional insert made in the liner and contacting the cup and the spherical head.

Figure 8.6 illustrates a cross-sectional view of the implanted endoprosthesis in the frontal plane. A metal or ceramic cylindrical insert 9 is installed along the axis of liner 2 inside the central angle α between the vertical axis of the head and the pelvic neck axis. The contact plane of the insert with the spherical head 7 is of the form of a concave sphere with diameter D equal to that of the spherical cavity in the polymer liner and the head. The insert diameter is $d < D$.

During loading of the endoprosthesis, insert 9 props by its upper face against cup 1 and by the shoulders of its cylindrical part against an annular recess in the liner, which hinders its upward shift. At abduction-adduction and extension-flexion of the femur, head 7 slides over the spherical surface formed by the internal surface of liner 2 and lower face of insert 9. Therefore, load on the endoprosthesis is spread over the contact areas "head 7 – liner 2" and "head 7 – insert 9". As a result, the weakest section of the endoprosthesis, the polymer insert, perceives just a part of the load, which retards its wear rate and enlarges lifespan of the endoprosthesis. The optimum load distribution in this junction is defined by the ratio *d/D* and angle α. The highest loads on the endoprosthesis operate over the vertical line and neck 6 axis, therefore the optimum position of insert 9 axis is inside the central angle α. The optimum criterion of the structure will be the balance between wear rates of the insert–head and liner–head junctions.

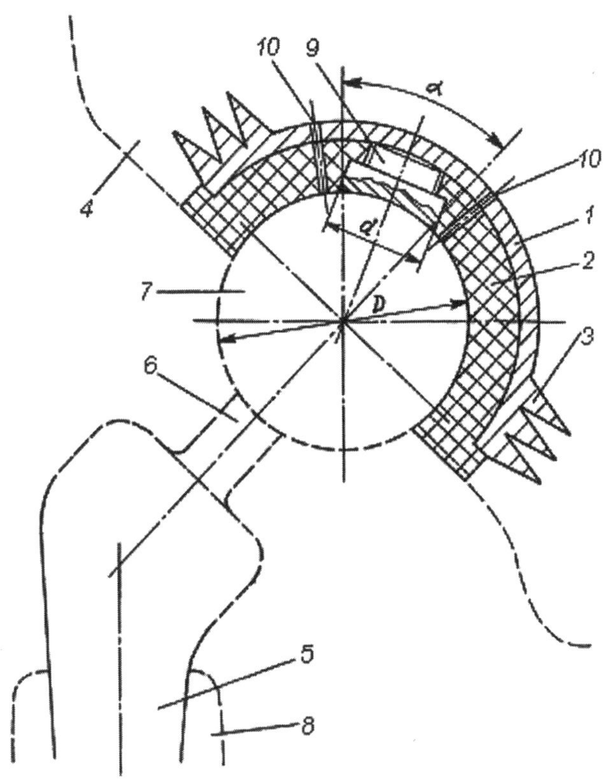

Fig. 8.6. Pelvic component of the hip endoprosthesis [30]: 1 – cup, 2 – liner, 3 – teeth, 4 – pelvic bone, 5 – stem, 6 – neck, 7 – head, 8 – femoral bone, 9 – insert, 10 - canal

The biological fluids that lubricate the friction pair are fed via canals 10 connecting the acetabulum and the clearance in the movable joint of the endoprosthesis. Uniformity of the lubrication is promoted by an annular clearance between liners 2 and inserts 9 as well as turning of the insert inside the liner socket. The annular clearance can be also used as a container of drugs (blood coagulation activating agents, antibiotics, and etc.).

The main advantage of proposed design consists in damping impact loads on the endoprosthesis by the polymer liner 2 when the larger portion of the load is carried by the wear-resistant pair head 7 – insert 9 protecting the liner from overloads and prompt wearing.

The total hip endoprosthesis [31] is to a certain degree simulating peculiarities of the functions of a natural joint. This prosthesis (Fig. 8.7) incorporates a *capitis femoris* ligament attached to the femoral head crest and inside a hole in the acetabulum centre. This flexible ligament has an artery feeding the head and the femoral neck from the upper buttock artery [32]. It does not restrict mobility of the joint within the anatomic limits but lowers the probability of dislocation. There is always a gap near the ligament between the head cartilage and acetabulum that contains the synovial fluid.

A design of endoprosthesis shown in Fig. 8.8 [31] has a pelvic component with metal cup 1 and polymer (UHMWPE) liner 2. The femoral component is made in the form of a metallic stem (the material is similar to cup 1) with a spherical head 7. Hollow 8 of the head has a spherical dome contacting the metal peg 9 located immovably in the liner.

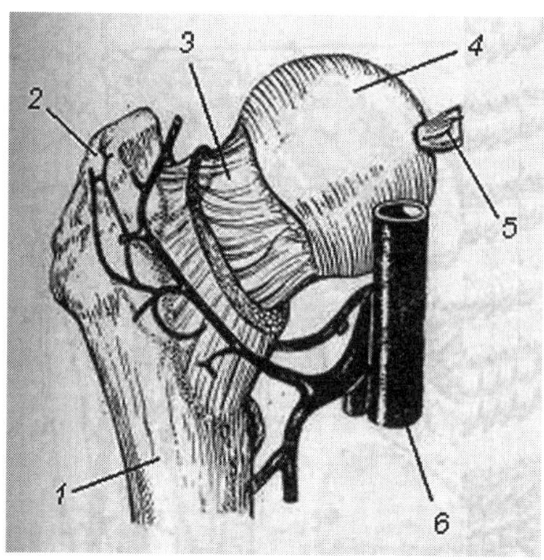

Fig. 8.7. Proximal femur: 1 – diaphysis, 2 – greater trochanter, 3 – neck, 4 – head, 5 – capitis femoris ligament, 6 - artery

The endoprosthesis design is shown in Fig. 8.8 [31]. Its pelvic component includes metal cup 1 and UHMWPE liner 2. The femoral element looks like a stem (the material is similar to the cup) having a spherical head 7. The head contains a hollow 8 with a dome contacting metal peg 9 immovably fixed in the liner.

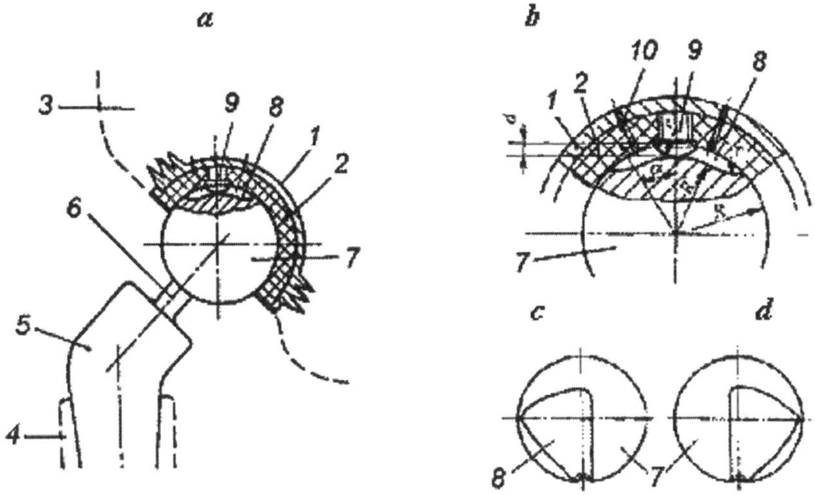

Fig. 8.8. Hip endoprosthesis [31]: *a* – cross-section of implanted endoprosthesis, *b* – fragment of immovable junction between head and liner, *c* and *d* – upper view of endoprosthesis head for left and right legs: 1 – cup, 2 – liner, 3 and 4 – pelvic and femoral bones, 5 – stem, 6 – neck, 7 – head, 8 – hollow, 9 – peg, 10 - canal

The peg on its spherical face is made of radius r. Its vertical axis passes through the upper point of the implanted liner. The peg props against cup *1* and the collars of its cylindrical part rest against an annular recess in the liner, which hinders it from shifting upwards. The peg's spherical end protrudes over the upper point of the dome of the implanted liner to a height $d < r$. The spherical hollow radius is $8R' = R - d$. The size of the hollow enables turning of the head in the liner by angle α relative to the vertical axis.

The shape of the head hollow is irregular oval. Its shape and size allow for the motions peculiar to natural hip joints, namely *max* flexion 120° and extension 13° in the sagittal plane; *max* abduction 90° in the frontal plane of the left and right limbs to the right and left [29]. The extreme points of each hollow shown in figures *c* and *d* are inclined to the vertical axis of the head in accordance with above-named angles.

During flexion or abduction of a joint, head 7 revolves in liner 2 till the rounded face of peg 9 leans against the wall of hollow 8. Such positioning with a normal mobility of the artificial joint reduces the risk of dislocations.

A larger part of the load on the implanted endoprosthesis is perceived by the peg 9 – head 7 contact, while the rest part is carried by the head 7 – liner 2 coupling. This geometry relieves the load from the liner, a most weak link of the endoprosthesis, retarding its wear rate significantly, and extending service life of the whole unit. Notice that the load on the metal pair peg 9 – head 7 is not as large as to accelerate wearing, and under optimal load distribution the wear rates of the metal–metal (peg–head) and polymer–metal (liner–head) joints are in fact similar.

Canals 10 supply the synovia and other biological fluids into the friction zone and provide lubrication of the endoprosthesis. The presence of fluids in hollow 8 averts dry friction and fatigue wearing of the polymer liner 2.

Hollow 8 also serves as a collector of wear debris formed during functioning. Removal of the abrasive wear particles alleviates wear rate of the liner. Accumulation of wear debris inside the endoprosthesis design assists in insulating them from the contact with surrounding tissues. These closed cavities envisaged in endoprostheses structures are also employed as vessels for medicinal preparations, including blood coagulation promoters.

The hip endoprosthesis [33] is based on the principles of the previous pair realizing the rolling friction.

This endoprosthesis includes a pelvic and femoral components fixed respectively in the pelvis and femur (Fig. 8.9).

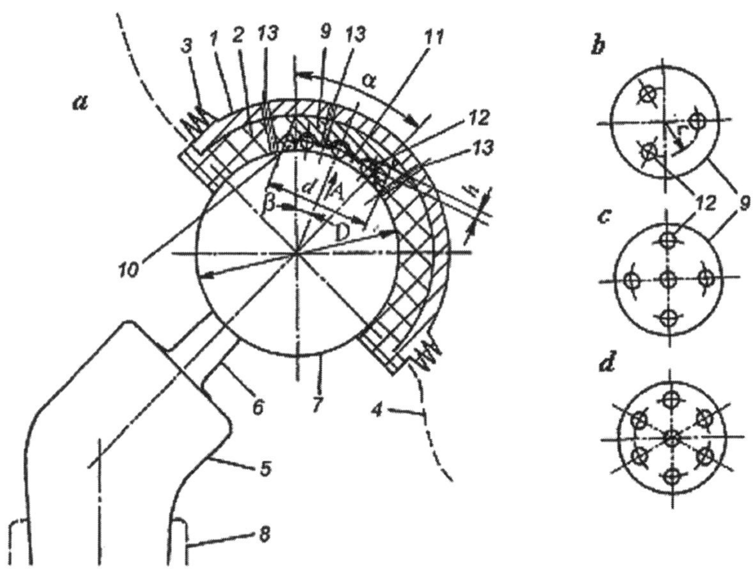

Fig. 8.9. Hip endoprosthesis [33]: ***a*** **– cross-section of implanted endoprosthesis with a frontal plane;** ***b, c*** **and** ***d*** **– faces of inserts 9 of different designs viewed along arrow A. 1 – cup, 2 – liner, 3 – teeth, 4 – pelvic bone, 5 – stem, 6 – neck, 7 – head, 8 – femoral bone, 9 – insert, 10 – collar, 11 – clearance, 12 – ball, 13 - canal**

A cylindrical metal insert 9 of the liner is located along the axis in the frontal plane of the body and inside the medial angle α between the vertical axis of the head and that of the neck 6. Its diameter $d < D$ is less than that of the head resting upon cup 1 and collar 10 in liner 2. Metal balls 12 are located in clearance 11 between insert 9 and head 7. The balls are fixed rotatable in the recesses made in the insert. Depth of recesses h allows for the balls to contact simultaneously the head and insert. Coaxial canals 13 are made in cup 1 and liner 2 connecting acetabulum with the clearance of the endoprosthesis friction pair.

When head 7 rotates in a loaded artificial limb, insert 9 revolves about its axis, and the balls roll over the head's surface. As a result, the total loading is distributed between the sliding pair head–liner and rolling pair head–balls. This is the latter one that provides low friction in the endoprosthesis. Thanks to rotation of insert 9 in liner 2 in the case of optimum number of balls (based on the mass and activity criteria of the patient) no wear paths are formed on the head. A design with three balls spaced at equal distances r from the insert centre at an angle of 120° to each other is presented in Fig. 8.9, *b*. Figure *c* illustrates a symmetrical positioning of four balls relative to a central fifth one. Figure d shows the same for six balls related to the seventh central one.

An optimum load distribution between the sliding and rolling pairs is defined by the d/D ratio and angle β inclination of insert 9 to the vertical axis. Maximum loads on the endoprosthesis are effective in vertical directions and along the axis of neck 6. Therefore, the location $\beta < \alpha$ of the insert is most favourable. The optimum criterion of the design is considered to be the parity of the wear rates of the sliding and rolling pairs.

This endoprosthesis [33] possesses the advantages of previously discussed design in damping impact loads, supply of biological fluids of the organism and injection of drugs into the operative wound.

Endoprosthesis of the knee [34] belongs to a sledge group of prostheses. Its structure differs by the presence of a polymer part serving as a knee meniscus. The natural meniscus is fixed by its external edge to the joint capsule, while its other thinning edge enters the clearance in the movable junction of the femoral and tibial bones. Flexibility of the capsule ensures independent motions in the horizontal plane for both medial and lateral menisci of the movable joint. The artificial meniscus allows for analogous movements of the endoprosthesis [34].

The main elements of the described endoprosthesis are illustrated in Fig. 8.10.

Bracket 1 (femoral component) has on its inner surface facets 2 for better fixing on the femoral plateau. A pair of rods 3 adds reliable fixation to the joint. The convex contact surface 4 of the bracket is of a variable curvature radius, which fits precisely the friction surface of the femoral plateau.

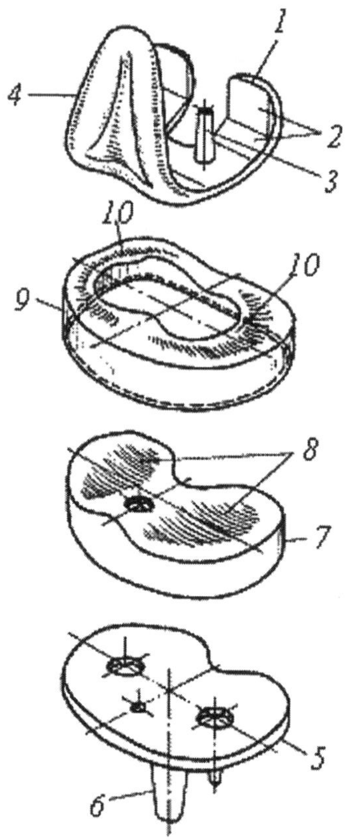

Fig. 8.10. Endoprosthesis of knee joint with artificial meniscus [34]

The tibial component of the endoprosthesis consists of plate 5 with stem 6 fixed intramedullary in the tibial bone.

Tray 7 has on its upper face grooves 8 whose curvilinear surface is congruent to surface 4 of the bracket. The tray is attached to plate 5.

The artificial meniscus is made as a thin-walled cylinder 9 enclosing the tray. Strip 10 is attached along the normal to the upper face inside the cylinder. It is shaped as a medial and laterial menisci of the knee.

The curvilinear surface 4 of the bracket slides over the congruent grooves 8 of the tray transferring on it the main load during motion. Strip 10 is inserted in the articulating artificial joint and is pinched between the bracket and tray on the nominal contact area periphery. The clearance fit of cylinder 9 in tray 7 and flexibility of cylinder's walls in combination with elasticity of strip 10 ensure independent shifting of strip portions in the clearances of the articulating pair. The motion kinematics of the bracket, artificial meniscus and tray during movement of the prosthesis is analogous to articulation of the femoral and tibial plateaus against lateral and medial menisci of a natural joint.

Strip 10 is made of UHMWPE, its gel or reinforced materials based on these binders.

The mentioned design does not complicate anyhow a classical sledge endoprosthesis but simulates the kinematics of frictional interactions in a natural knee containing menisci, and is devoid of any elements able to impede the endoprosthetic operation.

Above examples are one more proof to a continuing modification of metal–polymer joint endoprostheses and certain prospects in this direction at the level of designs. We may anticipate their further progress in conjunction with novel antifrictional materials that are forming neutral wear debris, non-metallic composites and advanced polymers as alternatives to UHMWPE.

8.4 TRENDS IN ENDOPROSTHETICS

The significance of maintaining healthy locomotor system of the organism is interrelated with raising actuality of endoprosthetics as an efficient means of treating pathological changes of human joints. Along with the needs of refining modern prosthetic designs, social concerns of endoprosthetics are connected with substantial lowering of their cost. These are the main stimuli of continuous perfection of endoprosthetics in response to scientific achievements. Modern trends in endoprosthetics are interrelated with profound understanding of the causes of joint diseases, improved methods of their diagnostics and therapeutics, striking results in genetic engineering, still new approaches to implantation and types of prostheses.

Pathogenesis of inflammations in joints is studied at present on the background of autoimmune reactions (directed against its own tissues). These reactions are characterized by the formation of auto antibodies for various auto antigens, particularly, for intrinsic in functional respect components of the nucleus and cytoplasm. As V. Nasonova, the Academician of the Russian AMS, has said, in a number of inflammatory rheumatoid diseases the infection and genetic predisposition to it play the role of etiological (causal) factors, and the major pathogenetic mechanism (the mechanism of disease developing) is the immune-mediated inflammation and autoimmunity. In this connection, we may anticipate that the use of bone autotransplants produced by genetic engineering methods and suppressing autoimmune reactions will become a customary trend in orthopaedics.

A promising trend in orthopaedics is the search for new infectious factors that initiate primarily chronic progressing diseases of joints. Of a paramount importance will be the discovery of the chief genes affecting the development of systemic rheumatoid sickness, osteoarthrosis and osteoporosis, which govern the quality and life-span of a broad category of population. The investigation findings will assign the main trend in genetic engineering and reconstructive orthopaedics.

Diagnostics of joint diseases is a starting point of the methodology of endoprosthetics.

The differentiated diagnostics of systemic diseases of the connectivum (immunodiagnostics) has become possible thanks to identification of auto antibodies reacting with auto antigens of cell nucleus and cytoplasm. Its results are accounted for at choosing the tactics of endoprosthetic operations including medicamental facilities and drugs preventing autoimmune reactions [34].

Radiographic studies of bones and joints present a valuable basis for evaluation of the state of rheumatoid arthritis, ankylosing spondylarthritis, osteoarthrosis inducing rapid destructive and erosive processes in joints. In the recent years comprehensive studies have been devoted to cartilage, its state and dynamics in the course of osteoarthrosis, and its treatment by non-steroid antiphlogistic medicines, many of which may damage cartilage at prolonged usage. For this purpose such methods as sonography and computer tomography are extensively employed today. Arthrosonography (ultrasonic echography) produces accurate estimates of cartilage condition in the arthrosic joint, visualizes the continuity of the bone head outline, which is requisite for early diagnostics of aseptic necrosis, and maintains a harmless dynamic control over treatment efficiency.

Densitometry (density measurement) occupies an important place among the diagnostic methods of the osseous tissue. This procedure assists in evaluating the osteoporosis degree and identifying bone illness, particularly necrosis, tumour, inflammation, metabolic osteopathy and so on [35].

Investigations of the synovia and nodules formations in joints play a significant role in differential diagnostics of microcrystalline arthropaties. The methods of polarization microscopy are used to identify microcrystalline structures in the synovial fluid for the case of acute arthritis or chronic exacerbation. They are a reliable means for identification of chondrocalcinosis, hydroxyapatite arthropaty and other diseases. The use of the electret-thermal analysis in studying liquid-crystalline and protein-polysaccharide constituents of synovia has been described earlier in chapter 7.4. Prognostication of joint diseases by above-named and other methods of diagnostics makes it possible to plan endoprosthetic operations with higher reliability.

Medicinal therapy as a complex method of treating joint diseases by non-steroid antiphlogistic medications (NAPM) and in specific cases by glucocorticosteroids in combination with cytotoxic and other drugs presents an alternative to endoprosthetics at certain stages of illness. Along with certain positive effects this kind of therapy may be accompanied by a set of adverse after-effects. Serious concern should be given here to gastropathy, which is an erosive-ulcerous affection of the upper section of the gastrointestinal tract closely related with the usage of non-steroid and anti-inflammatory medications. Such complications as the acute gastric bleeding, ulceration or perforation may sometimes be lethal for a group of elderly people suffering from rheumatoid arthritis. It is paradoxical that improved quality of people's life goes hand in hand with reduced longevity in connection with various complications of medicinal therapy of joint diseases. Understanding of pathogenesis of immune inflammation of joints and its mechanisms served as a theoretical substantiation of therapeutic treatment in the 21st century. The use of anticytokin antibodies as an important component of complex curing of the patients with resistant forms of rheumatoid arthritis presents a new trend in therapeutics of joints. One of the major principles of the modern therapy of joint diseases is becoming selectivity of the effect of non-steroid antiphlogistic medications.

Intraarticular injection of corticosteroid products (kenalogue, diprospane, hydrocartisone and other) is an efficient means of treating degenerative and dystrophic diseases of joints. At present they are substituted for the synovial endoprostheses being biological analogues of hyaluronate (Hyalgan, Orthovisk, Synvisk). Synvisk effects selectively the cartilage tissue, abates pain and restores mobility of joints. The synovial endoprostheses restore cartilage nutrition and represent perfect chondroprotectors, i.e. protect articular surfaces from fracture and retard osteoporosis [36]. Complex curing of joint diseases using named drugs gives a possibility to postpone the endoprosthetic operation for prolonged time (Fig. 8.11).

Genetic engineering and its achievements in creation of the bone and cartilage autotransplants seem to be a major reserve of joint endoprosthetics in the years to come. It is to be underlined that an ideal joint endoprosthesis is most likely the one produced by genetic engineering and grafted to the bone by transplantation methods. The main trends of genetic engineering set forth in chapter 4.5 will take a path of preventing immune pathologies.

The advances in molecular biology have now made it possible to address these problems at molecular level. By gene therapy, a defective gene is replaced with a normal or therapeutic gene. To be successful, the exact sequence and function of the specific gene must be understood, a vehicle for safe and efficient delivery of the gene into the cells must be located, and the expression of the gene should be well controlled. Until now, difficulties with efficient gene transfer and appropriate gene expression have still been an impediment. Besides, ethical problems of a carcinogenic and eugenic nature have arisen, but with gene transfer as drug delivery system, there is a great range of applications to acquired diseases [37].

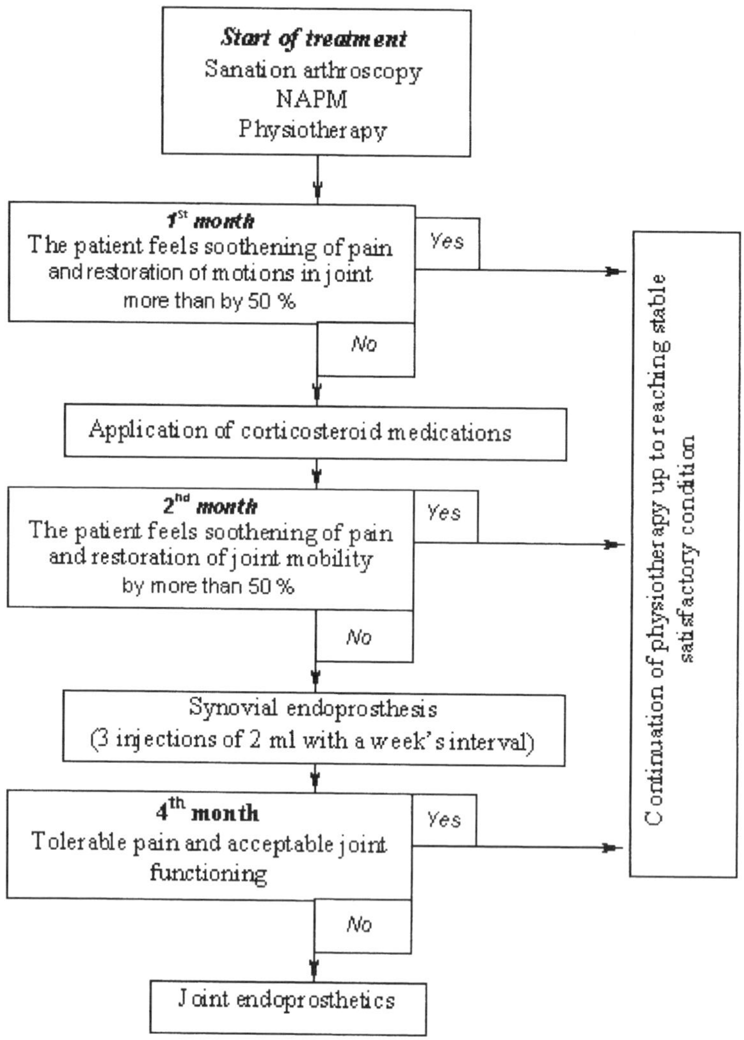

Fig. 8.11. Algorithm of treating arthrosis of the knee joint [36]

Modification of endoprostheses marches in step with their implantation methods. It is hard to believe that the approaches of genetic engineering will occupy a habitual everyday place in orthopaedists' practice in the nearest future. Evidently, during adaptation period the traditional joint endoprostheses will yield to demand and transfer to a qualitatively new level. Let's dwell in more detail on the possible means of attaining this level.

It may be anticipated that in spite of a protracted clinical history and apparent stability of the nearest and remote results of joint endoprosthetics based on metallic implants, the use of metals in endoprostheses will be contracting and the contact between metal parts and bone will be soon excluded. It is common knowledge that metals are a potential source of corrosion and metal stems form conducting external chains that may short-circuit sections of bony tissues carrying unlike in magnitude and sign bio potentials. A long-term contact between metal and bones leads inevitably to local necrosis of the bone. Metal wear debris are also extremely unfavourable for joint endoprostheses. They may be formed just as in the friction joints of endoprostheses so in the metal stem contact with cement mantle. Metal wear debris deposited in the joint capsule result in metallosis, and metal particles are often found in lungs and other organs of the patients with implanted endoprostheses. Meanwhile, surgeons of all times attempted to remove from patients' body various metal bullets, blades, fragments understanding their harmful affect for the health.

The analysis of works in biocompatibility of materials has shown that tribological aspects of biocompatibility, particularly the biocompatibility of wear debris of structural materials is insufficiently studied so far. Wear particles represent surface-active formations, which acquire high surface energy. This is the reason why their destructive effect on the bone–implant interface is realized as aseptic inflammatory reactions of the osseous tissue and osteolysis round the particulate-matter, leading to loosening of endoprostheses. Although we know the mechanisms of these reactions, there is not any comprehensive analysis of biocompatibility of structural materials and their wear debris, or a comparison of wear particles by their biocompatibility criterion. Unfortunately, the task on the development of antifrictional materials and friction joints that generate biocompatible wear debris has not been even posed in tribology up to now. The creation of customized frictional materials for endoprostheses whose wear products are friendly to surrounding tissues is an urgent problem of the materials science and tribology of medical spheres.

In this connection, composites as structural materials for endoprosthetics come first. High level of contemporary composite techniques is a warranty of reaching the strength of composite endoprostheses required to be on a par with metal ones. The use of ceramic and polymer binders in composites makes possible elimination of toxic effect of metals on the organism ensures corrosion resistance of endoprostheses and electric insulation of bone sections with unlike bioelectrical potentials. Composites favour the conditions for application of carbon materials in endoprostheses whose wear products are most biocompatible out of other structural materials. When contacting the bone, composite stems are less harmful for the osseous tissue as opposed to metals. By optimizing the choice of the binder, reinforcing elements and texture of the implant surface one may regulate adsorption of protein molecules and protein adhesion to the implant. In addition, composites are very convenient as drug and antimicrobial agent carriers, as well as sources of electrical and magnetic fields in the artificial limb. The presence of multifunctional components existing in intricate structurally stipulated interactions makes grounds for elaboration of smart composites able to respond in a preset form to the physiological environment of the organism or simulating some biological functions of a joint or its tissues.

Advanced design solutions of the friction joints open unexpected vistas in perfecting joint endoprostheses. We anticipate that their designs will realize a specific interaction between elements that imitate the biological structure and functions of joints or contain some sealing systems insulating the artificial joint from the rest systems of functioning. The former is a maximum task since we can not built-in the artificial limb into the biological environment of the organism without attracting the novelties in biology and genetic engineering. The latter task is attainable entirely based on the modern engineering level. The variants of its solution are discussed below. They are related, first of all, with (A) localizing wear debris within the bounds of the endoprosthesis, and (B) incompatible with the organism lubricating media.

A. Wear particles may be absorbed by endoprostheses components, e.g. artificial cartilage (Ch. 6) or contained in specific cavities insulated from the biological medium of the organism. This problem can be solved by the modern engineering means. For instance, in sealing machine parts we use similar elements called spacer gaskets. They may contain sealing elastic diaphragms, bellows and other shells able to isolate friction joints from the external media in the machinery like membrane pumps, sealed electric motors, magnetic and electromagnetic clutches and so on [38].
B. Sealing of the friction joint allows for the usage of any lubricating media for endoprostheses, incompatible for the organism included. Besides, very promising are endoprostheses that realize selective transfer at friction. This effect was first discovered by D.N. Garkunov and I.V. Kragelsky at copper alloy friction against steel in conditions of boundary lubrication (alcohol-glycerine mixture) that excludes copper oxidation. The phenomenon of selective transfer of copper from a solid copper alloy solution on steel and back is accompanied by the friction coefficient reduction. Note that the friction coefficient does not exceed in this case the value characteristic of the liquid friction. This results in a significant decreasing of the friction pair wearing and virtually to a wear-free sliding [39]. Such joint designs are atypical for the traditional mechanisms but may be a challenge for new approaches in endoprosthetics. Evidently, the first endoprosthesis design employing selective transfer at friction is the hip one [40].

Raised level of technical facilities in endoprosthetics and successes in genetic engineering will inevitably end in reassessing the surgical aspect of endoprosthetic strategy. Resection of bone epiphysis during preparation of the bone bed for the artificial joint is an irreparable injury for the bone as an autonomous formation and for the entire organism. The human osseous tissue is known [41] to experience the mechanical eigen-stresses whose magnitude depends on the bone section, skeletal elements, and changes with age. A complex field of spontaneous stresses equalized within the bone macro volume matches the bone model as a stressed structure. The external layer of the long bones experiences the compressive stresses, whereas the tensile ones occur in the bulk. The stress distribution gradient over a long bone is characterized by the growth of stresses in its medial part. Therefore, the bone deforms at resection under the effect of eigen-stresses. Each portion of the osseous tissue presents a source of the bioelectric potential, however, it has become a subject of biophysical investigations only lately (see 7.3). Along with disturbance of intrinsic for the bone stressed state, the resection alters the bio field parameters in joints and around them. So, along with doubtless profits, the operative techniques adopted in endoprosthetics today inflict drastic damages to the joint as a mechanically and electrically equilibrium organ.

The operation using autotransplants derived by the genic engineering methods is anticipated to restore the parameters of bioelectrical potentials of joints, while the mechanical stresses in such transplants are found in the equilibrium. An alternative decision, which can be realized in the nearest future, seems to be the strategy of repairing the articular end of the bone without its resection. This may be placing of artificial cartilage fragments, application of cartilage-simulating coatings, and treatment of the friction surface by specific medications and so on. This will undoubtedly necessitate much higher level of the medical techniques and medico-biological facilities for operations in joints.

Parallel to technical evolution in tumour surgery, limb function and life quality are being analyzed intensively. Although amputation remains an indispensable life-saving procedure in 10–30 % of cases, a better quality of life can usually be achieved following limb-saving procedures. In addition, patients in Europe and the Middle East demand limb-saving surgery due to traditional social reasons, even in case when the expectable function of the limb and life quality will not exceed the results of rehabilitation following ablative surgery. We believe, more and more modern diagnostic tools, highly effective chemotherapeutic agents and better surgical reconstructive techniques for limb-saving procedures will help to achieve a high-quality of life for the patients [42].

Documentation technology is an important part of the joint endoprosthetics system that assists in controlling the efficiency of implants design, lifetime of a prosthetic component and compares the results of operations by using a common scientific terminology.

In 1993 Sir Dennis Paterson wrote and editorial on the International Documentation and Evaluation System (IDES) [43]. He entitled the principles of IDES as consensus, hierarchical information, radiographic evaluation and acceptability. IDES was established by the International Society of Orthopaedic Surgery and Traumatology (SICOT) Standing Committee on Documentation and Evaluation, which was founded in 1990 by Prof. M.E. Muller as a chairman, and presented at the American Association of Orthopaedic Surgeons (AAOS) 61st annual meeting in 1994. The nomenclature used on the three IDES sheets for primary total hip arthroplasty, revision arthroplasty and follow-up is based on the consensus paper by the Hip Society, the SICOT Commission and the Task Force on Outcome Studies of the AAOS [44]. This consensus paper provided a terminology named CART (Clinical and Radiographic Terminology), in which each term, whether applying to a functional or radiographic parameter, was specifically defined to have a constant meaning. The initial impulse to create such a terminology was already given in 1985 when J. Galante [45] called for a uniform method of evaluation and reporting the results of hip-replacement surgery in order to compare the results on a common standard basis.

Today, the IDES represents one of the most valuable hip arthroplasty databases and documentation applications existing. During the 4 decades, extensive information about 50,000 primary and 12,000 revision operations, 77,000 follow-ups was collected. The IDES database provides clinically valuable and relevant information about treatment of hip diseases and long-term results and will contribute substantially to a further improvement of hip surgery and outcome research.

Based on experiences with the IDES application, a new generation documentation system with Internet technology was developed during the last 5 years. Data capture at source is possible for all users assigned to the documentation process, independent for each other and with different on-line and off-line data collection tools. These tools are interdependent, which means that all data are finally routed to the web interface for final submission, querying and analysis. That is why, the documentation process becomes highly flexible and dynamic, and can be adapted to the workflow of the respective department. Even questionnaires can be customized and extended beyond the essential data set, with an online question generator for individual research endeavours. An automated implant tracking and registration system with barcode technology (Secure Data Integration Concept, SEDICO) allows the direct identification of the implants used during surgery and offers an integrated order service for the implant manufacturers [46].

By making use of the offered tools, the orthopaedic community is enabled to collect and compare data more easily and accurately, perform clinical studies more transparently and therefore finally improve the quality and efficiency of medical treatments.

◈ ◈ ◈

The modern stage of endoprosthetics of joints was pioneered by the marvellous developments of J. Charnley. His designs have predetermined today's prosthetic structures and fixation methods, as well as the nomenclature of biocompatible materials. Accumulation of follow-up results taking place up till now has spurred new ideas on imparting certain biomechanical, biophysical and biochemical functions of natural tissues to endoprosthetic components. This concerns just as materials structure, so their ability to generate physical fields to compensate various violations of natural bioelectric potentials due to a pathology or surgical treatment. The next stage which is about to come will be most likely connected with successes in biology and genetic engineering. Understanding of the forthcoming perspectives in the medical and technical aspects of endoprosthetics seems highly actual for modification of traditional endoprostheses designs and operative maintenance.

References:

1. Weller S. Fifteen years of experience with the BICONTACT hip endoprosthesis system – the past, the present, the future. What has been achieved. *Int. Orthop.*, 2003, V. 27 (Suppl. 1), S2 – S6.
2. Effenberger H., Witzel U., Lintner F., and Rieger W. Stress analysis of threaded cup. *Int. Orthop.*, 2001, V. 25, p. 228-235.
3. Wilkinson J.M., Gordon A., and Stockley I. Experiences with the Plasmacup – early stability, wear, remodelling, and outcome. *Int. Orthop.*, 2003, V. 27 (Suppl. 1), S16 - S19.
4. Malchau H., Herberts P., and Ahnfelt L. Prognosis of total hip replacement in Sweden. Follow-up of 92,675 operations performed in 1978–1990. *Acta Orthop. Scand.*, 1993, V. 64, p. 497-506.
5. Muller M.E., Nazarian S., Koch P., and Schatzker J. The comprehensive classification of fracture of long bones. Berlin, Springer-Verlag, 1996, 201 pp.
6. Groot K. Bioceramic consisting of calcium phosphate salts. *Biomaterials*, 1981, V.1, p. 47–50.
7. LeGeros R.Z., and Daculsi G. *In vivo* transformation of biphasic calcium phosphate ceramics: Ultrastructural and physiochemical characterizations. In: Handbook of bioactive ceramics. VII. Florida, CRC Press, 1990, p. 17–28.

8. Daculsi G. New technology for calcium phosphate bioactive ceramics in bone repair. *Medical & Biological Engineering & Computing*, 1999, V.37, No. 2, p. 1598–1599.
9. Fokin V.A. Friction pairs for total hip endoprostheses and the problem of wearing. *Margo Anterior*, 2000, No. 4, p. 1–4.
10. Toni A., Sudanese A., Terzi S. et al. Ceramic in total hip arthroplasty. In: *Proc. of 2nd Symp. on Ceramic Wear Couple*, Stuttgart, 1997, p. 30–33.
11. Fisher J., and Dowson D. Tribology of total artificial joints. *Proc. Inst. Mech. Engrs., Part H,* 1997, V. 205 (H2), p. 73–79.
12. Unswarth A., Pearcy M.J., White E.F.T., and White G. Frictional properties of artificial joints. *Journal of Engineering in Medicine*, 1988, V.17, p. 101–104.
13. Auger D.D., Medley J.B., Fisher J., and Dowson D. A preliminary investigation of the cushion form bearing in artificial joints. In: Mechanic of coatings. Ed. by D. Dowson. Amsterdam, Elsevier, 1990, p. 264–269.
14. Oka M., Noguchi T., Kumar P., et al. Development of an artificial articular cartilage. *Clin. Mat.*, 1990, V. 6, p. 361–381.
15. O'Carroll S., Jin Z.M., Dowson D., et al. Determination of contact area in cushion form bearings for artificial hip joints. Proc. Instn. Mech. Engzs, Part H, 1990, V. 204 (H4), p. 217–223.
16. Dowson D., Fisher J., Jin Z.M., Auger D.D., Jobbins B. Design considerations in cushion form bearings for artificial hip joints. *Proc. Inst. Mech. Engrs., Part H,* 1991, V. 205 (H2), p. 59–68.
17. Lin F.F. A new model of human joint cartilage. *Problems in friction and lubrication*, 1974, No. 3, p. 164–170.
18. Sanders J. Biocompatible breakthrough. Georgia Tech, 1999, No. 5, p. 21.
19. USSR Patent 1061811, A 61 F 1/03. Endoprosthesis of knee joint. I.R. Voronovich, S.P., Kozlovsky, E.D. Beloenko, L.S. Pinchuk, et al., 1983.
20. Belarus Patent Application 20000995, C 08 J 9/26. L.S. Pinchuk, Zh.V. Kadolich, E.A. Tsvetkova, V.I. Nikolaev, E.D. Beloenko, 2001.
21. Belarus Patent Application 20000468, A 61 F 2/30. Insert for hip joint endoprosthesis. L.S. Pinchuk, E.A. Tsvetkova, Zh.V. Kadolich, V.I. Nikolaev, 2001.
22. Kadolich Zh.V., Pinchuk L.S., and Tsvetkova E.A. Mechanism of friction and wear of microporous UHMWPE. *J. Friction and Wear*, 2002, V. 23, No. 1, p. 71-74.
23. Kadolich Zh. Relaxation mechanism observed at microporous UHMWPE friction in endoprostheses. *Mat. III Symp. "Inzyneria ortopedyczna i protetyczna"*, Bialystok, 2001, p. 101-106.

24. Kadolich Zh.V. Physical modification polymer-metal joints improvement of wear resistance based on simulation of biophysical properties of natural joints. Ph.D. Thesis, Gomel, 2002.
25. Begun P.I., and Shukeilo Yu.A. Biomechanics. Manual for Higher School. St. Petersburg, Polytekhnika, 2000, 463 pp.
26. Belarus Patent 2673, A 61 F 2/30, A 61 F 27/00. Joint endoprosthesis. L.S. Pinchuk, E.A. Tsvetkova, V.I. Nikolaev, V.A. Goldade, 1999.
27. Belarus Patent Application 20010073, A 61 F 1/06. Hip joint endoprosthesis head. L.S. Pinchuk, Zh.V. Kadolich, E.A. Tsvetkova, et al., 2002.
28. Atic S.A.R. Unipolar endoprosthetics of the hip joint in elderly people (experimental and clinical grounding). Ph.D. Thesis, Minsk, 2004.
29. Belarus Patent Application 20020070, A 61 F 2/34. Acetabular component of hip joint endoprosthesis. L.S. Pinchuk, E.A. Tsvetkova, V.I. Nikolaev, V.A. Goldade, 2002.
30. Belarus Patent Application 20011135, A 61 F 2/34. Acetabular component of hip joint endoprosthesis. L.S. Pinchuk, E.A. Tsvetkova, V.I. Nikolaev, et al., 2004.
31. Belarus Patent Application 20011131, A 61 F 2/32. Total hip joint endoprosthesis. L.S. Pinchuk, E.A. Tsvetkova, V.I. Nikolaev, V.A. Goldade, 2004.
32. Mariev E.N., and Mallatt J. Human anatomy update. 3rd Ed., NY, Benjamin Cummings, 2002, 844 pp.
33. Belarus Patent Application 20011134. A 61 F 2/32. Hip joint endoprosthesis. L.S. Pinchuk, E.A. Tsvetkova, V.I. Nikolaev, 2002.
34. Beloenko E.D. Optimization of orthopedic and surgical treatment of rheumatoid arthritis based on the method of biophysical regulation of antifrictional properties of synovial medium in joints. Dr. Med. Sci. Thesis, Kiev, 1992.
35. Manual of bone densitometry measurements. Ed. by J.N. Fordham. Heidelberg, Springer-Verlag, 2000, 226 pp.
36. Clinical aspects of applying Synvisk preparation. Margo Anterior, 2002, No. 4, p. 5–8.
37. Li. H., Zou X., and Bunger C. Gene therapy and spinal disorder. *Int. Orthop.*, 2001, V. 25, p. 1-4.
38. Pinchuk L.S. Hermetology. Minsk, Nauka i Tekhnika, 1992, 216 pp.
39. Litvinov V.N., Mikhin N.M., and Myshkin N.K. Physico-chemical mechanics of selective transfer at friction. Moscow, Nauka, 1979, 188 pp.

40. Belarus Patent Application 20040362. A61 F 2/32. Total hip joint endoprosthesis. L.S. Pinchuk, E.A. Tsvetkova, V.I. Nikolaev, et al., 2005.
41. USSR Discovery No. 181. Phenomenon of generation of eigen-stresses in human and animal bones. V.I. Loshchilov, G.A. Nikolaev, and E.P. Babaev, 1976.
42. Szendroi M. Advances in orthopeadic oncology. *Int. Orthop.*, 2002, V. 26, p. 195-196.
43. Paterson D. The International Documentation and Evaluation System (IDES). *Int. Orthop.*, 1993, V. 16, p. 11-14.
44. Johnston R.C., Fitzgerald R.H., Harris W.H., et al. Clinical and radiological evaluation of total hip replacement. A standard system of terminology for reporting results. *J. Bone Joint Surg. (Am)* , 1990, V. 72, p. 161-168.
45. Galante J. The need for a standardized system for evaluating results of total hip surgery. *J. Bone Joint Surg. (Am)*, 1985, V. 67, p. 511-512.
46. Roder C., Eggli S., El-Kerdi A., et al. The International Documentation and Evaluation System (IDES) – 10 years experience. *Int. Orthop.*, 2003, V. 27, p. 259-261.

CONCLUSIONS

Summing it up, we would like to put emphasis on the main ideas that encouraged us to write this book. Although these topics have run all through the book, they could seem to recede into the background due to the abundance of technical and medico-biological details of endoprosthetics.

1. Vice-president of the European Federation of National Associations of Orthopaedics and Traumatology (EFFORT), Prof. Ervin Morcher (Switzerland), has expressively characterized social significance of endoprosthetics of joints. He states that in the last 30 years of the 20th century it was the orthopaedics that demonstrated the greatest growth in the number of surgeons and the amount of operations made by them. This is not only because of the expansion of joint diseases on the Earth and their effect on people's capacity for work, but also due to the demographic shifts that had occurred in the last quarter of the past century, first of all the increasing quantitative ratio of elderly people. The significance of endoprosthetics rises also in connection with the accidents, which number and gravity will scarcely drop in future. The preventive measures of the road and industrial injuries will most likely be compensated by the growth of sports and everyday traumas.

In this connection, it is actual to specify the notion "quality of life" and its evaluation from the standpoint of orthopedy. We have to define the criteria that correlate most accurately the relation between the patient's ages, degree of his activity and active life duration. On this base, we may reach better compliance between choosing the methods and the outcomes of curing joint pathology. By international efforts a unified approach to the estimation of treatment outcomes can be elaborated, which is strongly important for working-out both the policy of public health and the strategy of endoprosthetic surgery.

2. Endoprosthetics of joints is a multiprofile science-extensive field of orthopedy making use of the achievements in various scientific domains. Its evolution was intermingled with industrial, technological and medical developments, successes in anaesthesiology, transfusion, prophylactics of infectious and thromboembolic complications. Assimilation of innovations of technical physics has lead to the development of new image methods of treating joints and essentially refinement of diagnostics. Today, a successful outcome of the endoprosthetic operation is less dependent on the knowledge and mastery of a surgeon. Being a specialized field of orthopedy, endoprosthetics of joints has shown a tendency to further specialization into organs: hip, knee, shoulder joints and other. Besides, a technological specialization of endoprosthetics is observed connected with application of arthroscopy, microsurgery, and laser surgery and so on.

There is a situation when the decision-making in a number of urgent orthopedy tasks is impeded by dissociation between the professionals representing medical, biological, physical, chemical and engineering disciplines. As a result, the attainments accumulated in adjoining sciences are not united and so inaccessible to broad audience of researchers and practical surgeons. This served an impetus for writing this book where we tried to analyze modern trends of endoprosthetics and set out the position of the authors presenting different specialties in various aspects of the problem.

3. The current stage of endoprosthetics is influenced enormously by J. Charnley's ideas. The level of technical perfection of the modern implants is manifested in the best hip endoprostheses designs, including Charnley's Golden Standard and Muller's straight-stem cemented ones, Spotorno (CLS) uncemented prosthesis and other. The durability indices of the endoprostheses attained by now are a result of the uniqueness of the materials (Protasul metal alloys, cross-linked UHMWPE, titanium nitrides, etc.) and high technologies used today (isothermal forging producing parts almost devoid of residual stresses, laser technologies, plasma coating application and so on). The quantity of primary endoprosthetic operation in the world has already reached a steady level. The growth of the total amount of operations is because of the revision ones. Their main reason is the aseptic loosening of endoprostheses brought about by lysis of the osseous tissue in the bone–metal contact initiated by wear debris ingress into the interface.

From the other hand, the current stage of joint endoprosthetics can be characterized by an adverse phenomenon typical of all technology and science-intensive industries. It is displayed in the unauthorized copying of the best endoprostheses designs. Some companies imitate the developments of other firms without any serious biomechanical or clinical investigations and permission or payment of royalties to creators of the original constructions. Some of pirate firms make insignificant visual modifications. However, such insignificant variations in, e.g. lengthening, increasing height of ribs, curvature radius or other, without preceding clinical testing may result in unpredictable complications. The buyers of such products jeopardize the health of patients and the reputation of physicians.

4. The forthcoming aims of joint endoprosthetics consist primarily in ameliorating the existing types of prostheses. The evolutionary periods of the design, biomechanical and material testing (biocompatibility and tribology) transfer into a revolutionary stage of smart biomaterials and structures. They are imparted certain reactions to variations in the biological medium of joints and may correct this reaction by a feedback system. Biological cybernetics will be primarily employed in the feedback system of endoprostheses. The absence of signals from the zone of reflexes removed at total endoprosthetics of joints

(bone, cartilage, joint capsule, intraarticular ligaments) can be outweighed by the signals of the intelligent endoprosthesis taught to regulate the bone and muscle system actuating motions in the artificial limb.

A corner stone opening new perspectives to orthopedy, E. Morcher considers biomechanics. However far would orthopaedic surgery go in biomechanical direction, its progress is closely interrelated with biophysics, and on the first place with tribophysics, biochemistry, pharmacology, microbiology and immunology. According to our firm conviction, the future of orthopedy will be the junction with genetic engineering that will elevate endoprosthetics to a much higher level.

5. Like any other medical branch, orthopedy serves for the society and is subjected to varying economy, politics and ideology. Together with growing life cost and technical progress, expenses on the public health services have risen too. The society is rightfully expecting reimbursement of the invested means.

Osseous tissue, being the primary object of orthopedy, is a dynamic intricately organized system that experiences both renewal and decline during its lifetime. Any affects result in the changes of physiological, biomechanical, electrophysical and other parameters. So it is natural for the orthopaedists to raise the efficiency of surgery and accelerate rehabilitation of operated joints. Nevertheless, no one has managed to speed up natural processes of bone regeneration. This will hardly occur even after adoption of microbiological and genetic achievements. The present task of orthopaedists is not in changing natural physiological modes of regeneration, but to exclude pathological processes that hamper normal restoration of the operated bone. This is just the approach to solution of endoprosthetic problems, which is kept to in the present book.

Being limited by the volume of the book, we could only disclose the major engineering and medico-biological aspects of joint endoprosthetics. Such topics as simulation of biomechanical properties of the osseous tissue and ligaments, immunological mechanisms of implant vs. organism relations, protection of synovial capsule from wear debris and some other are left beyond the bounds of the manuscript. Although some estimation cited in the book are still disputable, they are meeting, in our view, modern trends in orthopedy. The authors believe that the present work will stimulate new research in the development of artificial joints.

SUBJECT INDEX